Mathematics Study Resources

Volume 18

Series Editors

Kolja Knauer, Departament de Matemàtiques i Informàtica, Universitat de Barcelona, Spain

Elijah Liflyand, Department of Mathematics, Bar-Ilan University, Ramat-Gan, Israel

This series comprises direct translations of successful foreign language titles, especially from the German language.

Powered by advances in automated translation, these books draw on global teaching excellence to provide students and lecturers with diverse materials for teaching and study.

Kolja Knauer • Ulrich Knauer

Discrete and Algebraic Structures

A Concise Introduction

Kolja Knauer
Departament de Matemàtiques i Informàtica
Universitat de Barcelona
Spain

Ulrich Knauer
Universität Oldenburg
Oldenburg, Germany

ISSN 2731-3824 ISSN 2731-3832 (electronic)
Mathematics Study Resources
ISBN 978-3-662-70562-9 ISBN 978-3-662-70563-6 (eBook)
https://doi.org/10.1007/978-3-662-70563-6

This book is a translation of the original German edition "Diskrete und algebraische Strukturen - kurz gefasst," 2nd edition, by Ulrich Knauer and Kolja Knauer, published by Springer-Verlag GmbH, DE in 2015. The translation was done with the help of an artificial intelligence machine translation tool. A subsequent human revision was done primarily in terms of content, so that the book will read stylistically differently from a conventional translation. Springer Nature works continuously to further the development of tools for the production of books and on the related technologies to support the authors.

Translation from the German language edition: "Diskrete und algebraische Strukturen - kurz gefasst" by Ulrich Knauer and Kolja Knauer, © Springer-Verlag Berlin Heidelberg 2015. Published by Springer Berlin Heidelberg. All Rights Reserved.

© The Editor(s) (if applicable) and The Author(s), under exclusive license to Springer-Verlag GmbH, DE, part of Springer Nature 2025

This work is subject to copyright. All rights are solely and exclusively licensed by the Publisher, whether the whole or part of the material is concerned, specifically the rights of translation, reprinting, reuse of illustrations, recitation, broadcasting, reproduction on microfilms or in any other physical way, and transmission or information storage and retrieval, electronic adaptation, computer software, or by similar or dissimilar methodology now known or hereafter developed.
The use of general descriptive names, registered names, trademarks, service marks, etc. in this publication does not imply, even in the absence of a specific statement, that such names are exempt from the relevant protective laws and regulations and therefore free for general use.
The publisher, the authors, and the editors are safe to assume that the advice and information in this book are believed to be true and accurate at the date of publication. Neither the publisher nor the authors or the editors give a warranty, expressed or implied, with respect to the material contained herein or for any errors or omissions that may have been made. The publisher remains neutral with regard to jurisdictional claims in published maps and institutional affiliations.

This Springer imprint is published by the registered company Springer-Verlag GmbH, DE, part of Springer Nature.
The registered company address is: Heidelberger Platz 3, 14197 Berlin, Germany

If disposing of this product, please recycle the paper.

Preface

This is the first English edition of our second German edition [49], "Diskrete und algebraische Strukturen—kurz gefasst," which followed the first German edition [48]. With this translation, we are essentially presenting a third edition of our work. Naturally, we seized this opportunity to correct errors and make several improvements.

This volume can be viewed as a textbook for students of Computer Science and Mathematics. It is designed to be an introductory guide for beginners but should also serve as a valuable compendium for more advanced learners who wish to revisit and reinforce their understanding of these concepts.

The editor provided us with a preliminary automated English translation, and we were quite impressed with the initial quality. However, we were also surprised by how much additional work was necessary to make the translation both accurate and readable. Since August 2023, we have been refining the text, and we sometimes wondered whether it would have been more efficient to wait for the translation software to improve—a natural thought in the rapidly evolving field of artificial intelligence.

Some traces of the original language remain. For instance, the long German sentences have been retained in many places. We endeavored to balance the readability of the English text with the original structure, which sometimes resulted in preserving certain Germanic expressions. For example, we kept the symbol K for fields, called Körper in German. Similarly, we use $\mathcal{U}$ for the independent sets of a matroid, reflecting the German term "unabhängig." We also use $\mathcal{K}$ for category in Chap. 12, according to the German word "Kategorie."

We also retained several mnemonics from the original text, referred to as "bridges for donkeys" or "Eselsbrücken" in German. We hope you find these touches both amusing and helpful.

We extend our heartfelt thanks to the publisher and especially to the directing editor, Dr. Andreas Rüdinger, for their immense help and numerous consultations during the translation process. Their support has been invaluable in bringing this English edition to life.

Barcelona, Spain Kolja Knauer
Oldenburg, Germany Ulrich Knauer

Contents

1 Fundamentals 1
   1.1 Languages 1
   1.2 Syntax and Semantics 3
   1.3 Statements and Truth Values 7
   1.4 Logical Identities 11
   1.5 Axiomatic Structure 13
   1.6 Normal Forms 14
   1.7 Working with Quantifiers 17
   1.8 Building a Theory 20
   1.9 Proof by Mathematical Induction 22

2 Sets and Counting 27
   2.1 Sets 28
   2.2 Operations on Sets 31
   2.3 Cardinality and Infinite Sets 36
   2.4 Counting 38
   2.5 Permutations and Combinations 40

3 Numbers and Their Representations 47
   3.1 The Set of Natural Numbers 47
   3.2 Extensions of Number Systems 51
   3.3 Numeral Systems 58

4 Relations 63
   4.1 Fundamentals 64
   4.2 Equivalence Relations 69
   4.3 Order Relations 75

5 Mappings 83
   5.1 Partial and Total Mappings 83
   5.2 Composition and Diagrams 88
   5.3 Jectivities and Inversion 92
   5.4 Homomorphism Theorem 97
   5.5 Cardinality of Sets 100

6	**Graphs**		109
	6.1	Basics	110
	6.2	Directed Graphs	111
	6.3	Undirected Graphs	114
	6.4	Representations of Graphs	117
	6.5	Operations with Matrices	121
	6.6	Structure Preserving Mappings	127
	6.7	Euler and Hamilton Problems	131
7	**Groupoid, Semigroup, Group**		139
	7.1	Inner Compositions	139
	7.2	From Groupoid to Group	145
	7.3	Compatible Relations	153
	7.4	Groups	158
	7.5	Product Structures	164
8	**From Semirings to Fields**		167
	8.1	From Semirings to Hypercomplex Systems	167
	8.2	Rings and Fields	171
	8.3	Universal Algebras	178
9	**Act, Vector Space, Extension**		183
	9.1	Outer Operations	183
	9.2	Vector Spaces	189
	9.3	Generating System, Basis, Linear (In)dependence	192
	9.4	Linear Mappings and Matrices	195
	9.5	Field Extensions	199
	9.6	Coding	203
10	**Rings and Modules**		207
	10.1	Rings	207
	10.2	Modules Over Rings	213
11	**Matroids**		217
	11.1	From Vectors, Graphs, and Fields to Matroids	217
	11.2	Further Axiomatizations	225
	11.3	Operations on Matroids	228
	11.4	Structure-Preserving Mappings	230
12	**Categories**		233
	12.1	Basic Concepts	234
	12.2	Special Objects and Morphisms	236
	12.3	Products and Coproducts	239
	12.4	Free	243
Bibliography			249
Symbols			253
Index			257

Fundamentals

In this chapter, we specify the linguistic level at which we will operate in the following chapters. This includes the language itself, i.e., the *vocabulary*, its formal structure, i.e., the *syntax*, and its meaning, i.e., the *semantics*. Additionally, this includes the structure of the thinking required here or the representation of the thought models, which is relatively simple using *formal logic*. *Proofs* serve as the basis for mathematical thought products. Therefore, in this chapter, we present the most important proof methods. We use terms like *set* or *element of a set* in a "naive" form, only in Chap. 2 will we provide their definitions.

On the Literature The book by Gersting [31] serves as a reference for this chapter and many of the following chapters. The German book *"Das ist o. B. d. A. trivial! (This is w.l.o.g. trivial!)"* by Beutelspacher [7] helps beginners navigate mathematical language, and gives advanced learners the "final touch". Similar literature in English includes "How to think like a mathematician" by Houston [36]. For further study of mathematical logic we recommend "Logic for Computer Scientists" by Schöning [67], originally in German, available as e-book in English. Introductions to logic can also be found in many books on set theory. Böhme [12] provides a brief description of propositional logic and discusses multi-valued, intuitionistic and modal logic, in addition to an introduction to fuzzy logic.

1.1 Languages

Natural language, subset, (finite) alphabet, letter, character, (empty) word, concatenation, (formal) language

Natural languages (e.g., German, Mandarin, English) are used for communication. They are the basis of our ideas and our thinking. And they change with the people who use them. We can express feelings and opinions through language, but also convey facts and rules. Often sentences in a natural language are ambiguous, sometimes their meaning even depends on the personalities who formulate them.

This ambiguity has its charm in everyday life, but is often perceived as an obstacle in scientific work. To better describe and analyze facts and conclusions, people began to develop *formal languages* around the end of the nineteenth century. Today, formal languages are used, among other things, to communicate with machines and computers (as *programming languages* such as *assembly*-languages, Python, C++ or Lisp) and to analyze mathematical structures (as in *logic*).

Although we analyze mathematical facts using logic and represent them in the language of *formal logic*, we communicate about them in our everyday language. In particular, statements in formal logic tend to become almost unreadable. We choose a compromise by explaining formulas and relating them to each other through text in everyday language. These explanatory texts do not come from a formal language, but still (if they are good) follow certain conventions. This shows that the authors know how this text would be translated into formulas. Here is an example that anticipates the material of this book. The formula

$$A \subseteq B \; :\Leftrightarrow \; [(\forall x) : \; x \in A \; \Rightarrow \; x \in B]$$

reads, in everyday language: "A set A is a **subset** of a set B if for all x one has: if x is an element of A, then x is an element of B ". The fact that the term *subset* is being defined here is expressed in the formula by the colon together with the double arrow and in the text by the bold italic font. A less good, but still correct, everyday language translation of the formula reads: "A set is a subset of another if all its elements are in the other". It is obviously more difficult to reconstruct the defining formula from the second formulation, which in turn is harder for beginners to understand than the first of the everyday language formulations.

Formal Languages

An **alphabet** is a non-empty set A. We call the elements of an alphabet **letters** or **symbols**. With A^* we denote the set of all **words** that can be formed by concatenating (= **concatenation of**) a finite number of (not necessarily different) symbols from A. It will prove convenient to allow the **empty word** ε, which consists of 0 symbols. An alphabet is called **finite** if it contains only a finite number of symbols. Any subset $L \subseteq A^*$ is called a **language** or more precisely a **formal language over** A.

Obviously, with this definition, a contentless aspect of a written language is captured. The definition abstracts from everything that makes a natural written language interesting, beautiful, and above all meaningful and purposeful. What remains is the level at which we communicate with machines. If we imagine a washing machine, it is not difficult to describe the formal language that the machine understands: these are the sequences of symbols, usually represented by switch positions, to which the machine responds. Selecting a wash cycle without turning on the machine would be a word that does not belong to the machine's language.

1.2 Syntax and Semantics

So we are dealing with the definition of a formal language at the level of a machine, which recognizes whether a word belongs to its language, without however being able to describe the meaningful reaction of the machine.

Examples 1.1

(1) The set of all finite 0, 1 sequences is a formal language. For example, the words 01001, 010101 and 1 are included, but not the sequence consisting of infinitely many zeros 00000....
(2) Classical Sanskrit has a description as a formal language, consisting of about 2000 basic words and a system of about 4000 derivation rules. This description dates back to Panini[1] around 400 BC.
(3) Programming languages are formal languages, consisting of basic words and derivation rules. They are less complex than Sanskrit.
(4) The set of all rational numbers in decimal notation is a formal language over the alphabet, which consists of the digits 0 to 9, the dot, the negative sign and a symbol for the period. The word "273.5" is in this language, but not "6.54.2". Real numbers that are not rational cannot be represented in this way as words of finite length.
(5) Assume we want to describe the written English language as a formal language. Then the *alphabet* would have to contain, in addition to the letters (a, A, b, B, ...), all punctuation marks and a space. Words would not only be sequences of symbols from the alphabet like *weather* or *beautiful*, but also *The weather is beautiful.*—In the terminology of formal languages, the entire text of a novel—provided it does not contain a single typographical error—would be one *word* of this language.

Exercise 1.2 (Braille) Search in the literature for the Braille font.[2]—Describe this as a formal language. Over which alphabet?

1.2 Syntax and Semantics

Syntax, decidable, recursive, enumerable, procedure, semantics, language of mathematics, logic, propositional logic, predicate logic

The *syntax* (study of sentence structure) of a language L determines which sequences of characters are in L and which are not. This can basically be done in two ways:

---

[1] Indian Sanskrit grammarian, around 400 BC.
[2] Louis Braille, 1809–1852, Paris, became blind at the age of 3, developed the dot writing system, named after him, in 1825.

- by a procedure that decides after a finite number of steps, whether a given word $w \in A^*$ is in L (symbolically: $w \in L$), the language is called ***decidable*** or
- by a procedure that generates exactly the words from L. That is, for each word $w \in A^*$, the procedure provides after a finite number of steps[3] either the statement "w is in L" or the statement "w is not in L" (symbolically: $w \notin L$ or $w \in A^* \setminus L$, spoken: A^* without L). Such a procedure is called ***recursive***. A language is therefore ***enumerable*** (recursively enumerable) if there is a procedure that generates all its elements and no others. Such a procedure does not have to terminate.

Such procedures apparently do not exist for the English language. Note that "enumerable" and "countable" (compare Sect. 5.5) are different terms.

Lemma 1.3 (Enumerability) *If A is a finite set, then A^* is enumerable.*

Proof Establish an order for the elements of A. Start with ε, then count the (finitely many) words of length 1 (= elements of A) in their order, then the (also finitely many) words of length 2 in alphabetical (= lexicographic) order, etc. □

Theorem 1.4 (Decidability) *A language L over a finite alphabet A is decidable if and only if both L and $A^* \setminus L$ (the set of all words over A that are not in L) are enumerable.*

Proof Let L be decidable. Enumerate A^* (which is possible due to Lemma 1.3) and decide for each word whether it belongs to L or not. In this way, both L and $A^* \setminus L$ can be enumerated.

If L and $A^* \setminus L$ are enumerable and it is to be decided whether $x \in A^*$ belongs to L or not, then enumerate L and $A^* \setminus L$ simultaneously until x appears in one of the two lists. This decides in which of the two sets x lies. □

From this theorem it follows directly that every decidable language is already enumerable. We therefore have to ask ourselves whether there are enumerable languages that are not decidable, since otherwise we would use two names for the same concept. For a proof that such languages exist, we refer to Gersting [31], Theorem 12.30.

To assure that the words of a language are not just meaningless strings of symbols, a *semantics* (theory of word meaning) is assigned to a formal language. Semantics therefore does not deal with the syntax of the relationships of the signs and words among themselves, but with the relationships between the words and what they denote.

---

[3] The condition that the procedure ends after a finite number of steps is fundamental here and leads to the definition of algorithms, see also Sect. 3.3 before Example 3.16.

1.2 Syntax and Semantics

We emphasize that the somewhat retrospective endowment of a formal language with meaning or content is an unnatural process. Actually, a language always first has a sense, a meaning. This applies to the written and even more to the spoken language, which historically comes first anyway. It also applies to body language, but also to the stanzas of the blackbirds, the songs of the whales, etc. It is noteworthy that written Chinese—in contrast to the European languages, for example—can be seen as a direct representation of the content: There is a multitude of Chinese languages—often misleadingly called dialects—whose speakers do not understand each other, but can express the same content through the same characters, thus being able to communicate in writing. The situation is the same with the numerical symbols $0, 1, 2, \ldots, 9$. Here too, the meaning is directly associated with the symbol, regardless of the language in which it is spoken. An interesting mix of formal language and direct content transmission is represented by *sign languages*, which could therefore serve in part for communication across national language boundaries. They have actually been developed and used for this purpose by Native American peoples. However, there are hundreds of different sign languages used by deaf people in various regions, and mutual understanding is just as difficult as with different spoken languages.

In recent decades, considerable progress has been made in the search for formal definitions of the semantics of programming languages. Research is being conducted in this field to effectively analyze what computer programs can do. A bon mot of unknown origin says: "Every program performs some task correctly – the question is just which one ..."

Language of Mathematics

People have always communicated about mathematics using everyday language. Formalizations were not created for their own sake, but to eliminate misunderstandings that arose in this communication.—When we say: "The number x is a prime number or it is the sum of two square numbers", it is essential whether this *or* includes the possibility that both statements are true, or whether the validity of one automatically excludes the other. Also in everyday life, it is advantageous to know what to expect from the statement "You don't paint my fence or I'll buy you a bicycle", but usually we clarify this by asking questions, or the meaning of the statements composed by *or* gives us an idea of how the sentence is probably meant. The use of the word *or* (and a few others) in mathematics will be clarified in the following and then no longer requires any questions or assumptions.

In order to be able to present complex facts clearly, it has proven useful to replace some words of everyday language with symbols. The most common of these symbols will be introduced here and in Chap. 2 on sets.

The exact definition of *or* or the use of the symbol " $\subseteq$ " are syntactic problems. We will mainly deal with these. The question of what a set is, what a statement is and under what circumstances we consider it to be true or false, are semantic problems. We will hardly touch on these.

Logic

Logic is the science of the laws and forms of thought. It deals with the question of whether certain conclusions are "logical" or not. Aristotle[4] is considered the founder of logic. He was the first to work with logical propositional formulas and consciously used variables.

This concept gives rise to all sorts of speculation. We quote some statements for clarification (without further source reference). A variable is a designation for an object that can take on different values from a set of elements. In other words, the term variable refers to a name for a blank place in a logical or mathematical expression. Equivalently, the term "placeholder" is also used. In mathematics, the terms "indeterminate" or "unknown" are also used in equations and polynomials. Both do not explain anything. We will return to this issue at a later point.

Mathematical logic began with Leibniz, but fell into oblivion and was not consistently developed until about the third of the nineteenth century. Boole, Peano, de Morgan, Frege, and in the twentieth century Russell and Whitehead set milestones in the development of mathematical logic.[5] In the meantime, logic has experienced a resurgence due to its diverse applications in computer science.

Propositional logic is the study of the connections between statements. Sections 1.3–1.5 are dedicated to it. In *predicate logic* (Sect. 1.7), the internal structure of the statements is described in addition: There, using *predicates* and *quantifiers*, it can be specified to which individuals a statement refers, for which it is true and for which it is not.

Usually, mathematical theories are not formally developed, but are subsequently analyzed and possibly clarified using logic. It then turns out which basic assumptions *(axioms)* are needed to derive all known facts *(theorems)* from them using given rules *(proofs)* (see Sects. 1.4 and 1.5).

Exercise 1.5 (Enumerations) Given the alphabet $A = \{0, 1\}$.

(a) Write down all four-letter words over the alphabet A.
(b) Describe at least two systematic methods, with which (a) can be solved.
(c) Is the set of all words over A enumerable?—If yes, give an enumeration, if no, justify this statement.

Exercise 1.6 (Languages) Consider the following procedure for generating a language L_1 over $A = \{0, 1\}$ and describe which words are in L_1.

---

[4] Ancient Greek philosopher and polymath, 384 to 322 BC.
[5] Gottfried Wilhelm Leibniz, German mathematician, 1646–1716; George Boole, English mathematician, 1815–1864; Guiseppe Peano, Italian mathematician in Turin, 1858–1932; Augustus de Morgan, English mathematician, 1806–1871; Friedrich Ludwig Gottlob Frege, German mathematician, 1848–1925; Bertrand Russell, English mathematician and social critic, 1872–1969; Alfred North Whitehead, English mathematician, 1861–1947.

(a) Start with the symbol ⋆.
(b) Replace ⋆ with 0⋆ (short: ⋆ → 0⋆).
(c) Replace ⋆ with 1 (short: ⋆ → 1).
(d) Apply (b) or (c) until a word from A^* is formed.

1.3 Statements and Truth Values

Truth value, statement, valency, statement variable, true, false, tertium non datur, undecidable, Fuzzy Logic, (unary, binary) connection, composition, components, propositional formula, negation, conjunction, disjunction, implication, equivalence, truth assignment, truth table, logically equivalent, tautology, contradiction, logically dual truth table/ propositional formula

Let there be a given language and a set with at least two elements, whose elements are called **truth values**. At most one truth value is assigned to each string of the language. Strings to which a truth value has been assigned are called **statements**. The number of truth values is called the **valency** of a logic.

As usual, we use a two-valued logic and denote the truth values with 1 **(true)** and 0 **(false)**. We represent statements by letters A, B, C, …, which we call **statement variables** and write $\omega(A) = 1$ or $\omega(A) = 0$, depending on whether the statement A is true or false.

An important basis of mathematical reasoning is the principle of the excluded third, Latin: **tertium non datur**, which states: Every statement is either true or false. In three-valued logics, **undecidable** or *indeterminate* is common as a third truth value alongside *true* and *false*. In *Fuzzy Logic* all real numbers between 0 and 1 are allowed as truth values, which allows a correspondingly fine gradation between *true* and *false*.

If we ask ourselves which sentences of the English language are statements and which are not, it makes sense to label "Berlin is on the river Elbe" as false and "Spring is followed by summer" as a true statement. On the other hand, we should not assign a truth value to "How are you?", "She is a good mathematician" and "It will rain on July 30, 2099" in a two-valued logic. In a three-valued logic, however, we could assign the truth value *indeterminate* to the last example and thus make the sentence "It will rain on July 30, 2099" a statement. A precise definition of the term "good mathematician" would also elevate the penultimate example from an expression of opinion of indeterminate truth value to a statement, if a specific person is meant.

We will examine what truth values result for so-called composite statements if we know the truth values of the components, instead of dealing with the (actually more interesting) questions that arise from the content of the sentences considered.

Composition of Statements

A *composition* of statements is a rule that assigns to a certain number of statements, the *components* of the composition, a new one. Its truth value is determined by the truth values of the components. We have *unary* compositions, which assign a new statement to a given one, like the *negation* (see below). And *binary* compositions assign a new statement to two given ones, like the *conjunction, disjunction, implication* and *equivalence* defined below.

We call statement variables and composed statement variables ***propositional formulas***. So, "she is 20 years old" is a propositional formula that only becomes a statement when the variable *she* is assigned.

Here, "assignment of variables" means: the variable is replaced by a (fixed) element from a given set.

For propositional formulas, which can be composed of statement variables, we write calligraphic capital letters like $\mathcal{A}, \mathcal{B}, \mathcal{C}, \ldots$.

Definitions of the Most Important Compositions

- The *negation* $\neg A$ ("not A") of a statement A is true when A is false, and false when A is true.
- The *conjunction* $A \wedge B$ ("A and B") of two statements A and B is true when A and B are both true. Otherwise, it is false.
- The *disjunction* $A \vee B$ ("A or B") of two statements A and B is false when A and B are both false. Otherwise, it is true.
- The *implication* $A \Rightarrow B$ of two statements A and B is false when A is true and B is false. Otherwise, it is true.
  There are many ways to formulate $A \Rightarrow B$ in words: "If A, then B" or "B follows from A" or "A implies B" or "A is sufficient for B" or "B is necessary for A". A is also called "premise", B "conclusion".
- The *equivalence* $A \Leftrightarrow B$ of two statements A and B is true when A and B have the same truth value. Otherwise, it is false.
  Ways of saying $A \Leftrightarrow B$: "A if and only if B", abbreviated "A iff B", or "A exactly if B", or "A is necessary and sufficient for B".

Truth Assignment

A *truth assignment* of a propositional formula is an assignment of truth values to its components, with the same truth value being assigned to the same statement variables.

A *truth table* lists the truth values of propositional formulas for every possible truth assignment of their variables in tabular form.

1.3 Statements and Truth Values

Example 1.7 (Truth Tables) In Table 1.1, the truth assignment of the statement variables is to the left of the double bar, that of the (composite) propositional formulas to the right:

Table 1.1 Truth tables

A	B	$\neg A$	$A \wedge B$	$A \vee B$	$A \Rightarrow B$	$A \Leftrightarrow B$
0	0	1	0	0	1	1
0	1	1	0	1	1	0
1	0	0	0	1	0	0
1	1	0	1	1	1	1

For the analysis of more complex logical expressions, the following (algorithmic) form of the truth table in Table 1.2 is useful:

Table 1.2 Algorithmic truth tables

(A	$\Leftrightarrow$	B)	($\neg$	(A	$\vee$	B)	$\vee$	$\neg($	$\neg A$	$\vee$	$\neg B$))
0	1	0	1	0	0	0	1	0	1	1	1
0	0	1	0	0	1	1	0	0	1	1	0
1	0	0	0	1	1	0	0	0	0	1	1
1	1	1	0	1	1	1	1	1	0	0	0
1	2	1	3	1	2	1	5	4	2	3	2

The numbers in the last row indicate a possible order in which the columns can be filled out. Columns with the same numbers can be filled out simultaneously. The columns with the number 1 must contain all possible combinations of the truth values of the statement variables. The column with the highest number (2 on the left, 5 on the right) gives the truth value of the corresponding propositional formula for each possible truth assignment of the statement variables.

In the above example, the truth values for the two considered propositional formulas coincide. This leads to the following definition.

Two propositional formulas $\mathcal{A}$ and $\mathcal{B}$ are called ***logically equivalent***, if they have the same truth value for every possible truth assignment, i.e. if their columns in the truth table are the same, possibly after reordering rows. We then write $\mathcal{A} = \mathcal{B}$.

A propositional formula $\mathcal{T}$, which is true for every truth assignment, is called ***tautology***. Since all tautologies are logically equivalent, we denote them with the same symbol **1** and write $\mathcal{T} = \mathbf{1}$, if $\mathcal{T}$ is a tautology.

A propositional formula $\mathcal{W}$, which is false for every truth assignment, is called ***contradiction***. We write $\mathcal{W} = \mathbf{0}$, if $\mathcal{W}$ is a contradiction.

We can now formulate the principle of the excluded third in various ways: $A \vee \neg A$ is a tautology, i.e. $A \vee \neg A = \mathbf{1}$, $A \wedge \neg A$ is a contradiction, i.e. $A \wedge \neg A = \mathbf{0}$. For statement variables A and B, $A \Rightarrow B$ is neither a tautology nor a contradiction, as the truth table in Example 1.7 shows: the column under $A \Rightarrow B$ contains both the truth value 0 and the truth value 1.

By considering the truth tables, we obtain

Lemma 1.8 (Equivalent Propositional Formulas) *The propositional formulas $\mathcal{A}$ and $\mathcal{B}$ are logically equivalent if and only if $\mathcal{A} \Leftrightarrow \mathcal{B}$ is a tautology.*

Logical Duality

Let T be a truth table. The ***logically dual truth table*** T^d is obtained from T by replacing each 0 in T with a 1 and each 1 in T with a 0. Obviously, $(T^d)^d = T$.

Let $\mathcal{A}$ be a propositional formula with the truth table T. The ***dual propositional formula*** $\mathcal{A}^d$ is defined by the logically dual truth table T^d.

Example 1.9 (Dualizing)

$$\left(\begin{array}{c|c} A & \neg A \\ \hline 0 & 1 \\ 1 & 0 \end{array}\right)^d = \begin{array}{c|c} A & (\neg A)^d \\ \hline 1 & 0 \\ 0 & 1 \end{array}$$

$$\left(\begin{array}{c|c|c} A & \neg A & A \wedge \neg A \\ \hline 0 & 1 & 0 \\ 1 & 0 & 0 \end{array}\right)^d = \begin{array}{c|c|c} A & \neg A & (A \wedge \neg A)^d \\ \hline 1 & 0 & 1 \\ 0 & 1 & 1 \end{array}$$

$$\left(\begin{array}{c|c|c} A & B & A \wedge B \\ \hline 0 & 0 & 0 \\ 0 & 1 & 0 \\ 1 & 0 & 0 \\ 1 & 1 & 1 \end{array}\right)^d = \begin{array}{c|c|c} A & B & (A \wedge B)^d \\ \hline 1 & 1 & 1 \\ 1 & 0 & 1 \\ 0 & 1 & 1 \\ 0 & 0 & 0 \end{array}$$

Using the truth tables in Example 1.9 and replacing the right sides with logically equivalent propositional formulas, we have proven:

Theorem 1.10 (Dual Propositional Formulas) *For all propositional formulas A and B we get the following:*

1. $(\neg A)^d = \neg A$.
2. $(A \wedge \neg A)^d = A \vee \neg A$ and $(A \vee \neg A)^d = A \wedge \neg A$.
3. $(A \wedge B)^d = A \vee B$ and $(A \vee B)^d = A \wedge B$.

Theorem 1.11 (Logical Dualizing) *If $\mathcal{A}$ and $\mathcal{B}$ are propositional formulas that only contain $\wedge$, $\vee$ and $\neg$ and possibly **0** or **1**, and if $\mathcal{B}$ is derived from $\mathcal{A}$ by replacing $\wedge$ with $\vee$ and $\vee$ with $\wedge$, as well as **1** with **0** and **0** with **1**, then $\mathcal{A}$ and $\mathcal{B}$ are logically dual to each other.*

Proof The claim follows from the logical duality of $\wedge$ and $\vee$, **0** and **1** as well as the fact that negation is dual to itself. □

Theorem 1.12 (Logical Duality and Equivalence) *If $\mathcal{A}$ and $\mathcal{B}$ are equivalent propositional formulas, then the two propositional formulas dual to $\mathcal{A}$ and $\mathcal{B}$ are also equivalent. If $\mathcal{A}$ is a tautology, then the statement dual to $\mathcal{A}$ is a contradiction and vice versa. Dualizing twice does not change a statement.*

Proof The proof follows from the definitions of duality and equivalence. □

Exercise 1.13 (Duality) Prove the following statements
(a) $(A \vee B)^d = A \wedge B$; (b) $(A \vee \neg A)^d = A \wedge \neg A$.

Exercise 1.14 (Formalization, Equivalence) Describe the following statements using compositions of statement variables.—Assuming that the statement in (a) is true, what then applies to the others?

(a) If rents are high, then many apartments are vacant or landlords make high profits.
(b) Landlords make high profits or rents are not high or many apartments are vacant.
(c) If only a few apartments are vacant, then landlords make high profits or rents are not high.
(d) If landlords do not make high profits, then many apartments are vacant or rents are not high.
(e) If many apartments are vacant, then rents are high and landlords make high profits.
(f) Rents are high and landlords make high profits and many apartments are vacant.

1.4 Logical Identities

> Logical identities, associativity, neutral elements, complements, commutativity, distributivity, equivalence, negation as involution, implication, de Morgan's rules, rule, law, axiom, parallel axiom, affine/projective plane, elliptic/hyperbolic geometry

We now list some basic pairs of equivalent propositional formulas. Their equivalence of course, requires a proof. This can conveniently be conducted here using truth tables. Such propositional formulas, whose validity has been proven for all statements A, B and C, we are called ***logical identities*** or *laws* or *rules*.

We use them for the transformation of propositional formulas, i.e., for the derivation of further identities. The five pairs of logically dual identities in Table 1.3 serve, as we will see later in Sect. 8.3, as axioms for a *Boolean Algebra*.

Associative laws allow notations without brackets, i.e., $A \vee B \vee C$ or $A \wedge B \wedge C$ for three statements A, B, and C. By repeatedly applying the associative laws, it can be shown that for finitely many statements, which are only composed with $\vee$

Table 1.3 Logical identities

Associativities		Neutral elements	Complements
$(A \vee B) \vee C = A \vee (B \vee C)$		$A \wedge 1 = A$	$A \vee \neg A = 1$
$(A \wedge B) \wedge C = A \wedge (B \wedge C)$		$A \vee 0 = A$	$A \wedge \neg A = 0$
Commutativities		Distributivities	
$A \vee B = B \vee A$		$A \vee (B \wedge C) = (A \vee B) \wedge (A \vee C)$	
$A \wedge B = B \wedge A$		$A \wedge (B \vee C) = (A \wedge B) \vee (A \wedge C)$	

Table 1.4 Further logical identities

Equivalence	Negation as involution
$A \Leftrightarrow B = (A \Rightarrow B) \wedge (B \Rightarrow A)$	$A = \neg \neg A$
Implication	Contraposition
$(A \Rightarrow B) = \neg A \vee B$	$(A \Rightarrow B) = \neg B \Rightarrow \neg A$
de Morgan's rules	
$\neg(A \wedge B) = \neg A \vee \neg B$	
$\neg(A \vee B) = \neg A \wedge \neg B$	

or only with $\wedge$, any permissible bracketing delivers the same result, that brackets in these cases are therefore superfluous. (The same applies to all other associative operations, e.g., the addition or the multiplication of real numbers.)

In Table 1.4 we provide some additional useful identities that facilitate the handling of negation, implication, and equivalence and can also be proven using truth tables (An operation that, when performed twice, is the identical operation is called ***involution***):

The distinction between *rules* and *laws*—in mathematics we speak of de Morgan's *rules,* but of the commutative *law,* the associative *law,* and the distributive *law*—has no logical but merely historical reasons: *laws* usually come from a time when people believed they were dealing with an unchangeable fact. When a propositional formula is called a *rule,* it becomes clear that under certain circumstances it may not be valid. Later, when people realized that they could define meaningful structures in which such laws did not apply, they began to use the word *axiom.* This is to clarify that it must be agreed upon which axioms are assumed to be valid. In Euclidean geometry, for example, the *parallel axiom* is assumed for *affine planes*:

> For every straight line g and for every point P not on this line, there is one and only one line parallel to g that passes through point P, i.e., a line through P that has no common point with g.

A very important role is also played by the *projective planes*, in which any two different lines intersect (i.e., the parallel axiom does not apply there). This is like on a sphere, our actual living space: the shortest connections between two points are segments of straight lines, and these are the great circles on the sphere. These all intersect! An exact formulation of this situation occurs in the *elliptic* or also *projective geometry*. A finite model is the so-called Fano matroid, see Sect. 11.1.This geometry consists of 7 points and 7 lines, all of which intersect pairwise. Our visual space, i.e., a finite circle disc, offers exactly the "opposite": for every straight line, through a point outside there are infinitely many lines that do not intersect the original line. This is a *hyperbolic geometry*.

Exercise 1.15 (Logical Identities) Prove the logical laws listed in this chapter.

Identify "Eat your spinach or you won't get any pudding" as an implication, what is A, what is B?

1.5 Axiomatic Structure

Axiomatic structure, (complete, contradiction-free) axiom system, axiom, derivation rule, modus ponens

Here we hint at what an **axiomatic structure** of propositional logic might look like. An **axiom system** consists of a series of selected tautologies, which are called **axioms**, and **derivation rules**, with which new tautologies can be developed from given ones.

In all areas of mathematics, axiom systems are used and often explicitly stated. A meaningful interaction with these is only possible if an axiom system of formal logic is presupposed. The choice of such an axiom system—as well as an axiom system of set theory—is usually made implicitly. However, this negligence will not cause us any problems. All logical identities that we have shown so far using truth tables can be derived from the three axioms (left) and the derivation rule (right), also known as **modus ponens**, shown in Table 1.5:

Table 1.5 Axioms and derivation rule of classical propositional logic

$A \Rightarrow (B \Rightarrow A)$	If S and $S \Rightarrow T$ are tautologies, then T is also a tautology, in short:
$[A \Rightarrow (B \Rightarrow C)] \Rightarrow [(A \Rightarrow B) \Rightarrow (A \Rightarrow C)]$	
$(\neg B \Rightarrow \neg A) \Rightarrow (A \Rightarrow B)$	$\dfrac{S \quad S \Rightarrow T}{T}$

The modus ponens corresponds to the tautology $(A \wedge (A \Rightarrow B)) \Rightarrow B$. This corresponds to our everyday understanding of an implication, since we want to conclude the validity of B from the implication $A \Rightarrow B$, but we can only do this logically correctly if we also presuppose the validity of A.

It can be shown that this axiom system is **consistent**, i.e., that no contradictions can be derived from it, and that it is **complete**, i.e., that all tautologies of propositional logic can be derived from it, provided they are only expressed with

¬ and ⇒. The other known compositions must then be defined using these two, e.g., $A \lor B := \neg A \Rightarrow B$.

1.6 Normal Forms

> Disjunctive/conjunctive normal form, disjunction term, existence of disjunctive normal forms, logical equivalence, disjunctive normal form and contradiction, conjunction term, clause, conjunctive normal form and tautology

To analyze propositional formulas, it is sometimes useful to bring them into a *normal form* using the identities from Sect. 1.4. The most important normal forms for propositional formulas are the disjunctive and the conjunctive normal form. The former is particularly suitable for identifying contradictions, the latter for identifying tautologies.

Disjunctive Normal Forms and Contradictions

A propositional formula $\mathcal{D}$ is called *disjunctive normal form* if it is of the form $\mathcal{D} = \mathcal{A}_1 \lor \mathcal{A}_2 \lor \ldots \lor \mathcal{A}_k$ and each propositional formula $\mathcal{A}_i$ ($1 \leq i \leq k$)—hereafter called *disjunctive term* or—*clause* is a conjunction of statement variables or negated statement variables.

For example, $(\neg A \land B) \lor (C \land A \land \neg D) \lor E$ is a disjunctive normal form, $\neg(A \land B) \lor (C \land D)$ is not, since one of the disjunctive terms is negated.

Disjunctive normal forms can be used for *implementing logical functions through circuits* with "and", "or" and "non" elements.

From a disjunctive normal form, we obtain others, logically equivalent forms, by adding disjunctive terms that are contradictions, or by swapping disjunctive terms among each other.

Theorem 1.16 (Existence of Disjunctive Normal Forms) *For every propositional formula $\mathcal{A}$ there exists a disjunctive normal form $\mathcal{D}_\mathcal{A}$ with $\mathcal{A} = \mathcal{D}_\mathcal{A}$.*

Proof Let $\mathcal{A}$ be a composition of the statement variables $A_1, A_2, \ldots, A_n$. Construct the truth table for $\mathcal{A}$.

If $\omega(\mathcal{A}) = 0$ in every row of the table, then $\mathcal{A}$ is a contradiction. Set, for example, $\mathcal{D}_\mathcal{A} := A_1 \land \neg A_1$. (This representation is not unique.)

Otherwise, arrange the 2^n rows of the truth table so that first the k rows are listed, in which $\omega(\mathcal{A}) = 1$, so $k \geq 1$. This looks like the table in Example 1.17.

We now denote the truth value of A_j in the i-th row of the truth table with $w_{i,j}$. In Example 1.17 we then have $w_{1,1} = 0$, $w_{2,1} = 1$, $w_{3,1} = 1$. We set for each i with $1 \leq i \leq k$:

1.6 Normal Forms

$$\mathcal{C}_i := \mathcal{C}_1 \wedge \cdots \wedge \mathcal{C}_n \quad \text{with} \quad \mathcal{C}_j := \begin{cases} A_j, & \text{if } w_{i,j} = 1, \\ \neg A_j, & \text{if } w_{i,j} = 0 \end{cases} \text{ for } 1 \leq j \leq n.$$

Then $\mathcal{C}_j$ has exactly in the j-th row of the truth table the truth value 1 and in all other rows (also in the rows $k+1$ to 2^n, if $k < 2^n$) the truth value 0.

In Example 1.17 we therefore get $\mathcal{C}_1 = \neg A_1 \wedge A_2 \wedge \neg A_3$

The following table represents the above construction in a general form.

Row	A_1	A_2	$\cdots$	A_n	$\mathcal{A}$	$\mathcal{C}_1$	$\cdots$	$\mathcal{C}_k$	$\mathcal{D}_\mathcal{A}$
1	$w_{1,1}$	$w_{1,2}$	$\cdots$	$w_{1,n}$	1	1	0	0	1
$\vdots$	$\vdots$	$\vdots$		$\vdots$	$\vdots$		$\ddots$		$\vdots$
k	$w_{k,1}$	$w_{k,2}$	$\cdots$	$w_{k,n}$	1	0	0	1	1

From this it follows that the disjunction $\mathcal{D}_\mathcal{A} := \mathcal{C}_1 \vee \cdots \vee \mathcal{C}_k$ of these k propositional forms in rows 1 to k has the value 1 and in all other rows the value 0, i.e., it is logically equivalent to $\mathcal{A}$. □

Example 1.17 (Truth Table, Disjunctive Normal Form) The adjacent table shows directly the structure of a realizing logical circuit. It contains exactly those rows of the truth table of the propositional form $\mathcal{A}$, in which $\omega(\mathcal{A}) = 1$. The following disjunctive normal form $\mathcal{D}_\mathcal{A}$ then also has the value 1 exactly in these three rows of its truth table:

A_1	A_2	A_3	$\mathcal{A}$
0	1	0	1
1	1	0	1
1	1	1	1

$$\mathcal{D}_\mathcal{A} = (\neg A_1 \wedge A_2 \wedge \neg A_3) \vee (A_1 \wedge A_2 \wedge \neg A_3) \vee (A_1 \wedge A_2 \wedge A_3).$$

Theorem 1.18 (Logical Equivalence) *Two propositional forms are equivalent if and only if they are both contradictions, or if their disjunctive normal forms after omitting all clauses, that are contradictions, only differ by the order of clauses and variables within the clauses.*

Proof Let $\mathcal{A}$ and $\mathcal{B}$ be propositional forms with disjunctive normal forms $\mathcal{D}_\mathcal{A}$ and $\mathcal{D}_\mathcal{B}$. Then $\mathcal{A} = \mathcal{D}_\mathcal{A}$ and $\mathcal{B} = \mathcal{D}_\mathcal{B}$ and thus the assertion follows using the commutative laws for disjunction and conjunction. □

Theorem 1.19 (Disjunctive Normal Form and Contradiction) *A disjunctive normal form is a contradiction if and only if every clause is a contradiction, i.e., if in each clause a propositional variable appears both negated and not negated.*

The following propositional form $(\neg A \wedge B \wedge \neg B) \vee (A \wedge C \wedge \neg A)$ which apparently is a contradiction exemplifies the situation.

Conjunctive Normal Form and Tautology

The logically dual concept (in the sense of $(\ )^d$) of the *disjunctive normal form* is the *conjunctive normal form*. A propositional form $\mathcal{A}$ is called **conjunctive normal form**, if it is of the form $\mathcal{A} = \mathcal{A}_1 \wedge \mathcal{A}_2 \wedge \ldots \wedge \mathcal{A}_k$ and each conjunction term $\mathcal{A}_i, 1 \leq i \leq k$,—hereafter also called *clause*—is a disjunction of propositional variables or negated propositional variables. For example, the propositional form $(\neg A \vee C \vee D) \wedge B \wedge A \wedge (C \vee F)$ is a conjunctive normal form, while $((A \vee B) \wedge (C \vee D)) \vee E$ is not.

We can also prove the existence of a logically equivalent conjunctive normal form $\mathcal{K}_\mathcal{A}$ for any propositional form $\mathcal{A}$. We start with the disjunctive normal form $\mathcal{D}_{\mathcal{A}^d}$ according to Theorem 1.16 and dualize it, thus forming $(\mathcal{D}_{\mathcal{A}^d})^d$.

Just as we can consider whether a disjunctive normal form is a contradiction, it can be determined whether a conjunctive normal form is a tautology. The logically dual theorem to Theorem 1.19 states:

Theorem 1.20 (Conjunctive Normal Form and Tautology) *A conjunctive normal form is a tautology if and only if each clause is a tautology, i.e. if in each of the clauses a propositional variable appears both negated and not negated.*

The propositional form $(A \vee B \vee C \vee \neg A) \wedge (D \vee \neg D) \wedge (C \vee \neg C)$ is therefore a tautology.

Logical equivalence of two propositional forms can also be demonstrated by comparing their conjunctive normal forms, dual to the procedure described in Theorem 1.18.

Exercise 1.21 (Tautology and Contradiction) Prove without using truth tables

(a) The propositional form $((\neg A \wedge (\neg B \wedge C)) \vee (B \wedge C) \vee (A \wedge C)) \wedge \neg C$ is a contradiction;
(b) The propositional form $((A \vee B) \wedge \neg(\neg A \wedge (\neg B \vee \neg C))) \vee (\neg A \wedge \neg B) \vee (\neg A \wedge \neg C)$ is a tautology.

Exercise 1.22 (Duality) Find the propositional form that is logically dual to the implication $A \Rightarrow B$

(a) by dualizing the truth table,
(b) by transformation and exploiting the dualities shown so far.

Exercise 1.23 (Conjunctive Normal Form) Derive a procedure for determining a conjunctive normal form from Theorem 1.16, using logical duality.

Exercise 1.24 (Transformation of Propositional Forms) Show the logical equivalences without using truth tables.

(a) $(A \vee B) \wedge (A \vee \neg B) = A$.
(b) $(A \wedge B) \vee (A \wedge \neg B) = A$.
(c) $A \Rightarrow (B \Rightarrow C) = (A \wedge B) \Rightarrow C$.

1.7 Working with Quantifiers

> Individual/predicate variable, individual domain, interpretation, for all, universal quantifier, there exists, existential quantifier, quantifier, quantified/free variable, exactly one, one and only one, restriction of the individual domain, negation of propositional forms with quantifiers, convergent sequence

Now, in brief, those concepts are introduced that are important when using predicates and quantifiers. A systematic treatment we leave to any book on logic (see, for example, Schöning [67]).

Predicates

In predicate logic, we examine the internal structure, especially the subject predicate relationships, in the statements of traditional logic. In addition to statement variables such as A, B, C, we also use **individual variables**, e.g. x, y, z, as well as **predicate variables**, e.g. $P(x)$, $Q(x, y, z)$, $R(x, y)$. The individual variables come from a **individual domain** to be determined, e.g. the natural numbers $\mathbb{N}$. A predicate describes a (logical) relationship between individuals. So a predicate is a relation on the set of individuals (see Chap. 4). Predicates can be true for all individuals, for some of them, or for none.—This depends on which interpretation we choose: An **interpretation** of an expression of predicate logic determines an individual domain and assigns to each predicate a suitable property of individuals.

Example 1.25 (Interpretations of Propositional Formulas) In Table 1.6 we provide various interpretations for the following propositional formula

$$A(x, y) := (P(x) \wedge P(y)) \Rightarrow Q(x, y).$$

Quantifiers

If $P(x)$ is a true statement for all x from the individual domain, then we write

Table 1.6 Interpretations of propositional formulas

Interpretation of $A(x, y)$		Truth value
Individual domain	Predicates	
All lines in a plane	$P(z) := z$ is parallel to a given line g. $Q(x, y) := x$ is parallel to y or $x = y$.	$A(x, y)$ is a tautology.
All whole numbers	$P(z) := z$ is odd, $Q(x, y) := x + y$ is odd.	$A(x, y)$ is true for all $x, y \in \mathbb{Z}$, of which exactly one is even, otherwise false.
All shoes	$P(z) := z$ is a shoe, $Q(x, y) := x$ and y are a pair of shoes.	$A(x, y)$ is true exactly for those shoes x and y that form a pair.
All natural numbers	$P(z) := z > -1$, $Q(x, y) := x + y < 0$.	$A(x, y)$ is a contradiction.

$$(\forall x) : P(x), \quad \text{or also} \quad \bigwedge_{x} P(x),$$

a true statement of predicate logic. We say: "*For all* x, $P(x)$ holds." or also "*For each* x, $P(x)$ holds." The symbols $\forall$ and $\bigwedge$ both symbolize the so-called *universal quantifier*.

The statement $(\forall x) : P(x)$ becomes false as soon as there is even a single x in the domain of individuals for which $P(x)$ does not hold.

If $P(x)$ is true for at least one x in the domain of individuals, we write

$$(\exists x) : P(x), \quad \text{or also} \quad \bigvee_{x} P(x),$$

a true statement. We say: "*There exists* x, for which $P(x)$ holds." The symbols $\exists$ and $\bigvee$ symbolize the *existential quantifier*. The statement $(\exists x) : P(x)$ is only false when $P(x)$ does not hold for some x in the domain of individuals, i.e., when $P(x)$ does not hold for all x in the domain of individuals.

The common term for existential and universal quantifiers is *quantifier*.

The variable that stands next to or under the quantifier (x in the above example) is called *quantified* variable (as opposed to the *free* variable).

In the following, we will always use the symbols $\forall$ and $\exists$ for quantifiers as they are very suggestive and as the quantified variables can be written next to them, which is often advantageous in continuous texts.

Sometimes we want to express not only that there is an element x with the property $P(x)$, but also that this element is the only one that has this property, i.e., that $P(x)$ is true for x and that $P(y)$ is false for all elements y of the domain of individuals with $y \neq x$. We then say: "There exists *exactly one* x, for which ...holds.", or also "There is *one and only one* x, for which ...holds." To formalize this succinctly, the syntax of predicate logic is extended in mathematical everyday language: We use the notations $(\exists ! x) : P(x)$ or $(\exists^{1} x) : P(x)$.

1.7 Working with Quantifiers

Often in the applications of logic, a *restriction of the domain of individuals* is useful and common. This is then conveniently specified directly at the quantifier:

$$(\forall x \in X) : [P(x) \wedge (\exists y \in Y) : Q(x, y)],$$

instead of $(\forall x) : [x \in X \Rightarrow [P(x) \wedge (\exists y) : (y \in Y \wedge Q(x, y))]]$.

Negation of Propositional Formulas with Quantifiers

The negation of a propositional formula in which quantifiers are used, is obtained by replacing each $\forall$ with $\exists$ and vice versa and then negating the predicates. Statement forms can be conveniently negated with de Morgan's rules if they have been previously transformed so that they only contain the connections $\neg$, $\wedge$ and $\vee$.

$$\neg[(\forall x) : P(x)] \text{ becomes } (\exists x) : \neg P(x)$$
$$\neg[(\exists x) : P(x)] \text{ becomes } (\forall x) : \neg P(x)$$

If multiple quantifiers occur, we negate these step by step, e.g.:

$$\begin{aligned}
&\neg[(\forall x) \quad (\exists y) \quad (\forall z) : (A \wedge P(x, y, z))] \\
=\ & (\exists x) \neg[(\exists y) \quad (\forall z) : (A \wedge P(x, y, z))] \\
=\ & (\exists x) \quad (\forall y) \neg[(\forall z) : (A \wedge P(x, y, z))] \\
=\ & (\exists x) \quad (\forall y) \quad (\exists z) : \neg[A \wedge P(x, y, z)] \\
=\ & (\exists x) \quad (\forall y) \quad (\exists z) : \neg A \vee \neg P(x, y, z).
\end{aligned}$$

Example 1.26 (Convergent—Divergent) In analysis, we encounter the concept of the *convergent sequence*, a popular example to practice the negation of statements with quantifiers:

A sequence $a_0, a_1, \ldots, a_n, \ldots$ of numbers in $\mathbb{R}$ is called **convergent to** $a \in \mathbb{R}$, if

$$(\forall \varepsilon > 0)\ (\exists n_0 \in \mathbb{N})\ (\forall n \geq n_0) : \quad |a_n - a| < \varepsilon.$$

That is, for every (no matter how small) positive number ε, there is an index n_0 from which the sequence values a_n differ from the limit a by less than ε.

The negation of this statement says that the sequence $(a_i)_{i \in \mathbb{N}}$ does not converge to the value a, if:

$$(\exists \varepsilon > 0)\ (\forall n_0 \in \mathbb{N})\ (\exists n \geq n_0) : \quad |a_n - a| \geq \varepsilon.$$

Exercise 1.27 (Deformalize) Formulate the last statement about non-convergence in colloquial language. (It gains in liveliness, and it is easier to see how the formula symbols serve as abbreviations for certain phrases.)

Exercise 1.28 (Formalize and Negate) Formalize and negate:

(a) All humans become brothers (What about women?).
(b) All natural numbers less than 10 are prime numbers.
(c) There are no even prime numbers.

1.8 Building a Theory

Conjecture, proof, theorem, main theorem, lemma, corollary, assumption, hypothesis, claim, inference, search for theorems, Goldbach's conjecture, inductive/deductive approach, counterexample, refutation of a conjecture, (in-)direct proof, proof by contraposition/contradiction

Conjecture, Proof, Theorem

If it is not known whether a statement is true or false, it is called a *conjecture*. A *proof* of a statement is a sequence of logical inferences that shows that the statement is a tautology. A tautology is called a ***theorem*** in mathematics. To characterize theorems according to their meaning, terms such as ***main theorem***, ***theorem***, ***lemma*** (= auxiliary theorem, which is supposed to help in the proof of one or more theorems) and *corollary* (= conclusion from the previously proven theorem) are available.—For the correct use of these terms and many other expressions in the language of mathematics, see, for example, [7, 36].

Assumption, Assertion, Conclusion

In mathematics, implications $\mathcal{P} \Rightarrow \mathcal{Q}$ are mainly considered, which are usually still provided with quantifiers. Here, $\mathcal{P}$ is called the ***assumption*** or premise (also: *hypothesis*). If $\mathcal{P} \Rightarrow \mathcal{Q}$ has not yet been proven, then $\mathcal{Q}$ is an ***assertion***, after the proof, $\mathcal{Q}$ is called a ***conclusion*** or conclusion from the assumption $\mathcal{P}$.

Example 1.29 (Logical Structure) Let's look at the following theorem. The terms used in it will be defined later; they are not important for understanding the logical structure of the theorem, which is what we are concerned with here.

> **Theorem.** *Let G be a group and U a subset of G. If for any two elements x and y from U, xy^{-1} is also in U, then U is a subgroup of G.*

We define the following statement variables

1.8 Building a Theory

$\mathcal{A} :=$ G is a group
$\mathcal{B} :=$ U is a subset of G
$\mathcal{C} :=$ $(\forall x, y \in U) : xy^{-1} \in U$
$\mathcal{D} :=$ U is a subgroup of G

and can now specify the logical structure of the theorem through the propositional formula $\mathcal{A} \wedge \mathcal{B} \wedge \mathcal{C} \Rightarrow \mathcal{D}$. If we consider $\mathcal{A}$, $\mathcal{B}$, $\mathcal{C}$ and $\mathcal{D}$ as predicate variables, we can more accurately represent the logical structure of the theorem:

$$(\forall G)(\forall U) : \quad (\mathcal{A}(G) \wedge \mathcal{B}(U, G) \wedge \mathcal{C}(U)) \Rightarrow \mathcal{D}(U, G).$$

In this form, it can be proven. However, if we also formally logically formulate the statements $\mathcal{A}(G)$ and $\mathcal{D}(U, G)$ based on the definitions of the terms *group* and *subgroup*, the expression becomes cumbersome.

Searching for Theorems

Conjectures are often obtained from experiments or from mathematical intuition. As an example we cite the famous (and still unproven) **Goldbach Conjecture**:[6]

Every even natural number greater than 2 is the sum of two prime numbers.

For example, $4 = 2 + 2$, $6 = 3 + 3$ and $8 = 5 + 3$.

The process of inferring from the specific to the general is called an ***inductive reasoning***. It is one of the standard methods used in all sciences to form theories. Theories derived in this way are always only conjectures that can be maintained until they are refuted. Inductive reasoning can never serve as proof.

In contrast to inductive reasoning, ***deductive reasoning*** derives new theorems from existing theorems or axioms through correct logical conclusions, which are then also proven.

Proof Methods

If we want to prove or refute a conjecture, we can choose from a number of different proof methods, which are presented below based on their logical structure.

- A *counterexample* to the implication $\mathcal{P} \Rightarrow \mathcal{Q}$ is a fact, i.e., a special case, in which $\mathcal{P}$ is true and $\mathcal{Q}$ is false. By providing a counterexample, the conjecture $\mathcal{P} \Rightarrow \mathcal{Q}$ is refuted, thus showing that $\mathcal{P} \Rightarrow \mathcal{Q}$ is not a theorem.
- A *direct proof* of $\mathcal{P} \Rightarrow \mathcal{Q}$ starts with the assumption that "$\mathcal{P}$ is true" and leads to the result "$\mathcal{Q}$ is true" through a series of correct logical conclusions.

---

[6] Christian Goldbach, German mathematician, 1690–1764.

- A *proof by contraposition* of $\mathcal{P} \Rightarrow \mathcal{Q}$ is a direct proof of $\neg \mathcal{Q} \Rightarrow \neg \mathcal{P}$. It is applied when it seems easier to derive new statements from the statement $\neg \mathcal{Q}$, than from the statement $\mathcal{P}$. This proof method is based on the logical identity $(\mathcal{P} \Rightarrow \mathcal{Q}) = (\neg \mathcal{Q} \Rightarrow \neg \mathcal{P})$.
- An *indirect proof (proof by contradiction)* of $\mathcal{P} \Rightarrow \mathcal{Q}$ is a direct proof of $\mathcal{P} \wedge \neg \mathcal{Q} \Rightarrow \mathbf{0}$. This method is useful when neither the statement $\mathcal{P}$ nor the statement $\neg \mathcal{Q}$ alone are suitable for deriving the desired new statements. We start an indirect proof with the assumption that $\mathcal{P}$ is true, and the assumption that $\neg \mathcal{Q}$ is also true, and infer a contradiction from this. Since no contradiction can follow from the true statement $\mathcal{P} \wedge \neg \mathcal{Q}$, we then know by the law of excluded third that the statement $\mathcal{P} \wedge \neg \mathcal{Q}$, and that means $\neg \mathcal{Q}$, was false. Often such proofs therefore end with the words: "...so the assumption was false and $\mathcal{Q}$ is true." This proof method is based on the following formula

$$[\mathcal{P} \Rightarrow \mathcal{Q}] = [(\mathcal{P} \wedge \neg \mathcal{Q}) \Rightarrow \mathbf{0}]$$

which can be proven as follows:
$((P \wedge \neg Q) \Rightarrow \mathbf{0}) = \neg(P \wedge \neg Q) \vee \mathbf{0} = \neg(P \wedge \neg Q) = \neg P \vee Q = P \Rightarrow Q$.

Absolutely to be avoided—although formally correct—are "indirect proofs", which derive the "contradiction" $\mathcal{Q}$ from $\mathcal{P} \wedge \neg \mathcal{Q}$ without using $\neg \mathcal{Q}$ or infer the "contradiction" $\neg \mathcal{P}$ without using $\mathcal{P}$.—These are disguised direct proofs or proofs by contraposition and should not called indirect. We recommend: **No false indirect proofs!** An example for which an indirect proof appears indispensable is the statement that $\sqrt{2}$ is irrational, see Theorem 3.9.

1.9   Proof by Mathematical Induction

Proof by mathematical induction, initial value, induction anchor, induction step, arbitrary but fixed, strong induction, recursive definition, initial conditions, solution of a recursion

The proof method of mathematical induction is not to be seen on the same level with the previously mentioned methods. It is a special form of direct proof, which exploits properties of natural numbers given by the *Peano axioms* in Sect. 3.1.

Induction proofs are suitable for statements that can be formulated as properties of all natural numbers greater than or equal to an *initial value* n_0, (which is often equal to 0 or 1). This are statements of the form

$$(\forall n \in \mathbb{N}_{n_0}) : \quad \mathcal{P}(n) \quad \text{with} \quad \mathbb{N}_{n_0} := \{ n \in \mathbb{N} \mid n \geq n_0 \}.$$

The proof by mathematical induction consists of two parts, the *induction anchor* (IA) and the *induction step* (IS). Each of these two parts can be proven by one of the methods presented above, i.e. each part is proven deductively. In the case of (IS), the induction hypothesis (IH) is used.

1.9 Proof by Mathematical Induction

We proceed as follows: The statement $\mathcal{P}(k)$ is first proven for an initial value n_0. Subsequently, we have to show how we come from the validity of this statement for an arbitrary but fixed[7] value k, which is greater than or equal to the initial value n_0, to its validity for the successor $k+1$. This is the induction step (IS), i.e. the implication $\forall n \geq n_0 : \mathcal{P}(k) \Rightarrow \mathcal{P}(k+1)$ is proven.

The induction hypothesis (IH) is not a proof step itself, but it is often very helpful to write it down.

(IA) Induction anchor: Prove $\mathcal{P}(n_0)$ for the initial value n_0.
(IH) Induction hypothesis: For an arbitrary but fixed value $k \in \mathbb{N}$ with $k \geq n_0$, assume $\mathcal{P}(k)$.
(IS) Induction step: Prove $\mathcal{P}(k+1)$.

Example 1.30 (Gaussian Sum Formula) We illustrate the procedure using the sum formula

$$\sum_{i=1}^{n} i := 1 + 2 + 3 + \cdots + n = \frac{1}{2}n(n+1)$$

for the first n natural numbers ($n \geq 1$). (This formula was found by the student Gauss,[8] allegedly when a teacher wanted to keep him busy with the task of adding the numbers from 1 to 100.)

(IA) For the initial value $n_0 := 1$, the formula is valid, since $1 = \frac{1}{2}1(1+1)$.
(IH) For a natural number $k \geq 1$, we assumed $\mathcal{P}(k)$: $1 + 2 + 3 + \cdots + k = \frac{1}{2}k(k+1)$
(IS) We now prove $\mathcal{P}(k+1)$, the claim is therefore $1+2+3+\cdots+k+(k+1) = \frac{1}{2}(k+1)(k+2)$. We calculate

$$\begin{aligned}
1 + 2 + 3 + \cdots + k + (k+1) &= \\
&= \tfrac{1}{2}k(k+1) + (k+1) &&\text{(as } \mathcal{P}(k) \text{ holds)} \\
&= \tfrac{1}{2}(k(k+1) + 2(k+1)) &&\text{(factor out } \tfrac{1}{2}\text{)} \\
&= \tfrac{1}{2}((k+2)(k+1)) &&\text{(factor out (k+1))} \\
&= \tfrac{1}{2}(k+1)(k+2) &&\text{(commutative law of multiplication).}
\end{aligned}$$

---

[7] The choice of an *arbitrary but fixed* element $k \in M$ may sound a bit nebulous. With *fixed* it is meant that we fix a $k \in M$, which then cannot be changed anymore. *Arbitrary* means that we can only use knowledge about this element k that we also have from any other element from the set M. If, for example, $M = \mathbb{N}$, then for an *arbitrary* $k \in M$ it cannot be assumed that k is divisible by 2, but it can be assumed that $k+1$ is also in M, and a *fixed* k is in any case $\neq k+1$.

[8] Johann Carl Friedrich Gauss, German mathematician, 1777–1855.

With this, $\mathcal{P}(k+1)$ is proven and by the principle of mathematical induction, $\mathcal{P}(k)$ follows for every natural number k with $k \geq 1$.

A variation is the so-called *strong induction*. Here we take advantage of the fact that at the point where we want to prove $\mathcal{P}(k+1)$, all previous statements $\mathcal{P}(n_0)$ to $\mathcal{P}(k)$ can be assumed to have been proven. That is, we now use the induction hypothesis (IH): Let $k \in \mathbb{N}$ be an arbitrary but fixed value and for all $i \in \mathbb{N}$ with $n_0 \leq i \leq k$, we have $\mathcal{P}(i)$.

Attention The induction anchor (IA) must always be proven first without using the induction hypothesis (IH)! Otherwise, the following can happen. We prove $2n = 3n$ for every natural number greater than 0. The formulation of (IH) is $2k = 3k$. Since $k \neq 0$ we divide by k and get $2 = 3$, which is (IA) for $n_0 = 1$. To prove (IS) we derive from (IH) that $2k + 2 = 3k + 3$, since $2 = 3$. Factoring out gives $2(k+1) = 3(k+1)$. With this, we have proven (IS). This would not have happened if we had first tried to prove (IA); we would have seen that (IA) is a contradiction.

Induction or Recursion?

Inductive approach means developing the next step from the previous ones. The simplest example of this are the natural numbers: We get a new one by adding one to an already existing one. In this way, we can reach any natural number if we start with zero. This is the basis of the proof principle of mathematical induction. The *recursive definition* is a reversal of induction: in a recursion we go backwards to a known value, in induction we go forwards from known values. An induction proof uses a recursive relationship in the induction step. Also a recursive definition requires already one or more known values. These correspond to the induction anchor and are called ***initial conditions***.

Actually, we do not see this distinction as very sharp.

A formalization of a *recursive definition* we do not give here.

A formula, with which each value from a recursive definition can be determined directly, i.e. without recourse to its predecessors, we call a ***solution to the recursion***, see the following examples and exercises!

Examples 1.31 (Solutions of Recursions)

(1) The recursive definition $s_n = s_{n-1} + n$ with the initial condition $s_0 = 0$ gives the sum of the first n natural numbers. A solution is $s_n = \frac{1}{2}n(n+1)$ (cf. Example 1.30).
(2) In the proof of Theorem 2.16, it is shown by mathematical induction that the recursive definition $P_{n+1} = 2 \cdot P_n$ with the initial condition $P_0 = 1$ has the solution $P_n = 2^n$ for every $n \in \mathbb{N}$.

(3) The number sequence 1, 2, 6, 24, 120, 720, ... is defined by the initial condition $f_1 = 1$ and the recursive definition $f_n = f_{n-1} \cdot n$. Apparently, $f_n = 1 \cdot 2 \cdot 3 \cdots n$ is a solution to this recursion.

That it is sometimes necessary to specify more than one initial value, and that the solution to a recursion is not always so easy to find, we will see in connection with the so-called Stirling numbers in Sect. 2.5

Exercise 1.32 (Square Numbers) Prove by mathematical induction:

$$\sum_{i=0}^{n} i^2 = \frac{1}{6}n(n+1)(2n+1).$$

Exercise 1.33 (Further Sum Formulas) Find a formula for each of the following sums and prove it by mathematical induction.

(a) $\sum_{i=0}^{n} i(i+1) = 0 \cdot 1 + 1 \cdot 2 + 2 \cdot 3 + \cdots n(n+1)$.
   Hint: Use $i(i+1) = i^2 + i$ and the formulas for $\sum i$ and $\sum i^2$.
(b) $k_n := \sum_{i=0}^{n} i^3 = 0^3 + 1^3 + 2^3 + \cdots + n^3$.
   Hint: Compare the sequences $1, 2, 3, \ldots, k_1, k_2, k_3, \ldots, \sqrt{k_1}, \sqrt{k_2}, \sqrt{k_3}, \ldots$

This formula was published by Faulhaber[9] in his algebra book (in Latin) 1631.

Exercise 1.34 (Dollars) Assume an infinite supply of 4- and 5-dollar coins. Prove by induction that any whole amount of dollars greater than 3 and different from 6, 7, and 11 can be formed by a combination of such coins.

---

[9] Johannes Faulhaber, Festungsbaumeister und Rechenmeister der Stadt Ulm (Fortress builder and arithmetic master of the city of Ulm), 1580–1635.

Sets and Counting

The approach in set theory is to group certain objects into *sets* and then make these sets and their relationship to each other the subject of consideration. The properties of the objects are of secondary interest here.

In doing so, concepts are used that are so self-evident for our thinking that the question arises why it is necessary to explain or define them and express them in symbols. Upon closer inspection, however, it turns out that the symbols can be useful abbreviations and a good prerequisite for precise argumentation based on the logic already provided.

Set theory, as we know it today, began at the beginning of the twentieth century, when it turned out that the *naive* handling of sets can lead to paradoxes. The most famous example of such a contradiction is *Russell's Antinomy*[1] which we will introduce in Sect. 2.3. Efforts to get rid of such paradoxes led to a developmental push in logic and set theory at the beginning of the twentieth century.

Axiom systems were developed, through whose analysis the problems could be partly eliminated, or at least adequately described. The most commonly used axiom systems today for set theory are based on that of Zermelo and Fraenkel,[2] usually supplemented by the *Axiom of Choice*, see also the corresponding footnote in Sect. 2.4. Its special role, however, is likely to be more historical than content-related. It has not yet been possible to establish the final axiom system for set theory. The main point of contention is whether certain axioms, which regulate the existence of or the handling of very large sets, should or should not be included in the Zermelo/Fraenkel axioms. For the application of sets within and outside of mathematics, these questions are hardly of importance. Nevertheless, it can be

---

[1] Bertrand Arthur William Russell, 3rd Earl Russell, British philosopher, mathematician and logician, 1872–1970.

[2] Ernst Zermelo, German mathematician, 1871–1953; Adolph Abraham Fraenkel, German mathematician, 1891–1965.

exciting, for example, to think about whether an additional axiom is desirable, due to which the real numbers form the smallest uncountable set, or not (keyword: *Continuum Hypothesis*), cf. Sect. 5.5.

Anyone who, encouraged by initial experiences in formal handling with sets and perhaps curious due to Russell's antinomy, turns to one of the books on set theory, will probably find that the results of our intuitive approach to set theory for a systematic structure are not significantly modified, but must be viewed from a different perspective to obtain a viable theory.

On the Literature A good mix of rigor and understandability on the subject of set theory is provided by the book by Hrbacek and Jech [38]. There, among other things, the advantages and disadvantages of various additional axioms are explained. It is suitable for self-study and also for reference. For counting, the second topic of this chapter, we mention the books by Aigner [3] and Jacobs [41], which include insights into the many facets of combinatorial questions as well as into graph theory and coding theory. Aigner particularly considers application-related questions, Jacobs compiles a list of famous sequences of whole numbers in the last chapter.

2.1 Sets

Set, defining property, element, Venn diagrams, universal set, equality of sets, empty set, (proper) subset, inclusion

The concept of *set* "appears to the mind so fundamental that we cannot hope to define it with the help of even more fundamental concepts," states Quine [64] in his book *Set Theory and its Logic*.—So if we define *set* as a collection, totality or summary of any objects, we have not made any progress because now *collection*, *totality* or *summary* would have to be defined.

We therefore refrain from giving such a "pseudo-definition". We all know examples of sets: the real numbers, the consonants of our alphabet, the pages of this book, the planets of our solar system.

We describe a set by specifying which elements it contains and which it does not. It is useful to imagine that the considered elements come from a *universal set*. This corresponds to the individual domain in predicate logic.

We write $a \in M$, if a is an element of the set M and $a \notin M$, if a is not an element of the set M.

A set can essentially be specified in two ways: On the one hand by enumerating its elements between curly brackets, e.g.

$$A = \{\text{red, yellow, green}\},$$

Fig. 2.1 Venn diagrams of
the sets $A := \{1, 2, 3\}$,
$B := \{2, 3, 7, 12\}$,
$C := \{4, 5\}$,
$D := \{12, 24, 8\}$

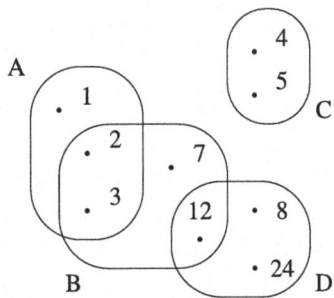

on the other hand by selection from an already defined set by means of a defining property, e.g.

$$M = \{x \mid x \text{ is a vector from } \mathbb{R}^2 \text{ and } x \text{ is a multiple of } (1, 1)\},$$

Read: "M is the set of all objects x (from the already defined set), for which ...". The set from which is selected can also be specified before the vertical bar, as follows:

$$M = \{x \in \mathbb{R}^2 \mid (\exists \lambda \in \mathbb{R}) : x = \lambda(1, 1)\}.$$

In both cases we have described the set M in the form $M = \{x \mid \mathcal{P}(x)\}$ using a predicate $\mathcal{P}(x)$, whose interpretation in the second representation is: "$(\exists \lambda \in \mathbb{R}) : x = \lambda(1, 1)$". We also call $\mathcal{P}(x)$ the ***defining property*** of the set M.

Venn[3] ***-Diagrams*** are suitable for a graphical representation of the relationships of sets to each other. They are also helpful in developing proof ideas, but they cannot replace a formal proof. In Austria, they are often called ***Mengenknödel (set dumplings)***. Figure 2.1 contains an example.

Equality and Subsets

Two sets M and N are called ***equal***, if they contain the same elements, i.e.:

$$M = N :\Leftrightarrow \quad ((\forall x) : x \in M \Leftrightarrow x \in N).$$

From this it follows: In proofs of set equality for an arbitrary x we must show the following two implications:

$$x \in M \Rightarrow x \in N \quad \text{and} \quad x \in N \Rightarrow x \in M.$$

---

[3] John Venn, English mathematician, 1834–1923.

Lemma 2.1 (Equal Elements) *One has $\{a\} = \{a, a\}$.*

Proof We have: $(\forall x): x \in \{a\} \Leftrightarrow x \in \{a, a\}$. $\square$

Generally, when we specify a set by listing its elements and an element appears multiple times, then this set is equal to the one that contains this element only once. For the case where an element should appear multiple times in a set, we provide a solution in Sect. 2.2: the individual copies of the same element are distinguished from each other by a formal trick. Another solution are so-called ***multisets***: for this one agrees to write the elements of the set multiple times in the curly brackets if necessary.

The set that contains no element is called ***empty set*** and is denoted by $\emptyset$, sometimes also by $\{\}$.

Theorem 2.2 (Uniqueness of the Empty Set) *There is only one empty set.*

Proof Let A and B be empty sets, then $(\forall x): x \in A \Rightarrow x \in B$, because an implication, whose left side is false, is itself true. Similarly, $(\forall x): x \in B \Rightarrow x \in A$, so $A = B$. $\square$

A set M is called ***subset*** of a set N (notation: $M \subseteq N$), if every element of M is also in N:

$$M \subseteq N :\Leftrightarrow ((\forall x): x \in M \Rightarrow x \in N).$$

Lemma 2.3 (Trivial Subsets) *For sets M, one has $\emptyset \subseteq M$ and $M \subseteq M$.*

Proof The statements $(\forall x): x \in \emptyset \Rightarrow x \in M$ and $(\forall x): x \in M \Rightarrow x \in M$ are obviously true. $\square$

A set M is called ***proper subset*** of a set N, if M is a subset of N, but not equal to N, i.e. if N contains at least one element that is not in M. We write $M \subset N$, if M is a proper subset of N. We point out that occasionally the symbol $\subset$ is used instead of $\subseteq$. Then the relation "is a proper subset of" can be expressed by the symbol $\subsetneq$. The symbol $\not\subseteq$ means: "is not a subset of". The relation "is a subset of" is also called ***inclusion***.

It is important to carefully distinguish the symbols $\in$ and $\subseteq$. The question of whether an object is an element or a subset of a set is not difficult to answer. Nevertheless, mistakes are often made which can have unpleasant consequences.

Example 2.4 (Elements and Subsets)

- $\{2, 3\} \subseteq \{4, 3, 2\}, 3 \in \{4, 3, 2\}$,
- $\{a\} \not\subseteq \{\{a\}, \{b\}, \{c\}\}$, but $\{\{a\}\} \subseteq \{\{a\}, \{b\}, \{c\}\}$ and $\{a\} \in \{\{a\}, \{b\}, \{c\}\}$ and $\{a\} \subseteq \{a, b, c\}$

- $\{x, y\} \in \{\{x, y\}, x, y, z\}$ and $\{x, y\} \subseteq \{\{x, y\}, x, y, z\}$,
- $\emptyset \subseteq \{a, b, c\}$, but $\{\emptyset\} \not\subseteq \{a, b, c\}$.

Lemma 2.5 (Transitivity of Inclusion) *For any three sets M_1, M_2 and M_3: $M_1 \subseteq M_2$ and $M_2 \subseteq M_3$ imply $M_1 \subseteq M_3$.*

Proof For the sets M_1, M_2 and M_3 let $M_1 \subseteq M_2$ and $M_2 \subseteq M_3$. To show: $(\forall x)$: $x \in M_1 \Rightarrow x \in M_3$. Let $x \in M_1$. Now $M_1 \subseteq M_2$ implies $x \in M_2$ and $M_2 \subseteq M_3$ implies $x \in M_3$. □

2.2 Operations on Sets

Difference set, intersection, union, symmetric difference, Cartesian product, complement, (ordered) pair, first/second component, power set, commutativity, associativity, distributivity, neutral element, de Morgan's rules, triple, r-tuple, component, r-th (I-th) power of a set, family/system of sets, partition, blocks of a partition, (disjoint) union

We recommend comparing the following definitions with those about operations of propositions from Sect. 1.3.

We present the most common operations of two sets in Table 2.1 and then give examples. In all definitions, A and B are sets.

If $A \subseteq B$, the difference set $B \setminus A$ is also called the ***complement of A in B*** and we write $\mathcal{C}_B(A)$ or $\overline{A}^B$, sometimes just $\mathcal{C}(A)$ or $\overline{A}$, if no confusion is expected.

The definitions of intersection and union give the motivation for the similarities of the symbols $\cap$ and $\wedge$ as well as $\cup$ and $\vee$.

The Cartesian[4] product $A \times B$ of A and B is thus the set of all *(ordered) pairs* (a, b) with *first component* $a \in A$ and *second component* $b \in B$. If $A = B = \mathbb{R}$, we also speak of the *Cartesian coordinate system*, see Sect. 4.1.

For $(a, b), (x, y) \in A \times B$ we define equality as follows:

$$(a, b) = (x, y) \Leftrightarrow a = x \text{ and } b = y.$$

Table 2.1 Operations on sets

Name	Definition	Speech
Difference set	$A \setminus B := \{x \in A \mid x \notin B\}$	A without B
Intersection	$A \cap B := \{x \mid x \in A \wedge x \in B\}$	A intersected with B
Union	$A \cup B := \{x \mid x \in A \vee x \in B\}$	A united with B
Symmetric Difference	$A \triangle B := (A \cup B) \setminus (A \cap B)$	A Delta B
Cartesian Product	$A \times B := \{(a, b) \mid (a \in A) \wedge (b \in B)\}$	A Cross B

---

[4] René Descartes, French philosopher and mathematician, 1596–1650.

If we consider the ordered pair (x, y) as an abbreviation for the set $\{x, \{x, y\}\}$, then this definition is superfluous, because it is then a consequence of the definition of set equality.

If for two sets M and N no element lies in both sets, i.e. if

$$(\forall x): \quad x \in M \Rightarrow x \notin N,$$

then these sets M and N are called *disjoint (to each other)*. Two sets M and N are disjoint if and only if $M \cap N = \emptyset$.

In the Venn diagram in Sect. 2.1 the sets A and D are disjoint, as are B and C, also C and D, and A and C. Although there is no element that lies in all three of the given sets A, B and D, we do not say that these three sets are disjoint. This term always refers only to two sets.—We define in general: n sets $M_1, M_2, \ldots, M_n$ ($n \in \mathbb{N}$) are called *pairwise disjoint*, if:

$$(\forall i, j \in \{1, 2, \ldots, n\} \text{ with } i \neq j) (\forall x): \quad x \in M_i \Rightarrow x \notin M_j.$$

As we see in the example in the Venn diagram in Sect. 2.1 $A \cap B \cap C \cap D = \emptyset$. From this we conclude that the formation of a common intersection is not suitable to describe the concept of pairwise disjoint sets.

Examples 2.6 (Set Operations)

(1) $\{a, \text{ber}, \text{witz}\} \setminus \{a, \text{ber}\} = \{\text{witz}\}$, but: $\{a, \text{ber}, \text{witz}\} \setminus \{\text{tz}\} = \{a, \text{ber}, \text{witz}\}$.
(2) $\{2, 4, 6, 8, \} \cap \{3, 6, 9, 12\} = \{6\}$.
(3) $\{\text{noble}, \text{helpful}, \text{good}\} \cap \{\text{helpful}, \text{noble}\} = \{\text{noble}, \text{helpful}\}$.
(4) $\{0, 2, 4\} \cup \{1, 2, 3, 5\} = \{0, 1, 2, 3, 4, 5\}$.
(5) $\{\{42\}, \{7, -68\}\} \cup \{\{42\}, 7\} = \{\{42\}, \{7, -68\}, 7\}$.
(6) $\{a, b\} \times \{0, 1, 2\} = \{(a, 0), (a, 1), (a, 2), (b, 0), (b, 1), (b, 2)\}$.

We have already noted at the beginning that the elements of a set can themselves be sets. An important "set of sets" is this:

The *power set* of a set M is the set $\wp(M)$, which has all subsets of M as elements. According to Lemma 2.3, we always have $\emptyset \in \wp(M)$ and $M \in \wp(M)$.

Example 2.7 (Powers) One has $\wp(\{1, 2\}) = \{\emptyset, \{1\}, \{2\}, \{1, 2\}\}$, $\wp(\{1\}) = \{\emptyset, \{1\}\}$ and $\wp(\emptyset) = \{\emptyset\}$, . However, $\wp(\{\emptyset\}) = \{\emptyset, \{\emptyset\}\}$.

Set Identities

For subsets A, B and C of set M, we have the following identities. Here, identity or more precisely set identity means that an equation form, such as $A \cup B = B \cup A$, is a true statement for all sets A and B. Note $\cong$ in the associativity of the Cartesian product means: The associative law holds if we consider the elements $((a, b), c) \in$

2.2 Operations on Sets

Table 2.2 Set identities

Associativities	Neutral elements	Complements
$(A \cup B) \cup C = A \cup (B \cup C)$	$A \cap M = A$	$A \cup \overline{A} = M$
$(A \cap B) \cap C = A \cap (B \cap C)$	$A \cup \emptyset = A$	$A \cap \overline{A} = \emptyset$
$(A \times B) \times C \cong A \times (B \times C)$		
Commutativity	Distributivity	
$A \cup B = B \cup A$	$A \cup (B \cap C) = (A \cup B) \cap (A \cup C)$	
$A \cap B = B \cap A$	$A \cap (B \cup C) = (A \cap B) \cup (A \cap C)$	
$A \triangle B = B \triangle A$		

$(A \times B) \times C$ and $(a, (b, c)) \in A \times (B \times C)$ as *essentially equal* and simply write (a, b, c). See Table 2.2.

The following proof of the first of the two distributive laws shows how these are related to the corresponding laws in Chap. 1.

Proof *(Distributivity)* Let A and B be subsets of a set M. For each element $x \in M$, one has

$x \in A \cup (B \cap C)$
$\Leftrightarrow (x \in A) \lor ((x \in B) \land (x \in C))$ (by definition of $\cup$ and $\cap$)
$\Leftrightarrow ((x \in A) \lor (x \in B)) \land ((x \in A) \lor (x \in C))$ (Distributivity)
$\Leftrightarrow (x \in A \cup B) \land (x \in A \cup C)$ (by definition of $\cup$)
$\Leftrightarrow x \in (A \cup B) \cap (A \cup C)$ (by definition of $\cap$).

□

The remaining logical identities from Sect. 1.4 can also be reformulated into true statements about sets. See Table 2.3.

As in Chap. 1, we can also omit brackets for unions, intersections of sets due to the associative laws. For Cartesian products, we do it according to the remarks made above. We write $A \cup B \cup C$, $A \cap B \cap C$ and $A \times B \times C$ for sets A, B and C, more

Table 2.3 More set identities for $A, B \subseteq M$

Equality	Complement as involution
$A = B \Leftrightarrow (A \subseteq B) \land (B \subseteq A)$	$A = \overline{\overline{A}}$
de Morgan's Laws	
$\overline{A \cap B} = \overline{A} \cup \overline{B}$	$A \subseteq B \Leftrightarrow \overline{B} \subseteq \overline{A}$
$\overline{A \cup B} = \overline{A} \cap \overline{B}$	

generally for subsets A_i of a set M:

$$\bigcup_{i=1}^{r} A_i := A_1 \cup A_2 \cup \ldots \cup A_r \quad \text{and even} \quad \bigcup_{i \in I} A_i$$

for an arbitrary index set I. Accordingly, we obtain the notations

$$\bigcap_{i=1}^{r} A_i := A_1 \cap A_2 \cap \ldots \cap A_r \quad \text{and} \quad \bigcap_{i \in I} A_i$$

as well as

$$\prod_{i=1}^{r} A_i := A_1 \times A_2 \times \ldots \times A_r \quad \text{and} \quad \prod_{i \in I} A_i.$$

In analogy to the notation for products of numbers, we also use additionally, if all factors A_i are equal to a set A,

$$A^r := \prod_{i=1}^{r} A_i, \quad \text{and even} \quad A^I := \prod_{i \in I} A_i$$

for the *r-th* or *I-th power of the set* A. The elements

$$(a_1, a_2, \ldots, a_r) \in \prod_{i=1}^{r} A_i$$

are called *(ordered) r-tuples* (*pairs* for $r = 2$ and *triples* for $r = 3$).

Family, Partition, Block

If I is a set and a set A_i is given for each $i \in I$, we say $(A_i)_{i \in I} = (A_i \,|\, i \in I)$ is a *family* or a *system of sets*. The round brackets express that we do not exclude that $A_i = A_j$ for $i, j \in I$ with $i \neq j$. In this case, we could not write $\{A_i \,|\, i \in I\}$, cf. Lemma 2.1. Nevertheless, it is possible to define a family of sets also as a set:

$$(A_i)_{i \in I} := \{(A_i, i) \,|\, i \in I\}.$$

In this way, it is ensured that for $i \neq j$ always $(A_i, i) \neq (A_j, j)$—even if $A_i = A_j$. Often, we still write $\{A_i \,|\, i \in I\}$.

If a set M is equal to the union of a set of pairwise disjoint, non-empty sets $\{A_i \,|\, i \in I\}$, then we say: $\{A_i \,|\, i \in I\}$, sometimes also $(A_i)_{i \in I}$, is a *partition*

2.2 Operations on Sets

or *decomposition* of M. The sets A_i, $i \in I$, are called **blocks of the partition**. We write

$$M = \dot{\bigcup}_{i \in I} A_i$$

and say that M is the **disjoint union**.

Example 2.8 (Partition) There are exactly five partitions of a three-element set $\{a, b, c\}$, namely:

$$\{\{a, b, c\}\}, \{\{a\}, \{b, c\}\}, \{\{b\}, \{a, c\}\}, \{\{c\}, \{a, b\}\} \text{ and } \{\{a\}, \{b\}, \{c\}\}.$$

The question of the number of partitions of an n-element set will be answered in Theorem 2.29.

Exercise 2.9 (Empty Set, Power Set) Determine $\wp(\{1, 2, 3\})$ and $\wp(\wp(\wp(\emptyset)))$. Prove $\emptyset \times M = \emptyset$ for any set M.

Exercise 2.10 (Characterization of Properties) Find for each of the following statements conditions on the sets A and B, so that the respective statement is true. Hint: In (a) by drawing Venn diagrams one (hopefully) arrives at the conjecture $(A \triangle B = \emptyset) \Leftrightarrow (A = B)$ and this should then be proven.

(a) $A \triangle B = \emptyset$ (b) $A \cup B = A$ (c) $A \cap B = A$
(d) $A \cup B \subseteq A \cap B$ (e) $A \cup \emptyset = \emptyset$.

Exercise 2.11 (Monotonicity) Which of the following statements are true for arbitrary subsets A, B, C of a given set M?—Proof or counterexample.

(a) $A \subseteq B \Leftrightarrow \overline{B} \subseteq \overline{A}$ (d) $A \subseteq B \Rightarrow A \cap C \subseteq B \cap C$
(b) $A = B \Leftrightarrow \overline{A} = \overline{B}$ (e) $A \subseteq B \Rightarrow A \setminus C \subseteq B \setminus C$
(c) $A \subseteq B \Rightarrow A \cup C \subseteq B \cup C$ (f) $A \subseteq B \Rightarrow A \triangle C \subseteq B \triangle C$.

Note: For (c) an elegant proof is possible with (a), (b) and de Morgan.

Exercise 2.12 (Intersection and Union) Show that for any sets A and B, $A \cap B \subseteq A \subseteq A \cup B$.

Exercise 2.13 (Sets in $\mathbb{R}^2$) The following sets are given:

$$A := \{(1, 2), (3, -2)\}, \quad C := \{(3\alpha, -2\alpha) \mid \alpha \in \mathbb{R}\},$$
$$B := \{(x, y) \mid 2x - y = 0\}, \quad D := \{(\alpha, 2\alpha) + (6\beta, -4\beta) \mid \alpha, \beta \in \mathbb{R}\}.$$

(a) Draw the sets in a coordinate system.
(b) Which of the following statements are true and which are false:

(α) $A \subseteq B$, (β) $A \subseteq D$, (γ) $B \cup C = D$, (δ) $\emptyset \in B$, (ϵ) $\emptyset \subseteq A$.

(c) Describe the following sets:

(α) $B \cap C$, (β) $B \triangle C$, (γ) $A \cap B$, (δ) $A \cap C$, (ϵ) $A \cap D$.

Exercise 2.14 (Partitions) Give all partitions of a 2-element and of a 4-element set.

Exercise 2.15 (Disjoint Union) Show for any sets A and B:

(a) $A = (A \cap B) \mathbin{\dot{\cup}} (A \setminus B)$;
(b) $A \cup B = (A \cap B) \mathbin{\dot{\cup}} (A \setminus B) \mathbin{\dot{\cup}} (B \setminus A)$.

2.3 Cardinality and Infinite Sets

Cardinality, cardinal number of a set, finite set, n-element set, axiom of separation, infinite sets, Russell's paradox, (proper) class

A set with finitely many elements is called a *finite set*. A finite set with n elements is called an *n-element set*. The **cardinality** (or **cardinal number**) of a finite set is the number of its elements.

Notations: card(M) or $|M|$ or #M.

Even for non-finite sets, there is the possibility to distinguish cardinalities. However, we can only satisfactorily formulate this distinction in Sect. 5.5 after introducing the concept of mapping.

Theorem 2.16 (Cardinality of the Power Set) *For every finite set M with* card(M) $= n$, *one gets* card($\wp(M)$) $= 2^n$.

Proof We prove by mathematical induction on n:
(BA) $n = 0$: The power set of the empty set has exactly $2^0 = 1$ element, namely the empty set itself (see Example 2.7).
(IH) If card(M) $= n$, then card($\wp(M)$) $= 2^n$.
(IS) Let A be a set with $n+1$ elements. We choose an element $a \in A$. The set $A \setminus \{a\}$ has n elements, so according to (IH) exactly 2^n subsets A_1, A_2, ..., A_{2^n}. These are also subsets of A (Transitivity!). Another 2^n subsets of A are $A_1 \cup \{a\}$, $A_2 \cup \{a\}$, ..., $A_{2^n} \cup \{a\}$. This tells us that A has *at least* $2 \cdot 2^n = 2^{n+1}$ subsets.

Now let B be any subset of A. There are two possibilities *(Tertium non datur!)*: either $a \in B$, or $a \notin B$.

2.3 Cardinality and Infinite Sets

1. Case: $a \notin B$. Then $B \subseteq A \setminus \{a\}$ is one of the sets A_i, $1 \leq i \leq 2^n$, counted in the first round.
2. Case: $a \in B$. Then $B \setminus \{a\}$ is one of the subsets A_i of $A \setminus \{a\}$ and $B = A_i \cup \{a\}$ was counted in the second round.

This shows that we have counted all subsets of A. $\square$

For finite sets we have thus shown, that the cardinality of the power set of a set is always greater than that of the set itself. This applies not only to finite, but even to infinite sets, as we will show in Theorem 5.33.

The **Axiom of Separation** states that for every set M and every predicate $\mathcal{P}(x)$ there exists a set P_M that contains exactly those elements $x \in M$ for which $\mathcal{P}(x)$ is a true statement. This means that we can form subsets for every set. This axiom is considered indispensable in mathematics. We use it here to prove the following theorem, which shows the limits of the concept of a set. It is summarized by Halmos[5] as "Nothing contains everything."

Theorem 2.17 ("Non-sets") *The totality of all sets is not a set.*

***Proof** (Indirect)* Let M be the totality of all sets, i.e., M contains all sets as elements. Assumption: M is a set. According to the Axiom of Separation, also

$$P_M := \{x \in M \mid x \notin x\}$$

is a set, as $x \notin x$ is certainly a predicate, and P_M is apparently a subset of M. Thus $P_M \subseteq M$. Since every set is an element of M, we have that $P_M \in M$.—We know from Example 2.4 that it is quite possible for a set to be both a subset and an element of another.

According to the law of the excluded third, either $P_M \in P_M$ or $P_M \notin P_M$. We examine both cases:

1. Case: $P_M \in P_M$ implies by definition of P_M because of $P_M \in M$, that $P_M \notin P_M$, which is a contradiction to $P_M \in P_M$.
2. Case: $P_M \notin P_M$ implies by definition of P_M, since we had assumed $P_M \in M$, that $P_M \in P_M$, a contradiction again.

Since both cases lead to a contradiction, our assumption, M is a set, must have been wrong. This proves the theorem. $\square$

The argument with which we derived this theorem is known as **Russell's paradox**. In popular form, it reads

> In a small village, there is a village barber. He shaves exactly the men of the village who do not shave themselves. – Who shaves the village barber?

---

[5] Paul Richard Halmos, Hungarian-American mathematician, 1916–2006.

This paradox is resolved by introducing the concept of the ***class***: Every set is a class. The totality of all sets is also a class. It is called a ***proper class*** since it is not simultaneously also a set. Analogous to set theory, a class theory can now be developed, in which similar laws apply as for sets. In particular, the totality of all classes is not a class itself ....

2.4 Counting

Principle of inclusion and exclusion, fundamental counting principle, independent events

We consider here some basic principles of counting.

A very elementary example is the *Dirichlet's pigeonhole principle*:[6]

If there are $n + 1$ pigeons in n pigeonholes, then there is at least one pigeonhole in which there are at least two pigeons.

Inclusion and Exclusion

We put the *principle of inclusion and exclusion* at the beginning. Here is a typical question:

How many numbers in the set $M := \{1, 2, 3, \ldots, 100\}$ are divisible by neither 4 nor by 6?

We set $A := \{n \in M \mid 4 \text{ divides } n\}$ and B analogously for 6. There are 25 numbers in M that are divisible by 4, so they are in A. And 16 numbers are divisible by 6, so they are in B. That makes a total of 41 numbers. But we have counted, for example, the 12 twice. Also all numbers divisible by 12, because these are precisely those that are divisible by 4 and by 6: a total of 8 numbers. We get $100 - (41 - 8) = 67$ numbers that are divisible by neither 4 nor 6.

Expressed in the language of sets, we have used the following relationship:

Lemma 2.18 (Cardinality of the Union) *For sets A and B we get*

$$|A \cup B| = |A| + |B| - |A \cap B|.$$

Proof Obviously, the cardinality of the disjoint union of sets is equal to the sum of their cardinalities. So we only need to write the set $A \cup B$ as a disjoint union. We use the formulas from Exercise 2.15:

$$A \cup B = (A \cap B) \ \dot{\cup} \ (A \setminus B) \ \dot{\cup} \ (B \setminus A) \quad \text{and} \quad A = (A \cap B) \ \dot{\cup} \ (A \setminus B),$$

---

[6] Johann Peter Gustav Lejeune Dirichlet, Belgian mathematician, 1805–1859.

2.4 Counting

Thus

$$|A \cup B| = |A \cap B| + |A \setminus B| + |B \setminus A| = |A \cap B| + |A| - |A \cap B| + |B| - |B \cap A|,$$

since obviously $|A \cup B| = |A| + |B|$ if and only if $A \cap B = \emptyset$. □

This counting principle, where multiple counts are subtracted, can be easily extended to more than two sets. We formulate it for three sets. Observe that there are elements subtracted multiple times, which then are added again.

Lemma 2.19 (Union of Three Sets) *For sets A, B and C one has*

$$|A \cup B \cup C| = |A| + |B| + |C| - |A \cap B| - |A \cap C| - |B \cap C| + |A \cap B \cap C|.$$

Proof We use the associative law to turn a union of three sets into a union of two sets, as well as the distributive law and Lemma 2.18 several times.

$$\begin{aligned}
|A \cup B \cup C| &= |A \cup (B \cup C)| \\
&= |A| + |B \cup C| - |A \cap (B \cup C)| \\
&= |A| + |B \cup C| - |(A \cap B) \cup (A \cap C)| \\
&= |A| + (|B| + |C| - |B \cap C|) \\
&\quad - (|A \cap B| + |A \cap C| - |A \cap B \cap A \cap C|).
\end{aligned}$$

Resolving the brackets and rearranging delivers the desired result. □

To please the friends of universally valid formulas: By means of mathematical induction, the following theorem can be proven. It generalizes our lemmas for $n = 2$ and for $n = 3$.

Theorem 2.20 (Principle of Inclusion and Exclusion) *For sets $A_1, A_2, \ldots, A_n$ one has*

$$|A_1 \cup A_2 \cup \ldots \cup A_n| = \sum_{1 \leq i \leq n} |A_i| - \sum_{1 \leq i < j \leq n} |A_i \cap A_j|$$

$$+ \sum_{1 \leq i < j < k \leq n} |A_i \cap A_j \cap A_k| - \sum_{1 \leq i < j < k < l \leq n} |A_i \cap A_j \cap A_k \cap A_l| + - \ldots$$

$$+ (-1)^{n-1} |A_1 \cap A_2 \cap \ldots \cap A_n|.$$

The Fundamental Counting Principle

Many of the questions considered in the following can be traced back to the *fundamental counting principle*. It states:

> If an event has k_1 possible outcomes and another event has k_2 possible outcomes, then the sequence of these two events has $k_1 k_2$ possible outcomes.

Obviously, this only applies if the outcome of the first event does not influence that of the second. We speak of ***independent events***, i.e., we have the cross product of two event sets.—By induction, the fundamental counting principle can also be used to determine the number of possible outcomes of the sequence of finitely many independent events.

Examples 2.21 (Independent Events)

(a) From three starters, five main courses, and two desserts, $3 \cdot 5 \cdot 2 = 30$ menus can be composed.
(b) If we throw a six-sided dice twice in a row, each throw has 6 possible results (= thrown number of eyes) and the two throws in sequence can yield 36 possible results.—Note that the results (2, 5) and (5, 2) are interpreted as different!
(c) If three cards are drawn in a row from a deck of 32 different cards, the first draw has 32 possible outcomes, the second has 31 (only 31 cards are available after the first draw) and the third has 30. In total, there are $32 \cdot 31 \cdot 30 = 29{,}760$ different outcomes.—Here too, the order in which the cards are drawn matters, i.e., we count triples and thus distinguish for example (7, 8, 9) from (8, 9, 7). The question of how many different "hands" of five cards there are, is answered below under the keyword *combinations*.

2.5 Permutations and Combinations

> Telephone number problem, lottery problem, mixing problem, permutations with/without repetition, combinations with/without repetition, n factorial, Stirling's approximation formula, binomial coefficients, binomial theorem, Stirling numbers

The following two problems we know in daily life:

Telephone number problem: How many k-digit telephone numbers (for example, composed of 10 digits) are there?
Lottery problem: How many combinations of k different numbers from a set of n numbers (e.g., 6 out of 49) are there?

A first glance shows: In the telephone number problem, the order matters and repetitions of digits are possible, in the lottery problem, the order does not matter and repetitions of numbers are not possible.

2.5 Permutations and Combinations

Table 2.4 Basic questions of counting

Basic questions of counting	With order *Permutations*	Without order *Combinations*
With repetitions	*Telephone number—Problem* $P^r(n, k) = n^k$ 2.24	$C^r(n, k) = \binom{n+k-1}{k}$ 2.28
Without repetitions	$P(n, k) = \frac{n!}{(n-k)!}$ 2.23	*Lottery—Problem* $C(n, k) = \binom{n}{k}$ 2.26

The criteria *order* and *repetition* can be combined in four different ways. Table 2.4 shows the relationships that yield four questions, it already contains the usual abbreviations and the answers as formulas, as well as references to the places where these are proven. *Permuting (lat.)* means swapping or rearranging. For example, when we shuffle a deck of cards, how many possible outcomes does this event have?—We ask more generally:

How many ways are there to arrange n objects in a row?

Here, we can argue the same way as in Example 2.21.

Now there are n choices for the first place, $n - 1$ choices for the second place, etc., for the penultimate place we have just two possibilities and for the last place one object remains.

So we have shown:

Theorem 2.22 (Factorial) *There are exactly*

$$n(n-1)(n-2)(n-3)\ldots 3 \cdot 2 \cdot 1 =: n!$$

ways to arrange n objects in a row, where we set $0! := 1$.

The abbreviated notation $n!$ for this product is read as n *factorial*. An approximation for $n!$ is provided by *Stirling's approximation formula*: $n! \approx \left(\frac{n}{e}\right)^n \sqrt{2\pi n}$, e is the *Euler's number*, π the *circle number*. Such a formula is of importance, as an exact calculation of $n!$ for large n is very extensive.

Now we can answer the more general question:

How many ways are there to select k objects from n different ones and arrange them in a row?

With $P(n, k)$ we denote the answer and call $P(n, k)$ the **number of k-permutations of an n-element set** or **of n objects**.

Theorem 2.23 (Permutations Without Repetition)

$$P(n, k) = n(n-1)(n-2)\cdots(n-k+1) = \frac{n!}{(n-k)!}.$$

Proof As before, we apply the fundamental counting principle to the k events of selecting one element each from the given set. We obtain the second equality sign with the definition of the factorial by calculating from right to left. □

With $P^r(n, k)$ we denote the **number of k-permutations of n objects with repetition**, which we asked for in the *telephone number problem*.

Theorem 2.24 (Permutations with Repetition)

$$P^r(n, k) = n^k.$$

Proof For each of the k objects to be selected, we have n possibilities. □

Combining means assembling. A popular example of a combination is the drawing of lottery numbers: It does not matter in which order the numbers were drawn. The general question is:

> How many ways are there to choose k objects from n?

In other words, what is the **number of k-combinations of an n-element set** or **of n objects**. We denoted this number by $C(n, k)$.

Lemma 2.25 (Permutations and Combinations)

$$C(n, k) \cdot k! = P(n, k).$$

Proof A k-permutation $P(n, k)$ of n objects can be understood as a k-combination of n objects, of which there are $C(n, k)$ possibilities—we do not yet know this number—followed by a permutation of the selected k objects (there are $k!$ possibilities for this). □

We can also interpret $C(n, k)$ as the number of k-element subsets of an n-element set.

From Theorem 2.23 and Lemma 2.25 we obtain by dividing by $k!$ and substituting the solution of the lottery problem:

Theorem 2.26 (Combinations without Repetition) *The number of k-combinations from an n-element set is*

$$C(n, k) = \frac{n(n-1)(n-2)\cdots(n-k+1)}{k(k-1)(k-2)\cdots 1} = \frac{n!}{k!(n-k)!} =: \binom{n}{k}.$$

For $k > n$ we set $\binom{n}{k} := 0$, then the statement follows also in these cases.

2.5 Permutations and Combinations

The expression $\binom{n}{k}$, read "*n choose k*", is called **binomial coefficient**. Binomial coefficients often appear, for example:

Theorem 2.27 (Binomial Theorem) *For real numbers a and b and $n \in \mathbb{N}$ we get*

$$(a+b)^n = \sum_{k=0}^{n} \binom{n}{k} a^k b^{n-k}.$$

Proof When expanding the expression $(a+b)^n$, for $k=0$ the term b^n appears once, namely when we choose the factor b from each of the n brackets and none of the factors a. In general, the term $a^k b^{n-k}, k \in \mathbb{N}$, occurs exactly $\binom{n}{k}$ times, since that is the number of ways to choose the factor a exactly k times from the n brackets $(a+b)$. □

In the combinations without order, each of the n objects could only be chosen once. If we imagine that we put the respective object back after the choice, the question arises about the number of k-**combinations of** n **objects with repetition,** denoted by $C^r(n,k)$ in analogy to $P^r(n,k)$ from Theorem 2.24.

For simplicity, let the considered n-element set be equal $M := \{1, 2, \ldots, n\}$. We can understand $P^r(n,k)$ as the number of all k-tuples with elements from M, i.e., as the cardinality of the set M^k.

Now, $C^r(n,k)$ differs from $P^r(n,k)$ only in that the order of the entries of the k-tuple is not considered: we choose one element—put it back, choose the next element—put it back, and so on until we have chosen k elements. The result of this process can be described as follows. We say for each element of M how often it was chosen. The sum of these frequencies is k, we make k choices. If we imagine n boxes—one for each element from M—then each result can be represented as a distribution of k balls on these n boxes. Figure 2.2 shows the situation for $n=7$ and $k=8$, where the objects 1, 1, 1, 3, 4, 4, 6, 6 (in any order) have been chosen. We put a ball in box 1 each time the 1 was chosen.

Between $P^r(n,k)$ and $C^r(n,k)$ there is no such simple relationship as between $P(n,k)$ and $C(n,k)$ (see Lemma 2.25), since the number of possible permutations of the selected objects depends on how many identical objects were selected.

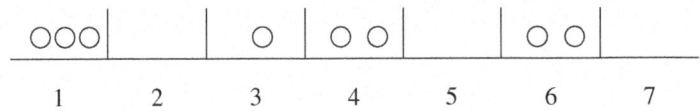

Fig. 2.2 The situation for $n=7$ and $k=8$

Theorem 2.28 (Combinations with Repetition) *The number of k-combinations of n objects with repetition is equal to*

$$C^r(n, k) = \binom{n+k-1}{k},$$

in particular, $C^r(n, k) = C(n+k-1, k)$.

Proof The above explanations have reduced the problem of determining $C^r(n, k)$ to counting the possibilities of dropping k balls into n boxes. The result of such an action can be represented as a sequence of k zeros (corresponding to the balls) and $n-1$ ones (corresponding to the partitions between the boxes), and each such 0, 1-sequence of length $n+k-1$ with exactly k zeros corresponds to the result of such an action. (In the example above, the 0, 1-sequence 0, 0, 0, 1, 1, 0, 1, 0, 0, 1, 1, 0, 0, 1 is represented.) According to Theorem 2.26, there are exactly $\binom{n+k-1}{k}$ such sequences, namely $C(n+k-1, k)$. □

Stirling Numbers

How many possibilities do 8 students have to divide themselves into 4 working groups?—Try an estimate!

It is difficult to solve this problem directly. Here, the principle of recursion helps. We denote the number of ways to distribute n students into k working groups by $S(n, k)$. The following initial conditions are obviously appropriate:

$$S(n, k) = 0 \text{ for } n < k \quad \text{and} \quad S(n, 1) = S(n, n) = 1 \text{ for } n \geq 1.$$

Now we consider how $S(8, 4)$ depends on numbers $S(n, k)$ with $n < 8$ or $k < 4$: If 7 students have already formed 3 working groups ($S(7, 3)$ possibilities), then an eighth person only has one option: she must work alone to form 4 working groups. If the seven have already formed 4 groups ($S(7, 4)$ possibilities), then the eighth person can join any of these groups ($4 \cdot S(7, 4)$ possibilities). In total we get:

$$S(8, 4) = S(7, 3) + 4 \cdot S(7, 4).$$

Here, the two numbers $S(7, 3)$ and $4 \cdot S(7, 4)$ are added, because it is not a sequence of events, but an alternative, we have a union of disjoint sets.—The same argument as in this special case leads to the general recursion formula

$$S(n, k) = S(n-1, k-1) + k \cdot S(n-1, k).$$

2.5 Permutations and Combinations

With the initial conditions $S(n, k) = 0$ for $n < k$ and $S(n, 1) = S(n, n) = 1$ for $n \geq 1$, the formula for natural numbers n and k with $n \geq k \geq 1$ gives the **Stirling**[7] **numbers of the second kind**. The number $S(n, k)$ answers the following question:

In how many ways can we divide an n-element set into k disjoint subsets?

A solution to this recursion was found by Stirling. We present it here without proof:

$$S(n, k) = \frac{1}{k!} \sum_{i=0}^{k} (-1)^{k-i} \binom{k}{i} i^n.$$

Apparently it may not be easy to solve a recursion.

With this formula we get the answer to the question posed at the beginning: There are no less than 1701 ways for 8 people to form 4 working groups!

We can now also answer the question from Example 2.8.

Theorem 2.29 (Number of Partitions) *The number of partitions of an n-element set is*

$$\sum_{k=1}^{n} S(n, k).$$

For completeness: **Stirling numbers** of the first kind are the coefficients $s(n, i)$ of the polynomial $k(k-1) \cdots (k-n+1) = \sum_{i=0}^{n} s(n, i) k^i$. Since they are not necessarily ≥ 0, they do not play a major role in counting problems.

Exercise 2.30 (Binomial Coefficients) For all $n, k \in \mathbb{N}$ prove that

(a) $\binom{n+1}{k} = \binom{n}{k} + \binom{n}{k-1}$.
 (This is a recursion formula for binomial coefficients.)
(b) $\binom{n}{0} + \binom{n}{1} + \binom{n}{2} + \ldots + \binom{n}{n} = 2^n$.
 (This formula can be proven directly or by mathematical induction.)
(c) $\binom{n+1}{k+1} = \binom{k}{k} + \binom{k+1}{k} + \ldots + \binom{n}{k}$.
 Hint: For any $k \in \mathbb{N}$, use induction on n:
 (IA) Show the formula for $n = k$,
 (IH) Assume that the formula holds for an $n_0 \in \mathbb{N}$ with $k \leq n_0$,
 (IS) Show the formula for $n_0 + 1$.
 Then the formula is proven for all $n, k \in \mathbb{N}$ with $k \leq n$.

---

[7] James Stirling, Scottish mathematician, 1692–1770.

Exercise 2.31 (Pascal's[8] Triangle) Look up the *Pascal's[8] Triangle* in the literature and compare its structure with the formulas from Exercise 2.30.

Exercise 2.32 (Binomial Theorem)

(a) Determine 2.01^6 using the binomial theorem.—How large is the error if only the first three terms are calculated?
(b) Why is the error smaller than in (a) if 1.99^6 is calculated?
(c) Is $1.0001^{10\,000} > 2$?

Exercise 2.33 (Counting)

(a) How many three-digit numbers < 600 can be formed from the digits 8, 6, 4 and 2?
(b) A commission of 3 people is to be chosen from a set of 5 Greens, 3 Reds and 4 Blues. In how many ways is this possible,
    (α) in general, (β) if at least one Blue must be included?
(c) For a party, 6 out of 14 possible people are to be invited. How many possibilities are there (α) in general, (β) if among the 14 there are two, one of whom does not come if the other comes, (γ) if there are two among them who only come together?

Exercise 2.34 (Partitions) Determine the number of partitions for sets with 0, 1, 2, 3, 4 and 5 elements.

Exercise 2.35 (Stirling Numbers)

(a) Make a table for $S(n, k)$ with $0 \leq n \leq 9$ and $0 \leq k \leq 4$.
(b) Find a formula for $S(n + 1, n)$ and one for $S(n, 2)$.
(c) Prove the formulas in (b) using a combinatorial interpretation ("number of partitions ...") and by mathematical induction using the recursion formula for the Stirling numbers.

Exercise 2.36 (Stirling's Approximation) Calculate Stirling's approximations for 1!, 10!, 100! and 1000! and compare with the exact results.

---

[8] Blaise Pascal, French physicist and mathematician, 1623–1662.

Numbers and Their Representations 3

Numbers are a self-evident part of our daily lives. We deal with them almost without thinking. Now, we will develop a mathematically deeper understanding of numbers. We will explore which of their properties we define—i.e., actually have to assume—and which properties then automatically apply—i.e., which we can prove, even if we will not conduct all proofs here.

In later chapters, sets of numbers with their *operations*, such as addition or multiplication, will serve as examples of *algebraic structures*.

On the Literature The topic of numbers is naturally of general interest. It also allows for historically interesting retrospectives; see, for example, Damerow et al. [18], Dedekind [20], and parts of Ebbinghaus et al. [24]. Especially for students and practicing teachers, there is a wide field of exciting questions. For a deeper introduction to mathematical questions about counting and enumerating, Klaua [47] is suitable.

3.1 The Set of Natural Numbers

Cardinal numbers, ordinal numbers, Peano axioms, categorical axiom system, monomorphic data type, successor, principle of mathematical induction, natural numbers $\mathbb{N}$, sum, associative law, commutative law, multiplication, powers, order on $\mathbb{N}$, well-ordering, distributive law

The natural numbers 0, 1, 2, 3, ... serve to count finite things *(cardinal numbers)* or to (linearly) order *(ordinal numbers)*, see for example Fig. 3.1.

Abstractly, natural numbers can be constructed using sets. This is a typical mathematical approach: At the beginning, we pretend as if we knew nothing. That is to say, more precisely, we only use the concepts of set and empty set, and thereby obtain the number 0 as the set ∅, which contains no element, the number 1 as a set

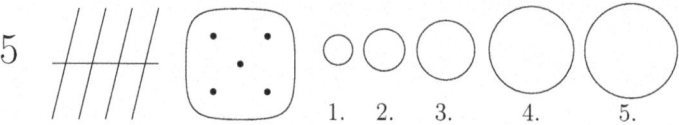

Fig. 3.1 Representations of Five, three times as cardinal number and once as ordinal number

that contains one element, the number 2 as a set that contains two elements, etc. Subsequently, we give these objects the names we are familiar with 0, 1, 2, ...:

$$
\begin{aligned}
0 &:= \emptyset & &= \emptyset \\
1 &:= \{0\} & &= \{\emptyset\} \\
2 &:= \{0, 1\} & &= \{\emptyset, \{\emptyset\}\} \\
3 &:= \{0, 1, 2\} & &= \{\emptyset, \{\emptyset\}, \{\emptyset, \{\emptyset\}\}\} \\
&\vdots & &\vdots
\end{aligned}
$$

However, it is also possible to define natural numbers by a suitable selection of their properties, an *axiom system*. The following comes from Peano and is the basis for the *proof by induction* described in Sect. 1.9.

The Peano Axioms

A set N is called *set of natural numbers*, if the following is true

(Peano1)   There is an element in N, we call it 0.
(Peano2)   For every $n \in N$ there exists exactly one $n^+ \in N$, the so-called *successor* of n.
(Peano3)   There is no $n \in N$ with $n^+ = 0$.
(Peano4)   If for n and m in N one has $n^+ = m^+$, then $n = m$.
(Peano5)   If a subset $A \subseteq N$ contains the element 0 and with every n also the successor n^+, then $A = N$ (*Principle of mathematical induction*).

It can be shown that there can essentially be at most one such set. This is why this axiom system is also called *categorical* or, as data type, *monomorphic*. Therefore, it makes sense that this set is represented by the special symbol $\mathbb{N}$.

In contrast, the axiom system for a group (see Chap. 7) is not categorical, as it is fulfilled by very different sets.

However, the Peano axioms contain no information about whether there is a set that has the properties (Peano1) to (Peano5). With Exercise 3.4 we can prove their existence relatively easily.

We will now proceed to prove properties of the set $\mathbb{N}$ using the axioms, without using the fact that such a set exists.

3.1 The Set of Natural Numbers

Addition, Multiplication and Order

The *sum of two natural numbers* is recursively defined by

$$(\text{Add1}) \; a + 0 := a \qquad \text{for all } a \in \mathbb{N},$$
$$(\text{Add2}) \; a + b^+ := (a+b)^+ \quad \text{for all } a, b \in \mathbb{N}.$$

If we set $1 := 0^+$, we get with (Add1) and (Add2) for each $a \in \mathbb{N}$:

$$a^+ = (a+0)^+ = a+1.$$

Lemma 3.1 (Sum) *The sum $a + b$ is defined for any two natural numbers a and b by (Add1) and (Add2).*

Proof Let $P(n)$ be the statement "For any $a \in \mathbb{N}$, the sum $a + n$ is defined". We prove by mathematical induction: $(\forall n \in \mathbb{N}) : P(n)$.

(IA) $P(0)$: For any $a \in \mathbb{N}$, $a + 0 = a$ (by (Add1))
(IH) $P(k)$ is valid for some $k \geq 0$, i.e., for all $a \in \mathbb{N}$ and some $k \geq 0$, $a + k$ is defined.
(IS)   Show $P(k+1)$, i.e. for all $a \in \mathbb{N}$, $a+(k+1)$ is defined. Indeed $a+(k+1) = a+k^+ = (a+k)^+$. According to (IH), $a+k$ is defined and according to (Peano2), $(a+k)^+$ is defined.

From (Peano5) it follows that $P(n)$ is true for every $n \in \mathbb{N}$. □

Theorem 3.2 (Associative Law of Addition) *For all $a, b, c \in \mathbb{N}$ one has $(a+b)+c = a+(b+c)$.*

Proof This proof can also be done by mathematical induction: So let $P(n)$ be the statement: "For any $a, b \in \mathbb{N}$, one has $(a+b)+n = a+(b+n)$".

(IA) $P(0)$: For any $a, b \in \mathbb{N}$, we get $(a+b)+0 = a+b = a+(b+0)$ by (Add1).
(IH)   For some $k \geq 0$, assume $P(k)$, i.e., for any $a, b \in \mathbb{N}$ and this $k \in \mathbb{N}$, we have $(a+b)+k = a+(b+k)$.
(IS)   Let $a, b \in \mathbb{N}$, we show that $(a+b)+(k+1) = (a+b)+k^+ = ((a+b)+k)^+$. The latter is equal to $(a+(b+k))^+ = a+(b+k)^+ = a+(b+k^+) = a+(b+(k+1))$ according to (IH), so we have $P(k+1)$.

According to (Peano5), $P(n)$ then applies for every $n \in \mathbb{N}$. □

We also define the *multiplication of natural numbers* recursively as follows:

$$(\text{Mult1}) \; a \cdot 0 := 0 \qquad \text{for all } a \in \mathbb{N},$$
$$(\text{Mult2}) \; a \cdot b^+ := (a \cdot b) + a \quad \text{for all } a, b \in \mathbb{N}.$$

In the following, we will mostly omit ·.
Powers of natural numbers can also be defined recursively:

(Pot1) $a^0 := 1$ for all $a \in \mathbb{N}$,
(Pot2) $a^{b^+} := a^b \cdot a$ for all $a, b \in \mathbb{N}$.

Corresponding statements as in Lemma 3.1 and Theorem 3.2 can now also be formulated and proven for multiplication, as well as all other usual calculation rules.

Now we define an *order* $\leq$ for $m, n \in \mathbb{N}$ as follows:

$$m \leq n :\Leftrightarrow \exists r \in \mathbb{N} \text{ with } m + r = n;$$

$$m < n :\Leftrightarrow m \leq n \text{ but } m \neq n.$$

While the following theorem may seem intuitively obvious, we will prove it to adhere to the rigor of our chosen axiomatic approach, which requires setting aside intuition.

Theorem 3.3 (Well-Ordering of $\mathbb{N}$) *Every non-empty subset of $\mathbb{N}$ has a smallest element, i.e., an element such that all (other) elements of the subset are larger (see Sect. 4.3).*

Proof We formulate the statement as follows:
$P(n)$: Every subset $A \subseteq \mathbb{N}$, which contains n, has a smallest element.
Now we prove $(\forall n \in \mathbb{N}) : P(n)$

(IA) $P(0)$: Every subset $A \subseteq \mathbb{N}$, which contains 0, has a smallest element, namely 0. This follows from the above given definition of the order on $\mathbb{N}$.

(IH) For some $n \in \mathbb{N}$ with $n \geq 0$ assume $P(n)$: Every subset $M \subseteq \mathbb{N}$, which contains n, has a smallest element.

(IS) We prove $P(n+1)$: Every subset $A \subseteq \mathbb{N}$, which contains $n+1$ has a smallest element. So let $A \subseteq \mathbb{N}$ be such a set. If n is also in A, we are lucky, because from (IH) it follows that A has a smallest element. We therefore only need to examine the case that $n + 1$ is in A, but not n. In this case we define a set B by $B := A \cup \{n\}$. Because the element n is in B, B has a smallest element b according to (IH), i.e. $b \leq x$ for all $x \in B$ and because $A \subseteq B$ it follows $(\forall x \in A) : b \leq x$. If $b \in A$, b is the sought smallest element of A. If not, then $b = n$ and we get $(\forall x \in A) : b < x$, so $n + 1$ is the smallest element of A.

$\square$

Exercise 3.4 (Natural Numbers Exist) According to the above mentioned construction of natural numbers as sets, define recursively:

$$0 := \emptyset \quad \text{and} \quad n^+ := n \cup \{n\}.$$

Show that the set of objects defined in this way $0, 0^+, (0^+)^+, \ldots$ satisfies the Peano axioms.

Exercise 3.5 (Commutative Law of Addition) Prove the commutative law of addition, i.e., show that for all $a, b \in \mathbb{N}$ one has

$$a + b = b + a.$$

Hint: Carry out two induction proofs: First show that $a + 0 = 0 + a$ for every $a \in \mathbb{N}$ and then prove that $a + 1 = 1 + a$ for every $a \in \mathbb{N}$.

Exercise 3.6 (Distributive Law) Prove that for all $a, b, c \in \mathbb{N}$ one has

$$c(a + b) = ca + cb.$$

Exercise 3.7 (Compatibility of Order) Prove that for all $a, b, c \in \mathbb{N}$ one has

$$a \leq b \quad \text{implies} \quad c + a \leq c + b \quad \text{and} \quad ca \leq cb.$$

3.2 Extensions of Number Systems

Equation, equation form, solution, integers $\mathbb{Z}$, (greatest common) divisor, coprime, prime number, Euclidean algorithm, division with remainder, rational numbers $\mathbb{Q}$, $\sqrt{2}$ is not rational, real numbers $\mathbb{R}$, irrational, $\sqrt{-1}$ is not real, imaginary unit i, imaginary, complex numbers $\mathbb{C}$, Gaussian number plane, Euclidean plane, algebraic number, algebraic equation of degree n, transcendental, rational $\Rightarrow$ algebraic, e and π are not algebraic, $\sqrt{-1}$ is algebraic, Fundamental Theorem of Algebra, algebraically closed

Equations

The expressions "$a + x = b$ with $a, b \in \mathbb{N}$" or "$2 + x = 3$" are usually called *equations*. Analogous to propositional forms, one should rather speak of *equational forms*, since it cannot be decided whether they are true or false. Formally, such an " equation" is not a statement. This also emphasizes that nothing can be said about the so-called *variable* x. Despite this formal, but quite illuminating, escapade, we will use the word " equation" as is common in mathematics.

Any number $c \in \mathbb{N}$ for which $a + c = b$ is called a *solution* of the equation $a + x = b$. We also say that c *satisfies* the equation $a + x = b$. Substituting a solution into the equation, makes it a true statement.

So the equation $a + x = b$ is actually a *question about the set* $\{x \in \mathbb{N} | a + x = b\}$. Is it empty or which elements does it contain?

In the following, we will extend the number system, starting from the set $\mathbb{N}$ of natural numbers, by adding further numbers until we can solve any *algebraic equation*. (The definition of an algebraic equation follows in the subsection "Algebraic and Transcendental Numbers.") We will implicitly assume that operations like addition and multiplication are defined in each number system as usual. In particular, this implies that restricting the operation on a larger number system to a smaller one yields the operation that was defined on the smaller one. Moreover, types of equations that can be solved in a smaller system can also be solved in its extensions. Proving this formally is not trivial. Compare, for example, Ebbinghaus et al. [24].

Integers

If a is a natural number $\neq 0$, then there is no $x \in \mathbb{N}$, such that $a + x = 0$. However, a solution x to this equation can be found among the negative integers. We set

$$\mathbb{Z} := \mathbb{N} \cup \{-a \mid a \in \mathbb{N}, a \neq 0\}$$

and call this set the ***set of integers***. If $\mathbb{Z}$ is defined in this way, it is quite cumbersome to define addition, multiplication and the order of integers in such a way that the statements from Sect. 3.1 remain valid for natural numbers (which also occur in $\mathbb{Z}$). We would have to define the sums $a + b$, $(-a) + b$, $a + (-b)$, $(-a) + (-b)$ for $a, b \in \mathbb{N}$ and likewise products and prove the corresponding properties of these operations.

Therefore, usually a different approach is taken: $\mathbb{Z}$ is defined as a set of equivalence classes on $\mathbb{N} \times \mathbb{N}$ (see Sect. 4.2, Exercise 4.28). Then addition, multiplication and the order of $\mathbb{N}$ can be easily transferred to $\mathbb{Z}$.

Once this is done, for every equation $a + x = b$ with $a, b \in \mathbb{Z}$ there is a solution $x \in \mathbb{Z}$, i.e. $(\forall a, b \in \mathbb{Z})(\exists x \in \mathbb{Z}): a + x = b$.

Next we observe: The equation $ax = b$ does not always have a solution in $\mathbb{Z}$.

We say a ***divides*** b or a ***is a divisor of*** b and write $a|b$, if a and b are integers and there is an $x \in \mathbb{Z}$ with $ax = b$. A number $p \in \mathbb{N}$ is called a ***prime number***,[1] if the only divisors of p in $\mathbb{Z}$ are the numbers $p, -p, 1$ and -1. The ***greatest common divisor of*** a ***and*** b is the largest integer t that divides both a and b. Notation: $\gcd(a, b) = t$. If $\gcd(a, b) = 1$, then a and b are called ***coprime***. For example, $\gcd(12, 8) = 4$ and $\gcd(9, 10) = 1$.

If $b \in \mathbb{Z}$ is not divisible by $a \in \mathbb{Z}$, a division with remainder can be performed.

Theorem 3.8 (Division of b by a with Remainder r) *If $b \in \mathbb{Z}, a \in \mathbb{N}, a \neq 0$, then there are uniquely determined numbers $z \in \mathbb{Z}$ and $r \in \mathbb{N}$ with $0 \leq r < a$, so*

---

[1] Very large prime numbers play an important role in coding issues. Whenever a larger prime number is found, it is always worth a news report. Keywords for your own research might be titanic, gigantic, mega or Mersenne prime numbers, (Marine Marsenne, French mathematician and catholic priest, 1588–1648).

3.2 Extensions of Number Systems

that

$$b = z \cdot a + r.$$

The numbers b and z are either both negative or both in $\mathbb{N}$. If b is smaller than a, then $z = 0$ and $r = b$.

Proof We first prove the theorem for $b \in \mathbb{N}$. Let

$$S := \{ b - ka \mid k \in \mathbb{N}, \, b - ka \in \mathbb{N} \}.$$

Because $b = b - 0a$ lies in $S \subseteq \mathbb{N}$, it follows that $S \neq \emptyset$. Now, according to Theorem 3.3, S has a smallest element, say $r := b - za$. Because r is the smallest element of S, it follows that $r < a$.

It remains to show that z and r are unique. If $b = z_1 a + r_1 = z_2 a + r_2$ with $0 \leq r_1, r_2 < a$, then $(z_1 - z_2)a = r_2 - r_1$. So if $r_1 = r_2$, then $z_1 = z_2$ because of $a \neq 0$. If, on the other hand, $r_1 \neq r_2$, say $r_1 < r_2$, then $0 < r_2 - r_1 \leq r_2 < a$. So $0 < (z_1 - z_2)a = r_2 - r_1 < a$ and $0 < z_1 - z_2 < 1$ as $0 < a$, a contradiction to $z_1, z_2 \in \mathbb{Z}$.

If now $b \notin \mathbb{N}$, then $b = (-1)y$ with $y > 0$. According to what has just been shown, we have a unique representation $y = za + r$ with $z \in \mathbb{N}, r \in \mathbb{N}$ and $0 \leq r < a$. This then implies $b = -za - r = -za - a + a - r = -(z+1)a + (a-r), 0 \leq a - r < a$. □

This theorem is the basis of the *Euclidean*[2] *Algorithm* (cf. Schöning [67]) for determining the greatest common divisor of two integers.

Rational Numbers

In order to solve equations of the form $ax = b$ for arbitrary $a, b \in \mathbb{Z}$, we must once again extend our number system.

We set

$$\mathbb{Q} := \{ \frac{a}{b} \mid a \in \mathbb{Z}, \, b \in \mathbb{N}, \, a, b \neq 0, \, \gcd(a, b) = 1 \} \cup \{ 0 \}$$

and call this set the **set of rational numbers**. With an appropriate definition of multiplication, there is now a solution $x \in \mathbb{Q}$ for every equation of the form

$$a \cdot x = b \text{ with } a, b \in \mathbb{Q} \text{ and } a \neq 0.$$

---

[2] Euclid of Alexandria, ca. 325–265 BC.

Just as integers can be defined as equivalence classes on $\mathbb{N} \times \mathbb{N}$, rational numbers can also be defined as equivalence classes—now on $\mathbb{Z} \times \mathbb{Z}$ (cf. Exercise 4.29 in Sect. 4.2).

We now show that there are also equations over $\mathbb{Q}$ that have no solution in $\mathbb{Q}$.

Theorem 3.9 ($\sqrt{2}$ is not rational) *The equation $x^2 = 2$ has no solution in $\mathbb{Q}$, i.e. $(\forall x \in \mathbb{Q}): x^2 \neq 2$.*

Proof This proof is conducted by contradiction. Assuming there were $\frac{a}{b} \in \mathbb{Q}$ with $(\frac{a}{b})^2 = 2$, then $a^2 = 2b^2$, so the number 2 would be a divisor of a (we use this as prior knowledge), i.e. there would be an element $\tilde{a}$ in $\mathbb{Z}$ with $a = 2\tilde{a}$. With this, we would then get from the assumption that $(\frac{2\tilde{a}}{b})^2 = 2$. This gives $4\tilde{a}^2 = 2b^2$ and as above, it would follow that 2 must also be a divisor of b, contradicting the definition of $\mathbb{Q}$ (the fractions are cancelled out, are irreducible!). Since a true statement cannot lead to a false one, it is thus shown that our assumption was false, and we have proven the theorem. □

Real Numbers

Now we extend the set of rational numbers to the set of real numbers, which, in particular, also contains the roots of non-negative rational numbers. However, this is not possible constructively. Intuitively, the ***set of real numbers*** $\mathbb{R}$ is the set of all points on the number line. We assume without further discussion that every real number can be represented as an (infinite) generally non-periodic decimal fraction, even though this is a very theoretical assumption and no computer can store such a number. No one can write infinite decimal fractions. The set of all points on the number line corresponds to the representation of real numbers by finite and infinite decimal fractions. A decimal fraction with period 9, of the form $\ldots k\overline{9}$ for $k \in \{0, 1, \ldots, 8\}$, is identified with $\ldots (k+1)$. For example, $0.\overline{9} := 1$, $3.764\overline{9} := 3.765$, and $-84.87\overline{9} := -84.88$.

We identify the rational numbers with the finite or periodic decimal fractions. In particular, $\mathbb{Q} \subsetneq \mathbb{R}$. Every non rational number is called ***irrational***.

Abstractly, $\mathbb{R}$ can be constructed, for example, by *Dedekind's cuts*, as a set of equivalence classes of *interval nestings* or of *Cauchy sequences*.[3]

The connection with the Cauchy sequences arises from the fact that the decimal fractions are actually convergent sequences.

Both constructions are beyond the scope of this book. For details you may check any book on Calculus.

---

[3] Richard Dedekind, German mathematician, 1831–1916; Augustin Louis Cauchy, French mathematician, 1789–1857.

3.2 Extensions of Number Systems

In $\mathbb{R}$ also every equation $x^n = a$ with $a \in \mathbb{R}$, $0 \leq a$, $n \in \mathbb{N}$ has a solution, i.e.

$$(\forall a \in \mathbb{R}, \ 0 \leq a)\,(\forall n \in \mathbb{N})\,(\exists x \in \mathbb{R}) : \ x^n = a.$$

But we still can't solve all equations because one has:

Theorem 3.10 ($\sqrt{-1}$ Is Not Real) *The equation $x^2 = -1$ has no solution in $\mathbb{R}$, i.e.* $(\forall x \in \mathbb{R}) : \ x^2 \neq -1$.

Since we have only hinted at a definition of real numbers, we cannot prove Theorem 3.10, which would otherwise not be difficult.

Complex Numbers

In order to finally be able to solve the equation $x^2 = -1$, we extend $\mathbb{R}$, this time constructively again.

We set $i := \sqrt{-1}$, the *imaginary unit*, defined by the property $i^2 = -1$, and thus obtain as *set of complex numbers*

$$\mathbb{C} := \{a + ib \mid a, b \in \mathbb{R}\}.$$

The multiples $\neq 0$ of i are called *imaginary numbers*.[4] Often $\mathbb{C}$ is also represented as $\mathbb{R} \times \mathbb{R} = \{(a, b) \mid a, b \in \mathbb{R}\}$ and called the *Gaussian number plane*.

Because $(a + ib) + (c + id) = (a + c) + i(b + d)$ we have in the Gaussian number plane a componentwise addition:

$$(a, b) + (c, d) = (a + c, b + d).$$

The multiplication is a bit more complicated: By multiplying out we get $(a+ib)(c+id) = ac + i(ad + bc) + i^2 bd = (ac - bd) + i(ad + bc)$ and therefore we set in the Gaussian number plane:

$$(a, b)(c, d) := (ac - bd, ad + bc).$$

Note that $\mathbb{R} \times \mathbb{R}$ with componentwise addition is called the *Euclidean plane*.

---

[4] Leibniz, in 1702, called imaginary roots of this kind a "fine and wonderful refuge of the divine spirit, almost a hybrid between being and non-being", see Ebbinghaus et al. [24]. There you will find a brief presentation of the "mental tortures" associated with the complex numbers.

Algebraic and Transcendental Numbers

Any number $c \in \mathbb{C}$, for which there exist $a_0, a_1, \ldots a_n \in \mathbb{Z}$ with $a_n \neq 0$, such that

$$a_n c^n + a_{n-1} c^{n-1} + \ldots + a_1 c + a_0 = 0,$$

is called *algebraic number*. The equation

$$a_n x^n + a_{n-1} x^{n-1} + \ldots + a_1 x + a_0 = 0$$

is called *algebraic equation of degree n*. So an algebraic number is a *solution of an algebraic equation*. Numbers that do not appear as a solution of an algebraic equation are called *transcendental numbers*.

The following theorem—despite its name—is usually not proven in lectures on algebra, since its proof needs analytical methods.

Theorem 3.11 (Fundamental Theorem of Algebra) *Every algebraic equation of degree $n \geq 1$ with coefficients $a_i \in \mathbb{C}$, $0 \leq i \leq n$, has at least one solution in $\mathbb{C}$.*

Because of this property, $\mathbb{C}$ is called *algebraically closed*.

Theorem 3.12 (rational $\implies$ algebraic) *Every rational number is algebraic.*

Proof If $q = \frac{a}{b}$ with $a \in \mathbb{Z}$, $b \in \mathbb{N}$, $b \neq 0$, then $bq - a = 0$. □

Theorem 3.13 ($\sqrt{-1}$ is algebraic) *There are algebraic numbers that are not real, e.g. the imaginary unit i.*

Proof The number i is a solution of the (algebraic) equation $x^2 + 1 = 0$. □

For a proof of Theorem 3.14 profound knowledge of Algebra is required. So we cannot give a proof here. It states that the two numbers e and π are transcendental. Surprisingly enough, it turns out that there exist "more" transcendental numbers than algebraic ones, cf. Theorem 5.28.

Theorem 3.14 (e and π are not algebraic) *The Eulerian[5] number e and the circle number π are transcendental real numbers.*

The Eulerian number e can be defined as the (infinite) sum

$$e := \sum_{k=0}^{\infty} \frac{1}{k!} = 1 + 1 + \frac{1}{2} + \frac{1}{6} + \frac{1}{24} + \ldots.$$

---

[5] Leonhard Euler, Swiss mathematician, 1707–1783.

3.2 Extensions of Number Systems

This is to be understood as a limit, since sums can only be finite. So one has to prove convergence and to determine the limit to obtain a numerical value. In addition, transcendence must be proven.

The circle number π can be defined as the circumference of the circle with diameter 1. However, this is not a method for calculation and it does not provide a proof of transcendence.

The background thereof is the prime example of an **unsolvable problem**, namely *squaring the circle*. That is: "Construct, using only compass and ruler (without measurement), a square with the same area as a given circle."

The problem arises from the fact that only rational numbers can be constructed with a compass and ruler. Since we know that π is not rational, the unsolvability is clear.

The other two unsolvable classical problems are *doubling the cube* and *trisecting an angle*. In both cases irrational numbers would have to be constructed.

We have now gathered all the terms and information to be able to draw the **picture of the number systems** (Fig. 3.2).

Exercise 3.15 (Prime Factorization) Use Theorem 3.8 to prove that in $\mathbb{Z}$ every number can be represented as a product of *prime numbers*, for example, $6 = 2 \cdot 3 = (-2) \cdot (-3)$.

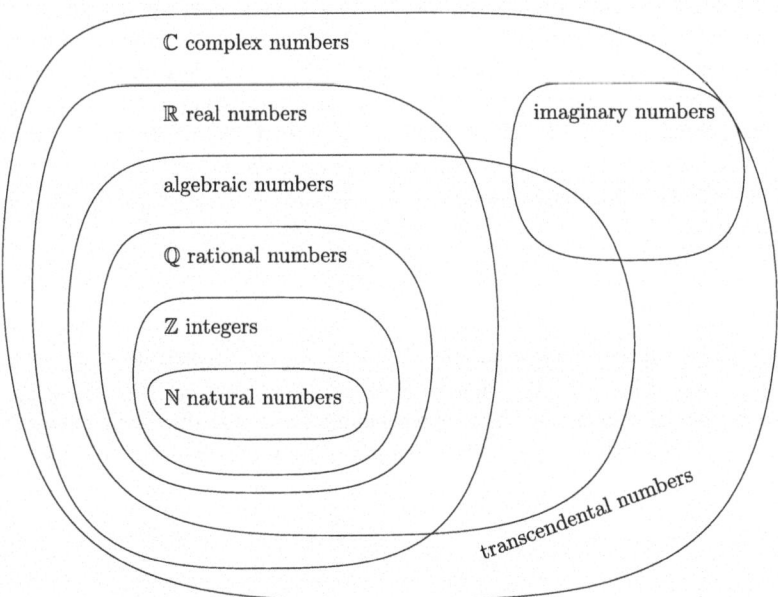

Fig. 3.2 Picture of the number systems

3.3 Numeral Systems

Base of a numeral system, binary numbers, dyadic numbers, dual numbers, octal numbers, hexadecimal numbers, representation of a number to base b, length of the representation, uniqueness

So far, this chapter has been about describing different types of number systems. Another problem is the *representation* of numbers. We usually represent numbers—due to the number of our fingers(?)—in the ***decimal system***, i.e., the numbers written side by side, usually called ***digits***, correspond to multiples of powers of ten.[6]
As examples, we consider

$$1576 = 1 \cdot 10^3 + 5 \cdot 10^2 + 7 \cdot 10^1 + 6 \cdot 10^0$$
$$230.75 = 2 \cdot 10^2 + 3 \cdot 10^1 + 0 \cdot 10^0 + 7 \cdot 10^{-1} + 5 \cdot 10^{-2}.$$

A completely different representation of numbers are the ***Roman numbers*** I, $II, III, IV, V, \ldots$. We can't calculate well with these, but they are still used today for numbering or for the representation of years. They are composed of the digits $I = 1$, $V = 5$, $X = 10$, $L = 50$, $C = 100$, $D = 500$, $M = 1000$. Adjacent digits are added, except the smaller one is to the left of the larger one, then it is subtracted ($IV = 4$). In this way, groups of four identical digits ($\neq M$) are avoided. By convention, at most one smaller digit is written in front of a larger one.

If we retain the idea underlying the decimal system, but replace the ***base*** 10 with another natural number $b \geq 2$, we get other numeral systems, of which the ***binary numbers*** (base 2), the ***octal numbers*** (base 8) and the ***hexadecimal numbers*** (base 16) have proven particularly useful for number representation in computers. In the hexadecimal system, the "digits" $0, 1, 2, \ldots, 9, A, B, C, D, E, F$ are used. Binary numbers are also called ***dyadic numbers*** or as ***dual numbers***.

In the following, the natural number $b \geq 2$ denotes the base of a numeral system. Consider

$$n = a_{k-1} b^{k-1} + a_{k-2} b^{k-2} + \ldots + a_1 b + a_0,$$

with $n \in \mathbb{N} \setminus \{0\}$ and natural numbers $a_0, a_1, \ldots, a_{k-1}$ where $0 \leq a_i < b$ for $0 \leq i < k$ and $a_{k-1} > 0$. Then we call the expression $a_{k-1} a_{k-2} \cdots a_1 a_0$ the ***representation of n to base b***. As a representation for 0, one always chooses $a_0 = 0$ with $k = 1$. The number k is called the ***length of the representation***. We also write

$$n = (a_{k-1} a_{k-2} \cdots a_1 a_0)_b.$$

---

[6] In some Mesoamerican and Inuit cultures numeral systems with base 20 developed, see Mayan and Kaktovik numerals.

3.3 Numeral Systems

Note that a sequence of digits can represent completely different numbers, when using different bases: For example, $1233_4 = 3 \cdot 4^0 + 3 \cdot 4^1 + 2 \cdot 4^2 + 1 \cdot 4^3 = 3_{10} + 12_{10} + 32_{10} + 64_{10} = 111_{10}$ but $1233_{16} = 3 \cdot 16^0 + 3 \cdot 16^1 + 2 \cdot 16^2 + 1 \cdot 16^3 = 3_{10} + 48_{10} + 512_{10} + 4096_{10} = 4659_{10}$.

To find the representation of a number for a given base, we look at how we can calculate it for $b = 10$. We describe two methods using examples and formulate the first as an **algorithm**. That is, we provide a method that produces the desired result (in this case, the decimal representation of n) for every input value (in this case n) after a finite number of steps.

Method 1 Repeated division with remainder by the largest possible power of the base b ("from the front").

Example 3.16 (From the Front)

(1) We calculate the digits of the number $n = 1576$ for its representation to base $b = 10$:

$$\begin{aligned}
10^3 &\le 1576 < 10^4 & 1576 &= \boxed{1} \cdot 10^3 + 576 \\
10^2 &\le 576 < 10^3 & 576 &= \boxed{5} \cdot 10^2 + 76 \\
10^1 &\le 76 < 10^2 & 76 &= \boxed{7} \cdot 10^1 + 6 \\
10^0 &\le 6 < 10^1 & 6 &= \boxed{6} \cdot 10^0 + 0
\end{aligned}$$

(2) Similarly, we calculate the digits of a representation of $n = 27_{10}$ to base $b = 5$:

$$\begin{aligned}
5^2 &\le 27 < 5^3 & 27 &= \boxed{1} \cdot 5^2 + 2 \\
& & 2 &= \boxed{0} \cdot 5^1 + 2 \\
5^0 &\le 2 < 5^1 & 2 &= \boxed{2} \cdot 5^0 + 0 & &\Rightarrow 27_{10} = 102_5
\end{aligned}$$

Algorithm for Method 1 Representation of $n \in \mathbb{N}$ to base $b \ge 2$.

 0. Input: n, b. If $n = 0$, $k = 1$, $a_0 = 0$, go to 3.
 1. Determine $k \in \mathbb{N}$ such that $b^{k-1} \le n < b^k$.
 2. For $i = k-1, k-2, \ldots, 1, 0$ repeat

   (a) Division with remainder: determine a, r such that
   $n = a \cdot b^i + r$ with $0 \le r < b^i$.
   (b) Set $a_i := a$ and $n := r$.

 3. Output: $a_{k-1} a_{k-2} \cdots a_0$.

We have $n = a_{k-1}b^{k-1} + a_{k-2}b^{k-2} + \ldots + a_1b + a_0$. This means $n = (a_{k-1}a_{k-2}\cdots a_0)_b$ is a representation of n to base b and $a_i < b$ for $0 \leq i < k$.

It is noteworthy that this method works without having n in a representation to any base, as long as we can decide which k to choose in Step 1 and which numbers a and r appear in Step 2(a). We can formally secure this using division with remainder.

Theorem 3.17 (Representation to Base b) *If $n, k \in \mathbb{N}$ with $b^{k-1} \leq n < b^k$, then there are uniquely determined numbers $a, r \in \mathbb{N}$ with $0 \leq r < b^{k-1}$, such that*

$$n = a \cdot b^{k-1} + r.$$

One has $0 \leq a < b$ and if $(a_{k-2}a_{k-3}\cdots a_0)_b$ is the unique representation of r to base b, then $(aa_{k-2}a_{k-3}\cdots a_0)_b$ is the unique representation of n to the base b of length k.

We exemplify this for $n = 101_{10}$. Now $101 = 1 \cdot 10^{3-1} + 1$, i.e. $a = 1, r = 1, k = 3$. Then $(a_1a_0)_{10} = (0, 1)$ is the representation of $r = 1$ to the basis 10. And indeed $(101)_{10} = (aa_1a_0)_{10}$.

Since we have assumed that every real number can be written as (infinite) generally non-periodic decimal fraction, non-integers can also be written with respect to any other base b as an infinite b-fraction.

Method 2 Repeated division with remainder by the base b. Here, the digits are generated in reverse order ("from the back").

Example 3.18 (From the Back) Again, we first calculate in (1) the digits of the number $n = 1576$ for its representation to base $b = 10$ and then apply the same procedure to determine in (2) the digits of $n = 13_{10}$ for its representation to the base $b = 2$:

$$
\begin{array}{ll}
(1)\ 1576 = 10 \cdot 157 + \boxed{6} & (2)\ 13 = 2 \cdot 6 + \boxed{1} \\
157 = 10 \cdot 15 + \boxed{7} & 6 = 2 \cdot 3 + \boxed{0} \\
15 = 10 \cdot 1 + \boxed{5} & 3 = 2 \cdot 1 + \boxed{1} \\
1 = 10 \cdot 0 + \boxed{1} & 1 = 2 \cdot 0 + \boxed{1} \quad \Rightarrow 13_{10} = 1101_2
\end{array}
$$

A representation of a number $n \in \mathbb{N}$ to base 1 is called **unary number representation**. Such a representation essentially consists of tally marks. To obtain unique unary representations for numbers, the algorithms in this chapter must be modified. It should also be noted that decimal fractions in the unary numeral system cannot simply be represented by adding a comma. The unary representation of a number n takes up significantly more storage space (namely n characters) than representations using larger bases such as 2 (namely $\log_2(n)$). Therefore it is important in theoretical

3.3 Numeral Systems

computer science how numbers are represented. This leads, for example, to the distinction between polynomial and pseudopolynomial runtime of an algorithm.

Exercise 3.19 (Different Number Representations) Search for representations of numbers from different cultures and times.

Exercise 3.20 (Algorithm "From the Back") Formulate an algorithm for Method 2.

Exercise 3.21 (Roman Numerals) Formulate a Roman representation with respect to a given set of digits $D \subseteq \mathbb{N}$ and find a method to build a representation for given n. Under which conditions is the representation unique?

Relations 4

Relations describe relationships between some or all elements of one or more sets. The exceedingly simple definition of a relation in Sect. 4.1 should not deceive; any kind of relationship and binding can be formally captured with it. In daily life, we encounter such relationships, e.g., between people (Hans is friends with Klaus, Katja is the daughter of Sibylle and Kurt) or between objects and properties (the pillow is soft, the heating is warm). In mathematics, relations describe relationships between numbers (2 is smaller than 5, 9 is divisible by 3) or other mathematical objects (the points P_1, P_2 and P_3 lie on a straight line, the triangles ABC and DEF are congruent). Also, predicates (Sect. 1.7) are relations. In Sect. 4.2, we examine *equivalence relations*. These allow decompositions of sets into disjoint subsets, i.e., they correspond to partitions (Chap. 2). *Order relations* (Sect. 4.3) generalize the notion of order among natural numbers that we established in Chap. 3 by allowing pairs of elements *incomparable*. Other types of relations are *mappings* and *partial mappings*. They play a fundamental role almost everywhere inside and outside mathematics. Therefore, we devote an entire chapter to them, Chap. 5.

There is a very effective set of tools that uses relations and lattices for concept analysis (in mathematics but also in other areas of life), see B. Ganter, R. Wille [30] and G. Stumme, R. Wille [69], as well as R. Wille [76]. Relational algebras and relational models in computer science extensively use relations and mappings.

On the Literature Relations and (partial) mappings are covered in many introductory books. For order relations, compare, for example, the book by Erné [27], which also deals with sets, numbers, relations, and mappings. The book by Davey and Priestley [19] establishes connections to computer science.

4.1 Fundamentals

Binary relation, empty relation, universal relation, diagonal, identity, n-ary relation, arity, unary, ternary, equality of relations, image, preimage, relation graph, coordinate system, complement, union, intersection, (right/left-)inverse, composition, associative law, commutative law, (ir)reflexive, (a/anti)symmetric, transitive, left/right unique, left/right total

A *binary relation between the set A and the set B* is a subset $\varrho \subseteq A \times B$ of the Cartesian product of A and B. Instead of $(a, b) \in \varrho$ we also write $a \varrho b$.

From Chap. 2 we know two special subsets of $A \times B$, which according to this definition are also relations: The relation $\varrho = A \times B$ which, is called *universal relation* and often denoted by ∇, called *Nabla*, and the relation $\emptyset$, called *empty relation*.

If $A = B$, then $\varrho \subseteq A \times A$ is called a *binary relation on* A. We set $\Delta_A := \{(a, a) \mid a \in A\}$ and call Δ_A the *diagonal* or *identity* on A.

Every relation σ between two different sets A and B we can also consider as a relation on a set M with $M \supseteq A \cup B$.

Relations can also represent relationships between the elements of more than two sets. The general definition is: An *n-ary relation of the sets* $A_1, A_2, \ldots, A_n$, ($n \in \mathbb{N}$, $n \geq 1$) is a subset $\varrho \subseteq A_1 \times A_2 \times \ldots \times A_n$. If $A_1 = A_2 = \ldots = A_n = A$, then ϱ is called an *n-ary* relation *on* A. The number n is called the *arity* of the relation. For $n = 1, 2, 3$ an n-ary relation is also called *unary, binary* or *ternary relation* respectively. If ϱ_i are n_i-ary relations on the set A, then $(A, (\varrho_i)_{i \in I})$ is called a *relational system* and the $|I|$-tuple $(n_i)_{i \in I}$ is called the *type* of the system.

Since relations are nothing more than special sets, two relations are called *equal*, if they are equal as sets.

We will restrict ourselves to binary relations in the following.

Let X and Y be sets and let $\varrho \subseteq X \times Y$. The *image of a subset A of X under* ϱ is defined by

$$\varrho[A] := \{ y \in Y \mid (\exists a \in A) : (a, y) \in \varrho \}.$$

Here we write $\varrho(a)$ instead of $\varrho[\{a\}]$, if the set $A = \{a\}$ contains only one element. Accordingly, we define the *preimage of a subset B of Y under* ϱ by

$$\varrho^{\leftarrow}[B] = \{ x \in X \mid (\exists b \in B) : (x, b) \in \varrho \}.$$

For $\varrho^{\leftarrow}[\{b\}]$ we write $\varrho^{\leftarrow}(b)$. Note that $\varrho^{\leftarrow}(b) = \{x \mid ((x, b) \in \varrho\}$ as well as $\varrho(a)$ can contain more than one element.

We now describe some ways to graphically represent binary relations, and explain these using three examples. We draw the *relation graph* of a binary relation ξ on a finite set M, by representing each element a of M as a point P_a in the plane $\mathbb{R}^2$ such that different elements correspond to different points. If $a \xi b$, then an arrow

4.1 Fundamentals

from P_a to P_b is drawn. If there is an arrow from P_a to P_b and another one from P_b to P_a, then these two points are also simply connected by a line. This procedure can also be applied to relations between two different sets as we will show in the example of the relation τ below.

On infinite sets we cannot draw relation graphs. But even here there is an appropriate way of graphical representation, which of course also works for finite sets. We draw a relation $\varrho \subseteq A \times B$ between sets A and B in a *coordinate system* with the A-axis and the B-axis perpendicular to it. An element $(x, y) \in \varrho$ is the corresponding point in the plane. The same principle is for function graphs. These are however somewhat more special, since there are no "stacked" points, i.e., different points with the same x-value cannot occur.

Example 4.1 (Relation Graphs) On the sets $\mathbb{R}, \mathbb{N}$, and $S := \{a, b, c, d, e, f, g, h, i, j\}$, $A := \{$Markus, Anja, Claudia, Hans, Susanne, Rainer$\}$ we consider the relations

$$\varrho := \{(x, y) \in \mathbb{R}^2 \mid x^2 + y^2 = 1\} \subseteq \mathbb{R} \times \mathbb{R},$$

$$\sigma := \{(f, e), (c, b), (h, a), (e, d) (f, g), (a, b), (d, c), (g, h)\} \subseteq S \times S,$$

$\tau := \{$(Susanne, 25), (Anja, 20), (Claudia, 22), (Markus, 22), (Hans, 24), (Rainer, 18)$\} \subseteq A \times \mathbb{N}$.

See Fig. 4.1 for the relation graphs of σ and τ and Figs. 4.2 and 4.3 for representations in the Cartesian coordinate system.

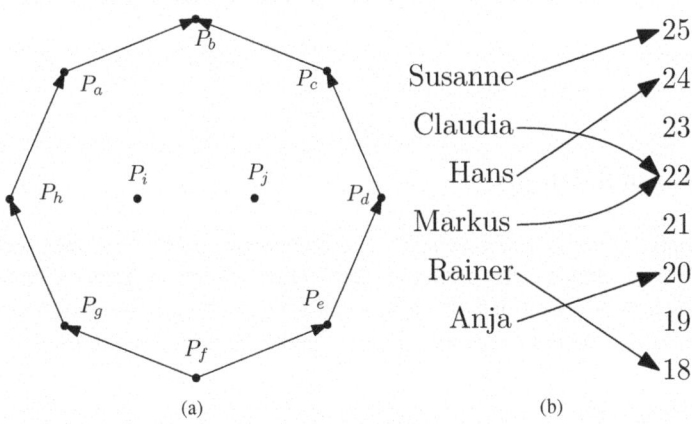

Fig. 4.1 (a) Graph of the relation σ. (b) Graph of the relation τ

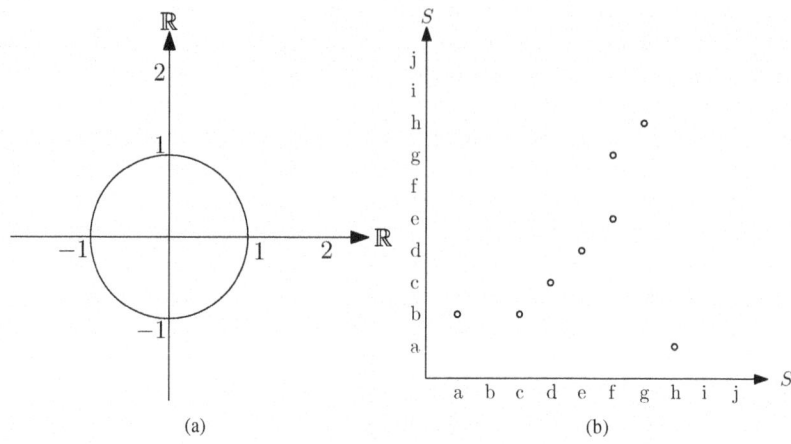

Fig. 4.2 (a) Relation $\varrho \subseteq \mathbb{R} \times \mathbb{R}$ in the (Cartesian) coordinate system. (b) Relation $\sigma \subseteq S \times S$ in the (Cartesian) coordinate system

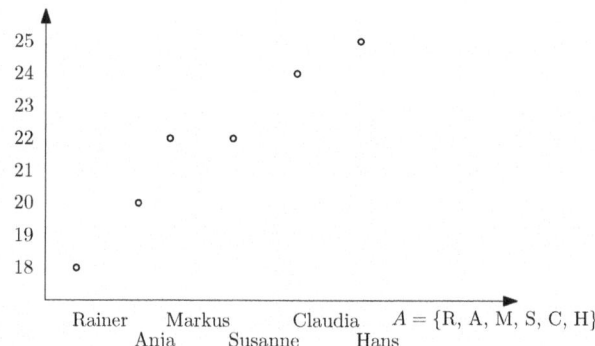

Fig. 4.3 Relation τ in the (Cartesian) coordinate system

Operations of Relations

We can apply operations of sets such as complement, union, and intersection to relations, since relations are particular sets. This is done in Table 4.1. We also provide two further operations of binary relations that are not possible for arbitrary sets, the *converse* and the *composition*. Let ϱ and σ be binary relations:

Table 4.1 Combinations of relations

Complement ϱ'	$a\varrho'b$	$:\Leftrightarrow \neg(a\varrho b)$
Union $\varrho \cup \sigma$	$a\,(\varrho \cup \sigma)\,b$	$:\Leftrightarrow a\varrho b$ or $a\sigma b$
Intersection $\varrho \cap \sigma$	$a\,(\varrho \cap \sigma)\,b$	$:\Leftrightarrow a\varrho b$ and $a\sigma b$
Converse $\varrho^{\leftarrow}$	$b\varrho^{\leftarrow}a$	$:\Leftrightarrow a\varrho b$
Composition $\sigma \circ \varrho$	$a\,(\sigma \circ \varrho)\,d$	$:\Leftrightarrow (\exists c): a\varrho c$ and $c\sigma d$

4.1 Fundamentals

Instead of the word converse, **reversal** or **oppositional** is also common. We read $\sigma \circ \varrho$ as "sigma after rho". If $\varrho \subseteq A \times B$ and $\sigma \subseteq C \times D$, then by definition a pair (a, d) is in the relation $\sigma \circ \varrho$, if $(a, d) \in A \times D$ and there exists $c \in B \cap C$ for which $a\varrho c$ and $c\sigma d$. If there is no such element c for any pair $(a, d) \in A \times D$, in particular, if $B \cap C = \emptyset$, then $\sigma \circ \varrho = \emptyset$. A relation $\sigma \subseteq A \times B$ is called **right-invertible**, if it has a **right inverse**, i.e., if there exists a relation $\varrho \subseteq B \times A$ such that $\sigma \circ \varrho = \triangle_B$; it is called **left-invertible**, if it has a **left inverse**, i.e. if $\lambda \subseteq B \times A$ exists, so that $\lambda \circ \sigma = \triangle_A$. A relation, which is both right- and left-invertible, is called **invertible**, a right-inverse of σ, which is also a left-inverse of σ, is called **inverse** of σ and is denoted by σ^{-1}.

Since relations are sets, we know from Sect. 2.2, that for union and intersection the associative law applies. For the composition we prove this in the following theorem.

Theorem 4.6 (Associative Law) *The composition of binary relations is an associative operation, i.e. for relations $\varrho \subseteq A \times B$, $\sigma \subseteq C \times D$ and $\tau \subseteq E \times F$ one has $\tau \circ (\sigma \circ \varrho) = (\tau \circ \sigma) \circ \varrho$.*

Proof First, from the definition of the composition of binary relations, it follows that $\sigma \circ \varrho \subseteq A \times D$, $\tau \circ \sigma \subseteq C \times F$ and thus both $\tau \circ (\sigma \circ \varrho)$ and $(\tau \circ \sigma) \circ \varrho$ are (possibly empty) subsets of $A \times F$.

For elements $a \in A$ and $f \in F$ now follows:

$$(a, f) \in \tau \circ (\sigma \circ \varrho)$$
$$\Leftrightarrow (\exists x \in D \cap E): \quad (a, x) \in \sigma \circ \varrho \text{ and } (x, f) \in \tau$$
$$\Leftrightarrow (\exists x \in D \cap E, y \in B \cap C): (a, y) \in \varrho, (y, x) \in \sigma \text{ and } (x, f) \in \tau$$
$$\Leftrightarrow (\exists y \in B \cap C): \quad (a, y) \in \varrho \text{ and } (y, f) \in \tau \circ \sigma$$
$$\Leftrightarrow \quad (a, f) \in (\tau \circ \sigma) \circ \varrho.$$

With this, the claim is proven. □

Example 4.7 (Composition Not Commutative) For $\varrho := \{(0, 1)\}$ and $\sigma := \{(1, 2)\}$, $\varrho \circ \sigma = \emptyset$ and $\sigma \circ \varrho = \{(0, 2)\}$.

Properties of Relations

Every relation σ between two different sets A and B can be considered as a relation on a set M with $A \cup B \subseteq M$. Table 4.2 describes possible properties of as binary relation ϱ on a set M, which we will need later, to define, for example, *order relations* or *equivalence relations*.

The properties in Table 4.3 are meaningful when we consider a binary relation ϱ between different sets A and B. We will use these terms to define *mappings*.

Table 4.2 Properties of relations

Reflexive	$(\forall a \in M)$:	$(a, a) \in \varrho$
Irreflexive	$(\forall a \in M)$:	$(a, a) \notin \varrho$
Symmetric	$(\forall a, b \in M)$:	$(a, b) \in \varrho \Rightarrow (b, a) \in \varrho$
Asymmetric	$(\forall a, b \in M)$:	$(a, b) \in \varrho \Rightarrow (b, a) \notin \varrho$
Antisymmetric	$(\forall a, b \in M)$:	$(a, b), (b, a) \in \varrho \Rightarrow a = b$
Transitive	$(\forall a, b, c \in M)$:	$(a, b), (b, c) \in \varrho \Rightarrow (a, c) \in \varrho$

Table 4.3 Unique and total

Left unique	$(\forall a_1, a_2 \in A, b \in B)$: $(a_1, b), (a_2, b) \in \varrho \Rightarrow a_1 = a_2$
Uniquely right	$(\forall a \in A, b_1, b_2 \in B)$: $(a, b_1), (a, b_2) \in \varrho \Rightarrow b_1 = b_2$
Left total	$(\forall a \in A)(\exists b \in B): (a, b) \in \varrho$
Right total	$(\forall b \in B)(\exists a \in A): (a, b) \in \varrho$

Exercise 4.10 (Composition of Right Unique Relations) If $f \subseteq A \times B$ and $g \subseteq C \times D$ are right unique relations, then $g \circ f \subseteq A \times D$ is also right unique, even for $B \cap C = \emptyset$, i.e., for $g \circ f = \emptyset$ as a relation, or $g \circ f = \theta$ as a mapping.

Exercise 4.11 (Presentation) How can a binary relation be presented by a table or a matrix?

Exercise 4.12 (Relations) Let A, B, C and D be sets.

(a) How many binary relations between two sets A and B are there, if $|A| = m$ and $|B| = n$?
(b) Determine the composition $\varrho \circ \sigma$ of the relations $\varrho := A \times B$ and $\sigma := C \times D$.

Exercise 4.13 (Kinship) On a set A of women, the following relations are defined:

$a\sigma b :\Leftrightarrow a$ is sister of b,
$a\tau b :\Leftrightarrow a$ is daughter of b.

(a) Describe the kinship relationship of a to b, for the following cases in the style of "$a\varrho b \Rightarrow a$ is cousin of b."

(α) $a\sigma^{\leftarrow} b$ (ϵ) $a(\sigma \circ \sigma)b$
(β) $a\tau^{\leftarrow} b$ (ζ) $a(\sigma \circ \tau)b$
(γ) $a(\tau \circ \tau)b$ (η) $a(\tau \circ \sigma)b$
(δ) $a(\tau \circ \tau)^{\leftarrow} b$ (θ) $a(\sigma \circ \tau)^{\leftarrow} b$

(b) Why do we have only "$\Rightarrow$" in (γ), (ϵ), (η) and (θ) and not "$\Leftrightarrow$"?

Fig. 4.4 The relation ϱ

(c) Which of the properties *(ir)reflexive, (anti)symmetric, transitive* do the relations σ and τ have?

Exercise 4.14 (Complement, Converse, Composition) The area occupied by the letter F in Fig. 4.4 is a subset ϱ of the depicted rectangle $A \times B$, thus a binary relation. – What do ϱ', $\varrho^{\leftarrow}$ or $\varrho \circ \varrho$ out?

Exercise 4.16 (Inverse Relations) Let $A := \{a, b, c, d\}$ and $B := \{1, 2\}$.

(a) Find a relation $\sigma \subseteq B \times A$ with at least two right inverses ϱ_1 and ϱ_2. – What properties must a binary relation have in order to have exactly one right inverse?
(b) Find a relation $\sigma \subseteq A \times B$ with at least two left inverses λ_1 and λ_2. – What properties must a binary relation have in order to have exactly one left inverse?
(c) Show: A relation $\sigma \subseteq X \times Y$ is invertible if and only if $\sigma^{\leftarrow}$ is the only right and left inverse of σ.
(d) How could the concept of (one-sided or two-sided) invertibility be generalized to a *partial* invertibility? – Why is it not sensible to denote the relation $\sigma^{\leftarrow}$ with σ^{-1}?

Exercise 4.17 (Relations in Everyday Life) Search in your everyday environment for sets with relations that clearly illustrate the differences between the defined properties.

4.2 Equivalence Relations

Equivalence relation, equivalence class, partitions, factor set, quotient set, representative (system), residue class, congruent modulo n, compatibility relation, construction of $\mathbb{Z}$ from $\mathbb{N}$, $\mathbb{Q}$ from $\mathbb{Z}$, transitive closure

Equivalence relations serve to group elements of a set into so-called equivalence classes. In order to be able to do this, a relation must have special properties:

A binary relation on a set M is called an ***equivalence relation***, if it is reflexive, symmetric, and transitive. The alphabetical order $(r-s-t)$ should help not to forget these properties.

Examples 4.18

(1) The relation "is parallel to or equal to" is an equivalence relation on the set $\mathcal{G}$ of all lines, for example in the plane $\mathbb{R}^2$.
(2) The relation "is perpendicular to" is not an equivalence relation on the set $\mathcal{G}$: it is symmetric, but neither reflexive nor transitive.
(3) For any set A, the all-relation $\nabla_A := A \times A$ on A, and the diagonal $\triangle_A := \{(x, x) \mid x \in A\}$ are equivalence relations on A.
(4) The empty relation $\emptyset$ is only an equivalence relation on the empty set, it is not reflexive on a non-empty set.

Equivalence Classes and Partitions

Every relation ϱ on a set M provides subsets of M: the images $\varrho(x) = \{y \in M \mid x\varrho y\}$ of $x \in M$ under ϱ. If ϱ is an equivalence relation, these subsets have particularly beautiful properties, which we describe in Corollary 4.20.

Lemma 4.19 (Images under a Relation) *Let $\varrho \subseteq M \times M$.*

1. *If ϱ is reflexive, then*
    (a) $(\forall x \in M): \quad x \in \varrho(x)$.
    (b) $(\forall x, y \in M): \quad \varrho(x) = \varrho(y) \Rightarrow x \in \varrho(y)$.
    (c) $(\forall x, y \in M): \quad \varrho(x) \cap \varrho(y) = \emptyset \Rightarrow x \notin \varrho(y)$.
2. *If ϱ is transitive, then $(\forall x, y \in M): \quad x \in \varrho(y) \Rightarrow \varrho(x) \subseteq \varrho(y)$.*
3. *If ϱ is symmetric and transitive, then*
    (a) $(\forall x, y \in M): \quad x \in \varrho(y) \Rightarrow \varrho(x) = \varrho(y)$.
    (b) $(\forall x, y \in M): \quad x \notin \varrho(y) \Rightarrow \varrho(x) \cap \varrho(y) = \emptyset$.

Proof

1. (a) is clear and (b) and (c) follow immediately.
2. If ϱ is transitive and $x \in \varrho(y)$, then for $z \in \varrho(x)$ due to $y\varrho x$ and $x\varrho z$, it follows that $y\varrho z$.
3. If ϱ is transitive and $x \in \varrho(y)$, then with 2. follows that $\varrho(x) \subseteq \varrho(y)$. For each $z \in \varrho(y)$ from $y\varrho z$ and $y\varrho x$ follows that $x\varrho z$, due to symmetry and transitivity. So $z \in \varrho(x)$, and therefore we have equality in (a). We prove (b) by contraposition: If $z \in \varrho(x) \cap \varrho(y)$, we get $y\varrho z$ and $z\varrho x$, so $y\varrho x$, i.e. $x \in \varrho(y)$. □

Corollary 4.20 (Equivalence Relations) *If ϱ is an equivalence relation on a set M, then*

4.2 Equivalence Relations

1. $(\forall x \in M): \quad x \in \varrho(x)$.

2. $(\forall x, y \in M): x \in \varrho(y) \Leftrightarrow \varrho(x) = \varrho(y)$.

3. $(\forall x, y \in M): x \notin \varrho(y) \Leftrightarrow \varrho(x) \cap \varrho(y) = \emptyset$.

The Corollary states that every element of the set M lies in a set $\varrho(x)$. Moreover, two sets $\varrho(x)$ and $\varrho(y)$ are either equal or have no element in common. This means that these sets form a *partition* of M. Compare Sect. 2.2. We therefore call $\varrho(x)$ the **equivalence class** or **class of x by ϱ** and instead of $\varrho(x)$ we also use the notation $[x]_\varrho$ for the equivalence class belonging to $x \in M$.

If, conversely, an arbitrary partition $M = \dot\bigcup_{i \in I} M_i$ of M is given, then by $\varrho := \{(x, y) \in M \times M \mid (\exists i \in I): x, y \in M_i\}$ an equivalence relation ϱ is defined. Its classes are exactly the blocks of the given partition.

This leads to

Theorem 4.21 (Main Theorem on Equivalence Relations) *Let M be a non-empty set. The different classes of an equivalence relation on M form a partition of M. The blocks M_i of a partition $M = \dot\bigcup_{i \in I} M_i$ of M are the classes of the (uniquely determined) equivalence relation ϱ on M, defined by*

$$x \varrho y :\Leftrightarrow (\exists i \in I): x, y \in M_i.$$

Proof Only the uniqueness of ϱ remains to be shown. If two equivalence relations ϱ and σ have the same classes, then $\varrho = \sigma$, since for all $x, y \in M$ one has:

$$(x, y) \in \varrho \Leftrightarrow x \varrho y \Leftrightarrow y \in [x]_\varrho = [x]_\sigma \Leftrightarrow x \sigma y \Leftrightarrow (x, y) \in \sigma.$$

□

Factor Sets and Representative Systems

Now we consider the equivalence classes of M with respect to an equivalence relation ϱ as elements of a set. We denote this set by M/ϱ and call it **factor set of M by ϱ**. Often one also finds the term **quotient set**.

If $M = \dot\bigcup_{i \in I} M_i$ is a partition of M and one chooses any $m_i \in M_i$ for each $i \in I$, then m_i is called a **representative** of the block M_i. The set $\{m_i \mid i \in I\}$ is called a **representative system** of this partition.[1]

Every element of an equivalence class can be chosen as a representative of this class. Often, however, the equivalence classes contain elements that stand out due

---

[1] It is not obvious that one can simultaneously choose a representative from infinitely many blocks. In mathematics, this guaranteed by the already mentioned Axiom of Choice.

to special properties. We can refer to such elements as "beautiful" and choose them as representatives. The "beautiful" lines in the set $\mathcal{G}$ of all lines of $\mathbb{R}^2$, could be considered to be the lines through the origin $(0, 0) \in \mathbb{R}^2$. These form a representative system for the classes of the relation $\varrho :=$ "is parallel to or equal", i.e. for the elements of the factor set $\mathbb{R}^2/_\varrho$.

Residue Classes

On the set $\mathbb{Z}$ of integers, we define for each $n \in \mathbb{N}$ the equivalence relation " $\equiv$ mod n" by

$$x \equiv y \bmod n \quad :\Leftrightarrow \quad n \mid (x - y),$$

where $\mid$ is the divisor relation (from Sect. 3.2) i.e.

$$x \equiv y \bmod n \quad \Leftrightarrow \quad (\exists z \in \mathbb{Z}) : zn = x - y \quad \Leftrightarrow \quad x - y \in n\mathbb{Z},$$

where we set $n\mathbb{Z} := \{ nz \mid z \in \mathbb{Z} \}$.

If $x \equiv y \bmod n$, we say x **is congruent to** y **modulo** n (instead of equivalent). This expresses that this relation is compatible with the structure of $\mathbb{Z}$, i.e. with addition and multiplication. This will be discussed in Chap. 7, Lemma 7.2, see also Exercise 4.27. The classes with respect to $\equiv \bmod n$ are called **residue classes modulo** n.

For $n = 0$ we get the equality relation $\Delta_\mathbb{Z}$, i.e. all equivalence classes are single-element, so each class corresponds exactly to one element of $\mathbb{Z}$. For $n = 1$ we get the universal relation $\nabla_\mathbb{Z}$, i.e. there is only one single equivalence class, namely all of $\mathbb{Z}$.

A common representative system for the residue classes modulo n are the remainders if divided by n, i.e. the numbers $0, 1, 2, \ldots, n - 1$.

Notation For the factor set of $\mathbb{Z}$ by $\equiv \bmod n$ we write $\mathbb{Z}/_{n\mathbb{Z}}$, $\mathbb{Z}_n$ or also $\mathbb{Z}/_n$. Instead of $x \equiv y \bmod n$ also $x \equiv y\,(n)$ is common. And the equivalence class of x is denoted by $[x]_n$ or $\overline{x}$, if it is clear which n is meant. A simple calculation shows that $x \equiv y \bmod n$ for non-negative x and y exactly if they have the same remainders if divided by n.

Example 4.22 ($\mathbb{Z}_3$) For $n = 3$, the set $\mathbb{Z}_3$ contains the elements

$[0]_3 = \overline{0} = \{\ldots, -6, -3, 0, 3, 6, \ldots\}$,
$[1]_3 = \overline{1} = \{\ldots, -5, -2, 1, 4, 7, \ldots\}$,
$[2]_3 = \overline{2} = \{\ldots, -4, -1, 2, 5, 8, \ldots\}$.

4.2 Equivalence Relations

These are sets themselves. So $\mathbb{Z}_3 = \{\bar{0}, \bar{1}, \bar{2}\}$ is a finite set, while $\mathbb{Z}$ and all three residue classes modulo 3 are infinite. Of course, $\{9, -8, 11\}$ is also a representative system for the residue classes modulo 3, although it seems not very reasonable.

Exercise 4.23 (Partitions) Give the equivalence relations for the following partitions of the set $\{1, 2, 3, 4, 5, 6\}$ and draw the relation graphs

(a) $\{\{1, 2\}, \{3, 4\}, \{5, 6\}\};$ (b) $\{\{1, 2, 3, 4\}, \{5\}, \{6\}\}.$

Exercise 4.24 (Properties of Binary Relations)

(a) Show for every binary relation ϱ on a set M: ϱ is transitive $\Leftrightarrow \varrho \circ \varrho \subseteq \varrho$.
(b) Draw the graphs of binary relations on the set $\{1, 2, 3\}$, which are r–s–¬t, r–¬s–t, ..., ¬r–¬s–t, ¬r–¬s–¬t. Here, ¬t stands for *not transitive* and so on.

Exercise 4.25 (Compatibility Relations) A relation ϱ is called *compatibility relation*, if it is reflexive and symmetric.

(a) Let $M :=$ {bit, coffee, night, computer, exercise} and for $x, y \in M$ let $x \varrho y :\Leftrightarrow$ x and y have at least one letter in common.
Show that ϱ is a compatibility relation on M and draw its relation graph.
Show that ϱ is not transitive, thus not an equivalence relation.
(b) How could one generalize the concept of equivalence class, to define a *compatibility set*?
Instead of the partition $M = \bigcup_{i \in I} A_i$ of M into equivalence classes $A_i, i \in I$ a "cover" $M = \bigcup_{i \in I} V_i$ of M through compatibility sets $V_i, i \in I$, should be achieved.

Exercise 4.26 (Divisibility) Examine the divisor relation for the following sets. Is it (ir)reflexive, (anti)symmetric or transitive.
(α) $\mathbb{N}$, (β) $\mathbb{Z}$, (γ) $\mathbb{Q}$, (δ) $\{2, 3, 4, 6\}$.

Exercise 4.27 (Addition $\bmod n$) Show that the modulo-relation defined by $x \equiv y \bmod n \Leftrightarrow n | (x - y)$ on $\mathbb{Z}$ is an equivalence relation for any $n \in \mathbb{N}$ and that for all $x_1, x_2, y_1, y_2 \in \mathbb{Z}$ one has:

$$x_1 \equiv x_2 \bmod n \text{ and } y_1 \equiv y_2 \bmod n \Rightarrow x_1 + y_1 \equiv x_2 + y_2 \bmod n.$$

Exercise 4.28 (Construction of $\mathbb{Z}$ from $\mathbb{N}$) Show that
$(x, y)\nu(u, v) :\Leftrightarrow x + v = u + y$ $(x, y, u, v \in \mathbb{N})$ defines an equivalence relation on $\mathbb{N} \times \mathbb{N}$. Write out the following classes:
$[(1, 2)]_\nu$, $[(7, 9)]_\nu$, $[(3, 1)]_\nu$ and $[(5, 5)]_\nu$.

Exercise 4.29 (Construction of $\mathbb{Q}$ from $\mathbb{Z}$) Show that $(x, y) \sim (u, v) :\Leftrightarrow xv = yu$ $(x, y, u, v \in \mathbb{Z}, y, v \neq 0)$ defines an equivalence relation on $\mathbb{Z} \times (\mathbb{Z} \setminus 0)$. Write out the following classes: $[(0, 1)]_\sim$, $[(1, 2)]_\sim$, $[(-2, 1)]_\sim$ and $[(5, 5)]_\sim$.

Exercise 4.30 (The Transitive Closure of a Binary Relation) The *transitive closure* ϱ^+ *of a binary relation* ϱ on a set M is the (smallest with respect to inclusion) relation that contains ϱ and is transitive. Set $\sigma := \varrho \cup \varrho^2 \cup \ldots \cup \varrho^{|M|}$ and show that every transitive relation τ with $\varrho \subseteq \tau$ also contains σ. The proof that σ is transitive is somewhat cumbersome. We then get that $\sigma = \varrho^+$.

Consider the binary relation

$$\varrho := \{(1, 2), (1, 4), (2, 3), (2, 5), (4, 2), (4, 5), (5, 6), (6, 3)\}$$

on the set $\{1, 2, 3, 4, 5, 6\}$.

(a) Draw the relation graphs for ϱ^k, $k \in \mathbb{N}$, where $\varrho^1 := \varrho$, $\varrho^{k+1} := \varrho^k \circ \varrho$.
(b) Which of the powers ϱ^k, $k \in \mathbb{N}$, are transitive?
   Is $\varrho \cup \varrho^2 \cup \ldots \cup \varrho^k$ transitive for any $k \in \mathbb{N}$?

Exercise 4.31 (Coin Denominations) Is it possible to find a currency system with only two types of coins (worth a and b Cents) so that any sum $n \geq 30$ Cents can be paid with it? Can one furthermore choose a and b so that n is exactly the "magic limit", i.e., 29 Cents cannot be paid with a- and b-Cent coins? If yes, what different solutions (a, b) are there?

The following steps should lead to a complete solution of this problem.

1. Concrete examples
   (α) What amounts can be paid if we choose
       $(a, b) = (5, 10), (4, 6), (3, 8)$?
   (β) For each $(a, b) = (3, 7), (3, 8), (5, 6), (5, 7)$, draw a coordinate system, where the horizontal axis carries the integer multiples xa of a and the vertical axis carries the integer multiples yb of b. Plot the values $xa + ya$ at the point (xa, yb) (at least for $-3 \leq x \leq b, 0 \leq y \leq a + 2$).
       See Fig. 4.5 for the example: $(a, b) = (2, 3)$.

   - Where are the payable amounts in the new currency system?
   - Where are the non-payable amounts?
   - From which amount n are all higher amounts payable? (Create a table for a, b, n.)

   (γ) Mark in the diagrams from (β) the residue classes mod a and answer for each of the pairs (a, b)—if possible also for a general pair (a, b)—the following questions:

4.3 Order Relations

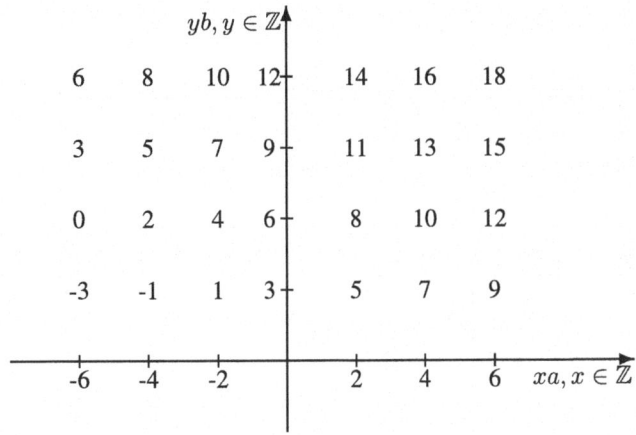

Fig. 4.5 Coin denominations

- What is the largest non-payable amount in each residue class and where can it be found in the coordinate system?
- What is the largest non-payable amount for this pair (a, b) and where is it located in the coordinate system?

2. General solution: Let $a, b \in \mathbb{N}$ with $\gcd(a, b) = 1$.
   (α) Show that $\{\, 0, b, 2b, 3b, \ldots, (a-1)b \,\}$ is a representative system for the residue classes mod a. Hint: Show that for $0 \le i, j \le a-1$, and $p, q, r \in \mathbb{N}$ one has:

   $$ib = pa + r \text{ and } jb = qa + r \quad \Rightarrow \quad i = j,$$

   because this means that all numbers ib ($0 \le i \le a-1$) lie in different residue classes. – Argument for the proof: Because of $\gcd(a, b) = 1$ one has

   $$(\forall u, v \in \mathbb{Z}) \; au = bv \Rightarrow a \mid v, \; b \mid u \,.$$

   (β) What is the largest non-representable amount of money in the residue class of ib? – What is the largest non-representable amount of money overall?
   (γ) Justify that with this the task is solved.

4.3 Order Relations

Order relation, partial order, quasi-order, partially ordered set, poset, total (= linear) order, chain, incomparable, covered by, divisor relation, component-wise/lexicographic order, Hasse diagram, x-downwards, minimal/maximal element, minimum/maximum, smallest/largest element, (largest/smallest) lower/upper bound, infimum/supremum, lower/upper semilattice, lattice, inf/sup-semilattice, complete (upper/lower) lattice

A binary relation on a set M is called **order relation** or **(partial) order**, if it is reflexive, antisymmetric and transitive. If it is only reflexive and transitive, it is called a *quasi-order*.

Order relations are often represented by the symbol $\leq$ (or by $\subseteq$, if the elements of M themselves are sets). We say: "$(M, \leq)$ is an **ordered** or **partially ordered set**" abbreviated *poset*. An order relation $\leq$ on a set M is called **total** or **linear order**, if

$$(\forall x, y \in M): \quad x \leq y \text{ or } y \leq x.$$

In this case, the poset $(M, \leq)$ is also called a **chain**: Any two of its elements are comparable.

If $(M, \leq)$ is a poset, we denote the complement of the relation $\leq$ **less than or equal to** by $\not\leq$ **not less than or equal to** and its converse by $\geq$ **greater than or equal to**. Moreover, we use the following relations:

less than: $\quad a < b :\Leftrightarrow a \leq b$ and $a \neq b$,

incomparable: $\quad a \| b :\Leftrightarrow a \not\leq b$ and $b \not\leq a$,

is covered by: $\quad a \lessdot b :\Leftrightarrow \begin{cases} a \leq b \text{ and for all } c \in M \text{ one has:} \\ a \leq c \leq b \Rightarrow a = c \text{ or } c = b \end{cases}$

with $a, b \in M$. The complement and converse of $<$ are analogously denoted with $\not<$ and $>$.

Examples 4.33 (Orders)

(1) Every set of sets is a poset with respect to the inclusion $\subseteq$. In this way, we even get representatives for all possible posets, as Theorem 4.37 will show.
(2) The divisor relation defined in Sect. 3.2 is an order relation on the set $\mathbb{N}$ of natural numbers, i.e., for $a, b \in \mathbb{N}$. It is not an order relation on $\mathbb{Z}$ as a whole, as it is not antisymmetric: $2 \mid -2$ and $-2 \mid 2$, although $2 \neq -2$.
(3) Let M be a set of judo fighters, then

$$a \prec b \quad :\Leftrightarrow \quad a \text{ defeats } b \text{ in competition}, \quad a, b \in M$$

defines a binary relation $\prec$ on M, which is usually not an order relation: It is generally neither reflexive, nor antisymmetric or transitive.
(4) The sets $\mathbb{N}, \mathbb{Z}, \mathbb{Q}$ and $\mathbb{R}$ with their natural order are chains.

Example 4.34 (Orders on the Cartesian Product) If $(A, \leq_1)$ and $(B, \leq_2)$ are posets, then

4.3 Order Relations

$$(a_1, b_1) \leq_k (a_2, b_2) \quad :\Leftrightarrow \quad a_1 \leq_1 a_2 \text{ and } b_1 \leq_2 b_2$$

defines an order relation on the Cartesian product $A \times B$, which we call the **componentwise order**.

Note: The complex numbers $\mathbb{C} \cong \mathbb{R} \times \mathbb{R}$ with this order do not form a chain.
Through

$$(a_1, b_1) \leq (a_2, b_2) \quad :\Leftrightarrow \quad a_1 <_1 a_2 \text{ or } (a_1 = a_2 \text{ and } b_1 \leq_2 b_2)$$

another order relation is defined on $A \times B$, the so-called **lexicographic order**. It is a total order if $\leq_1$ and $\leq_2$ are total order relations.

Hasse Diagram

The **Hasse diagram** of a poset $(M, \leq)$ is obtained by representing each element a of M as a point P_a in the plane $\mathbb{R}^2$. Now different elements a and b correspond to different points P_a and P_b, such that the second coordinate of P_a is smaller than that of P_b if $a < b$. That is P_b lies "higher" than P_a. If $a \lessdot b$, we connect P_a and P_b by a line.

The Hasse diagram of a poset $(M, \leq)$ thus contains fewer edges than the relation graph of $\leq$, as we only connect those points whose corresponding elements are in the relation $\lessdot$.

Example 4.35 (Power Sets) In Fig. 4.6 we show the power set $P_3 := \wp(\{1, 2, 3\})$ and a subset M of the power set $P_7 := \wp(\{1, 2, \ldots, 7\})$, both ordered by inclusion.

We will now show how every ordered set can be understood as a set of sets, ordered by inclusion.

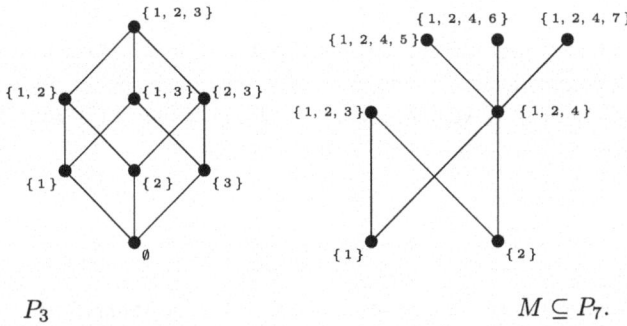

Fig. 4.6 Inclusion orders

Theorem 4.37 (Representation of Ordered Sets) *For every poset $(M, \leq)$ there is a subset of the power set $\wp(M)$, ordered by inclusion, which has the same Hasse diagram as $(M, \leq)$.*

Proof Let $(M, \leq)$ be given. In the Hasse diagram of M each $x \in M$ corresponds to a point $P_x \in \mathbb{R}^2$, as described. We now assign to each $x \in M$ a subset "x downwards" of M, defined by $\downarrow x := \{m \in M \mid m \leq x\}$. We show $x \leq y \Leftrightarrow \downarrow x \subseteq \downarrow y$.

On the one hand, $x \leq y$ implies that for all $m \in M$ one has $m \in \downarrow x \Rightarrow m \leq x \leq y \Rightarrow m \in \downarrow y$. Thus $\downarrow x \subseteq \downarrow y$.

On the other hand, $\downarrow x \subseteq \downarrow y$ implies $x \in \downarrow x \subseteq \downarrow y$. Thus $x \leq y$.

Now it is easy to see that $x \lessdot y \Leftrightarrow \downarrow x \subset \downarrow y$. This proves the claim. (Analogous to $\lessdot$, the symbol $\subset\!\!\!\cdot$ is used for the relation "is covered by " for $\subseteq$.) □

Special Elements of a Poset

Let $(M, \leq)$ be a poset and A a subset of M, carrying the same order. In Table 4.4 we define:

Table 4.4 Special elements of a poset

Element	Designation	Condition
$a \in A$	a *minimal element of* A	$(\forall x \in A) : x \leq a \Rightarrow x = a$
$b \in A$	a *maximal element of* A	$(\forall x \in A) : b \leq x \Rightarrow b = x$
$a \in A$	min A, the *minimum of* A, the *smallest element of* A	$(\forall x \in A) : a \leq x$
$b \in A$	max A, the *maximum of* A, the *largest element of* A	$(\forall x \in A) : x \leq b$
$m \in M$	a *lower bound of* A	$(\forall x \in A) : m \leq x$
$m \in M$	an *upper bound of* A	$(\forall x \in A) : x \leq m$

Note that neither lower nor upper bounds of A must lie in A! If a lower bound of A is in A, it is automatically the minimum of A. The same applies to upper bounds and the maximum of A. If we denote the set of lower bounds of A with A^{lb} and the upper ones with A^{ub}, we can formally express this by

$$a = \min A \Leftrightarrow a \in A \cap A^{lb},$$
$$a = \max A \Leftrightarrow a \in A \cap A^{ub}.$$

If there is a **greatest lower bound of** A, we call this the **infimum of** A, in short: inf A. Correspondingly the minimum of A^{ub} is called the **smallest upper bound of**

4.3 Order Relations

Fig. 4.7 Poset A

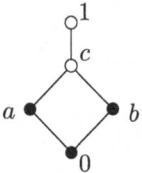

A or also the **supremum of** A, in short: sup A, so

$$\inf A := \max A^{lb} \quad \text{and} \quad \sup A := \min A^{ub}.$$

Example 4.39 (Minimax) Figure 4.7 shows a poset M with a subset A marked by black circles.

Here we get

a and b are maximum elements of A,
1 is the maximum element of M,
0 is the minimum element of A and of M,
A has no greatest element, 1 is the greatest element of M,
0 is the smallest element of A and of M,
$\sup(A) = c$, $\sup(M) = 1$, $\inf(A) = \inf(M) = 0$,
$A^{ub} = \{1, c\}$, $M^{ub} = \{1\}$, $A^{lb} = M^{lb} = \{0\}$.

Theorem 4.41 (Large and Small)

$$\textit{The greatest element} \quad \text{is} \quad \begin{cases} \textit{maximal} \\ \textit{the supremum} \end{cases} \quad \text{which is} \quad \textit{an upper bound,}$$

Further implications are not valid in general. Correspondingly, we get:

$$\textit{the smallest element} \quad \text{is} \quad \begin{cases} \textit{minimal} \\ \textit{the infimum} \end{cases} \quad \text{which is} \quad \textit{a lower bound.}$$

Proof Using the above example, we see that the not given implications do not apply:

The element a is maximal in A, but not the greatest element, supremum or upper bound of the set, c is the supremum of A, but not a maximal or even greatest element of this set and 1 is an upper bound, but neither the supremum nor a maximal element of A.

Now we prove the claimed implications: From the definition follows that the greatest element $g \in A$ is maximal in A. It is the supremum, since it lies in $A \cap A^{ub}$, and for every $z \in A^{ub}$ one has $g \leq z$, since $g \in A$. Therefore $g = \min(A^{ub})$.

That the smallest upper bound is an upper bound follows from the definition.

The corresponding implications for the "small ones" are proven analogously. □

Semilattices and Lattices

A poset $(M, \leq)$ is called a *lower semilattice* or *inf-semilattice* if for any two elements $x, y \in M$ the infimum $\inf\{x, y\}$ exists in M. Correspondingly, $(M, \leq)$ is called an *upper semilattice* or *sup-semilattice*, if for any two elements $x, y \in M$ the supremum $\sup\{x, y\}$ exists in M. A poset, which is both a lower and an upper semilattice, is called a *lattice*.

Since $\inf\{x, y, z\} = \inf\{\inf\{x, y\}, z\}$, the infimum also exists in a lower semilattice for any three-element set. With complete induction it is shown that even every finite subset of a lower semilattice has an infimum.

If every (in particular infinite) subset of a lower semilattice has an infimum, then it is called a *complete lower semilattice*. Correspondingly, a *complete upper semilattice* is defined. For a *complete lattice* we require that every subset has both an infimum and a supremum. Since a finite (semi)lattice only has finite subsets, it is always complete.

We will return to lattices in Example 8.20. There we will see that infimum and supremum can be interpreted as operations.

Examples 4.42 (Lattices)

(1) The sets $\mathbb{Q}$ and $\mathbb{R}$ of rational and real numbers, each with their natural order, are lattices, but not complete lattices: In $(\mathbb{Q}, \leq)$ the set of all rational numbers that are less than π has no supremum. In $\mathbb{R}$ it has a supremum (namely π), but no infimum. Closed intervals in $\mathbb{R}$, such as $[0, 1]$, are complete lattices.
(2) The power set $\wp(M)$ of any set M, ordered by inclusion, is a complete lattice, in which the supremum of a set (of subsets of M) is the union of the subsets and the infimum is their intersection.
(3) In Example 4.35, P_3 is a lattice, but M is neither an upper nor lower semilattice, since for example the set $\{\{1\}, \{2\}\}$ has neither a supremum (although two upper bounds) nor an infimum (not even a lower bound).
(4) In Example 4.39 we see that M is a lattice, A is a lower semilattice, but not an upper semilattice, so it is not a lattice.

Exercise 4.43 (Orders on Equivalence Relations)

Let N be a 4-element set.

(a) Give the equivalence classes for all equivalence relations on N.
(b) Find equivalence relations $\varrho_0, \varrho_1, \varrho_2, \varrho_3, \varrho_4$ on N, which fulfill the inclusions $\varrho_0 \subseteq \varrho_1, \varrho_1 \subseteq \varrho_2, \varrho_1 \subseteq \varrho_3, \varrho_2 \subseteq \varrho_4, \varrho_3 \subseteq \varrho_4$ shown in Fig. 4.8, and give their graphs.
(c) Define the relation $\varrho \subseteq M \times M$ on the set

$$M := \{\text{KUR, GUT, RAT, STAR, LOS, BAU}\}$$

4.3 Order Relations

Fig. 4.8 Inclusions of equivalence relations

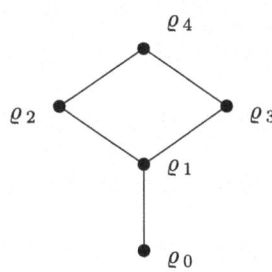

by

$x \sigma y :\Leftrightarrow x$ and y have at least one letter in common.

(α) Draw the relation graph of σ and prove or disprove that σ is reflexive, symmetric or transitive.

(β) Is there a (with respect to inclusion) smallest equivalence relation ϱ with $\sigma \subseteq \varrho$? Is there a (with respect to inclusion) largest equivalence relation τ with $\tau \subseteq \sigma$? Discuss these questions through graphical representations and explanations.

Exercise 4.45 (Equivalence and Order) Provide a set M with a relation $\varrho \subseteq M \times M$ such that ϱ is both an order and equivalence relation on M. How many such examples are there?

Exercise 4.46 (Hasse Diagrams) Let A be the set of all divisors of the integer m, ordered by the relation $|$ ("divides"). Draw Hasse diagrams for
(a) $m = 2$; (b) $m = 6$; (c) $m = 12$; (d) $m = 30$; (e) $m = 45$; (f) $m = 210$.

Exercise 4.47 (Largest and Smallest) The Hasse diagram of the set $M = \{a, b, c, d, e, f\}$ shown in Fig. 4.9 defines an order σ on M.

(a) Which of the following statements are true:
$a \sigma c, c \sigma c, b \sigma c, d \sigma a, b \sigma f, c \sigma f, e \sigma b$?
(b) State the maximum and minimum as well as the largest and smallest elements of (M, σ), if they exist.
(c) State all upper and lower bounds as well as the supremum and infimum for the following subsets of (M, σ): $\{b, c\}, \{b, e, f\}, \{c, d, e\}, \{a, b, c\}$.

Exercise 4.49 (Sets and Order) Let A and B be subsets of a set M. Prove (for each subset X of M):

Fig. 4.9 Hasse diagram

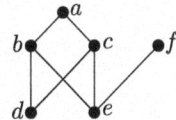

Fig. 4.10 A partially ordered set $(M, \leq)$

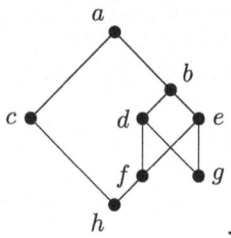

(a) $X \subseteq A$ and $X \subseteq B \Rightarrow X \subseteq A \cap B$.
(b) $A \subseteq X$ and $B \subseteq X \Rightarrow A \cup B \subseteq X$.
(c) For A and B, there always exists a supremum and an infimum in $(\mathcal{P}(M), \subseteq)$.

Exercise 4.50 (Ordered Sets) The Hasse diagram in Fig. 4.10 represents a partially ordered set $(M, \leq)$.

Consider the subsets of M:
$A := \{d, e, f, g, h\}$, $B := \{b, c\}$, $C := \{b, d, e, f\}$, $D := \{a, c, e\}$.

(a) Which of the following statements are true and which are false?
$a \leq b$, $f \leq g$, $h \leq b$, $g \leq a$, $g \leq c$, $d \leq e$, $f \leq e$.
(b) Draw the Hasse diagrams of the sets A, B, C and D with the inherited order relations from $(M, \leq)$
(i.e., $x \leq y$ in $A \subseteq M \iff x, y \in A$ and $x \leq y$ in M).
(c) Provide the sets P^{ub} and P^{lb} of the upper and lower bounds of P as well as, if available, sup P and inf P for $P = A$, B, C and D.

Exercise 4.52 (Componentwise and Lexicographic Order)

(a) Let $A := \{0, 1, 2\}$ be equipped with a linear order $\leq$. Draw the Hasse diagrams of $(A, \leq)$, $(A \times A, \leq_k)$ and $(A \times A, \leq_l)$.
(b) Let the set $B := \{u, v\}$ be provided with the order $\triangle_B = \{(u, u), (v, v)\}$. Draw the Hasse diagrams of $(B, \triangle_B)$, $(A \times B, \leq_k)$ and $(A \times B, \leq_l)$.
(c) Define the lexicographic order on $A_1 \times A_2 \times \ldots \times A_n$ for any $n \in \mathbb{N}$ and arbitrary sets $(A_1, \leq_1)$, $(A_2, \leq_2)$, ..., $(A_n, \leq_n)$. Show that this is a linear order if the order relations $\leq_1, \leq_2, \ldots, \leq_n$ are linear. Is the converse true?

Mappings 5

A photo is an image of some detail of reality. It is not identical to reality, but it represents it sufficiently accurately for many purposes. For example, it is usually possible to distinguish a person from others using passport photos. However, confusions are also possible, since the passport photo does not represent many characteristics of the person. On the other hand, there are situations in which information (such as the appearance of a person) can only be passed on using an image: the original itself would be completely unsuitable for books or magazines.

It is similar with mappings in mathematics. They can be used to represent complicated mathematical objects through simpler ones. We have experienced this, for example, when we represented the set $\mathbb{Z}$ through the set of its equivalence classes with respect to $\equiv$ mod 3 and these again through a set $\{0, 1, 2\}$ of representatives. Here, the set $\{\ldots, -5, -2, 1, 4, 7, \ldots\}$ is mapped to the equivalence class $\overline{1}$ and this again to the representative 1. One can say that a mapping creates a model.

In many areas of mathematics, mappings are also called *functions*. This is also the case in physics. There they are mostly used to describe dependencies of various processes. For example, a function can express that the distance covered is proportional to the elapsed time, i.e., that the object under consideration moves at a constant speed: each point in time is assigned the location where the object is at that moment.

Mappings are also very useful in computer science. A frequently mentioned and much studied example is the output of a computer program as an image of the input.

5.1 Partial and Total Mappings

Partial/total mapping, equality of (partial) mappings, image, preimage, source, domain, target, range, mapping rule, surjective, identity, identical mapping, constant mapping, lower/upper Gaussian bracket, integer function, absolute value function, distance function, metric, metric space, embedding, injection, projection, union/Cartesian product of mappings

Formally, for sets A, B we would like to define a mapping as an assignment of elements from B to elements from A. Usually this is written as $f : A \to B$. But what is an assignment? From the desired result, it turns out that it is a special relation. For the following definitions, we therefore set $f \subseteq A \times B$ and convince ourselves that the definitions do exactly what is desired.

We start with the following more general concept of a partial mapping. A triple (f, A, B) is called a *partial mapping from A to B*, (sometimes) written $f : A \multimap B$, if for each $a \in A$ there exists **at most one** $b \in B$ with $(a, b) \in f$. In this case e write $f : a \mapsto b$ or $f(a) = b$. In other words:

$$\forall a \in A : \ |f(a)| \leq 1.$$

A partial mapping (f, A, B) is called a *total mapping from A to B*, written $f : A \to B$, if for each $a \in A$ there exists **exactly one** $b \in B$ with $(a, b) \in f$. In other words:

$$\forall a \in A : \ |f(a)| = 1.$$

In both cases we use *mapping (from A to B)* and write $f : A \to B$.

We see that a partial mapping f is a right unique relation, a total mapping, in addition is a left total relation.

So a mapping (f, A, B) consists of three components, namely the sets A, B and $f \subseteq A \times B$. To test for equality one has to check that all three components agree:

Two mappings (f, A, B) and (g, C, D) are *equal*, if $A = C$, $B = D$ and $f = g$ apply.

Example 5.1 (Relation Graph) We use the representation of the relation graph of the relation f to illustrate essential features of partial mappings, total mappings and non-mappings (Fig. 5.1).

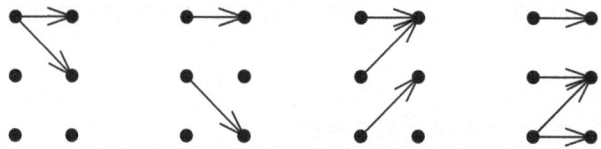

Fig. 5.1 From left to right: no mapping, partial mapping, mapping, no mapping

5.1 Partial and Total Mappings

If (f, A, B) is a (partial) mapping, then we call

$f[A]$	*image* of A, also written as $f(A)$, $\operatorname{Im} f$ or $\operatorname{ran} f$,
$f^{\leftarrow}[B]$	*preimage* of B, also written as $\operatorname{dom} f$,
A	*domain*, also: *source*,
B	*range*, or *codomain*, sometimes also: *target*
f	*mapping rule* of (f, A, B).

The definitions of image and preimage are already known from relations, see Sect. 4.1. Domain and range are not always mentioned, so we speak of the (partial) mapping f. We use the same symbol for mapping and mapping rule.

A partial mapping (f, A, B) is total, if $f^{\leftarrow}[B] = A$. If $f[A] = B$, then (f, A, B) is called a *surjective* mapping, see Sect. 5.3. The relation f is in this case right total. If the relation f is left unique, then the mapping (f, A, B) is called *injective*, see also Sect. 5.3.

Be careful with the empty set Suppose that (f, A, B) is a partial mapping with $A \neq \emptyset$ and $f \neq \emptyset$, then f is not the empty relation (see Sect. 4.1). Now from the definition of partial mapping, it follows that $B \neq \emptyset$. On the other hand, $(\emptyset, A, B)$ is always a partial mapping, which is total exactly if, $A = \emptyset$. Conversely, if $A = \emptyset$, then from the mapping rule $f \subseteq A \times B$ automatically $f = \emptyset$.

A remark on language: When a mapping (which is a special relation) is represented in a coordinate system in the previously described way (see Sect. 4.1), then one also speaks of the *function graph*. We do not use this terminology here: Graphs as mathematical structures, as will be defined in Chap. 6, correspond exactly to what we know as relation graph. They are not the representation of a function (mapping, relation) in the coordinate system.

Examples 5.2 (Partial and total mappings) We consider the following triples $F_i = (f_i, A_i, B_i)$ with $f_i \subseteq A_i \times B_i$ for $1 \leq i \leq 7$:

$$F_1 = (\{(a, 1), (b, 1)\}, \{a, b, c\}, \{1, 2, 3\}),$$
$$F_2 = (\{(a, 1), (b, 1), (c, 3)\}, \{a, b, c\}, \{1, 2, 3\}),$$
$$F_3 = (\{(a, 3), (b, 1), (c, 2)\}, \{a, b, c\}, \{1, 2, 3\}),$$
$$F_4 = (\{(a, 1), (b, 1)\}, \{a, b\}, \{1, 2, 3\}),$$
$$F_5 = (\{(a, 1), (c, 3)\}, \{a, c\}, \{1, 2, 3\}),$$
$$F_6 = (\{(1, a), (1, b), (3, c)\}, \{1, 2, 3\}, \{a, b, c\}),$$
$$F_7 = (\{(1, a), (3, c)\}, \{1, 2, 3\}, \{a, c\}).$$

The relation f_6 is not right unique, so F_6 is not a (partial) mapping. The triples F_1 and F_7 are partial mappings, F_2, F_3, F_4 and F_5 are even total mappings. For the mapping rules apply: $f_6 = f_2^{\leftarrow}$, $f_7 = f_5^{\leftarrow}$, $f_1 = f_4$. But $F_1 \neq F_4$, since $A_1 \neq A_4$.

Examples 5.3 (Special Mappings)

(1) For any set M, the relation *Diagonal* (see Example 4.18) is in particular a mapping, which is called the *identity* id_M *on M* or as *identical mapping of M*:

$$\mathrm{id}_M : \begin{cases} M \to M \\ x \mapsto x. \end{cases}$$

(2) For every set M and each $m \in M$, we denote by c_m the **constant mapping**, which maps all elements $x \in M$ to the same element m:

$$c_m : \begin{cases} M \to M \\ x \mapsto m. \end{cases}$$

(3) Each real number $x \in \mathbb{R}$, written as a decimal number $x = x_0, x_1 x_2 x_3 \ldots$ with $x_0 \in \mathbb{Z}$ and $x_i \in \{0, 1, 2, 3, \ldots, 9\}$ for $i = 1, 2, 3, \ldots$, is assigned the **greatest integer** $\leq x$ (*rounding down*) or the smallest integer $\geq x$ (*rounding up*), by

$$\lfloor \ \rfloor : \begin{cases} \mathbb{R} \to \mathbb{Z} \\ x \mapsto \lfloor x \rfloor := \begin{cases} x_0, & \text{if } x \in \mathbb{Z} \text{ or } x \geq 0, \\ x_0 - 1, & \text{if } x < 0 \text{ and } x \notin \mathbb{Z}, \end{cases} \end{cases}$$

the so-called *lower Gauss bracket* (*floor function*), also written as [ ], and by

$$\lceil \ \rceil : \begin{cases} \mathbb{R} \to \mathbb{Z} \\ x \mapsto \lceil x \rceil := \begin{cases} x_0, & \text{if } x \in \mathbb{Z} \text{ or } x \leq 0, \\ x_0 + 1, & \text{if } x > 0 \text{ and } x \notin \mathbb{Z}, \end{cases} \end{cases}$$

the so-called *upper Gauss bracket* (*ceiling function*) also written as { }. These mappings are also called *integer functions*.

(4) The *absolute value function* assigns a non-negative real number to each real number:

$$| \ | : \begin{cases} \mathbb{R} \to \mathbb{R} \\ x \mapsto |x| := \begin{cases} x, & \text{if } x \geq 0, \\ -x & \text{otherwise.} \end{cases} \end{cases}$$

5.1 Partial and Total Mappings

(5) Let M be a set. A mapping $d : M \times M \longrightarrow \mathbb{R}^+ = \{x \in \mathbb{R} | x \geq 0\}$ is called a *metric (distance function) on* M, if

(Met1) $d(x, y) = 0$ exactly if $x = y$,
(Met2) $d(x, y) = d(y, x)$,
(Met3) $d(x, y) + d(y, z) \geq d(x, z)$ (*triangle inequality*).

If d is a metric on M, then the pair (M, d) is called a *metric space*.

(6) The *distance function*

$$d : \begin{cases} \mathbb{R}^2 \times \mathbb{R}^2 \to \mathbb{R}^+ \\ (a, b) \mapsto \sqrt{(b_1 - a_1)^2 + (b_2 - a_2)^2} \end{cases}$$

is a metric. It assigns the distance $d(a, b)$ to each pair of points $a := (a_1, a_2)$ and $b := (b_1, b_2)$ in the Euclidean plane.

(7) If A and B are sets with $A \subseteq B$, the *embedding* or *injection* of A into B is defined by

$$\iota : \begin{cases} A \to B \\ x \mapsto x \end{cases}$$

So (ι, A, B) is the extension of the range A of (id_A, A, A) to B and also the restriction of the source B of (id_B, B, B) to $A \subseteq B$.

In particular, there is always the *k-th injection*

$$\iota_k : \begin{cases} A_k \to \bigcup_{i \in I} A_i \\ x \mapsto x \end{cases}$$

into the union of sets $A_i, i \in I$, if $k \in I$.

(8) As a counterpart to the k-th injection ι_k, for each Cartesian product of sets $A_i, i \in I$, for $k \in I$ the *k-th projection* is defined as follows:

$$\pi_k : \begin{cases} \prod_{i \in I} A_i \to A_k \\ (a_i)_{i \in I} \mapsto a_k. \end{cases}$$

(9) An analogue to the embedding of a subset into its "superset" is the *projection of the set B onto its subset A*

$$\pi : \begin{cases} B \to A \\ x \mapsto x, \end{cases}$$

which is generally only a partial mapping. The partial mapping (π, B, A) is thus the restriction of the range of (id_B, B, B) to A.

(10) For every family ($f_i : A_i \to B_i \mid i \in I$) the **union of mappings** $\dot\bigcup f_i$ is defined by the rule

$$\dot\bigcup f_i : \begin{cases} \dot\bigcup A_i \to \bigcup B_i \\ a \mapsto f_i(a), \text{ if } a \in A_i, \end{cases}$$

This provides a mapping between unions of sets.

(11) For every family ($f_i : A_i \to B_i \mid i \in I$) the **Cartesian product of mappings** $\prod f_i$ is defined by the rule

$$\prod f_i : \begin{cases} \prod A_i \to \prod B_i \\ (a_i)_{i \in I} \mapsto (f_i(a_i))_{i \in I}, \end{cases}$$

This provides a mapping between product sets.

5.2 Composition and Diagrams

Composition, argument, associativity, non-commutative, (commutative) diagram, restriction, extension,

For mappings (f, A, B) and (g, C, D) the composition $g \circ f$, is a non-empty mapping if ran $f \subseteq$ dom g. Note that dom $g = C$, if g is a total mapping. So for mappings (f, A, B) and (g, C, D) with ran $f \subseteq$ dom g (in particular if $B \subseteq C$) set

$$g \circ f : \begin{cases} A \to D, \\ x \mapsto (g \circ f)(x) := g(f(x)). \end{cases}$$

The mapping

$$(g, C, D) \circ (f, A, B) := (g \circ f, A, D)$$

is called the **composition of** (f, A, B) **and** (g, C, D). We read it as "g after f" since first f is applied to $x \in A$ and then g is applied to $f(x)$.

Since every mapping rule is a binary relation, we can identify this composition as a composition of relations. From Exercise 4.10 in Sect. 4.1 we know: Right uniqueness is preserved by composition.

$$(x, y) \in (g \circ f) \Leftrightarrow (\exists z \in B)\ f(x) = z \text{ and } g(z) = y \Leftrightarrow : y = g(f(x)).$$

For the peculiarities of the composition of partial mappings, that are not total, compare Exercise 5.6.

5.2 Composition and Diagrams

Attention It is also common to write the composition $g \circ f$ as fg. In this case, the *arguments*—these are the elements to which the mapping rule is applied—are written to the left of the mapping, so instead of $(g \circ f)(x)$ one writes $x(fg)$. We will stick with the first notation.

As the composition of binary relations is not commutative, the composition of mappings is not either: The relations in Example 4.7 are right-unique, so they are partial mappings. The example shows that their composition is not commutative. As a special case of the composition of binary relations, we also obtain the associativity of the composition for mappings:

Theorem 5.4 (Associativity) *The composition of mappings is associative:*
$(h \circ g) \circ f = h \circ (g \circ f)$, *provided the compositions are defined.*

As we have already seen, we can graphically represent mappings in the same way as binary relations: we can draw the graph of a mapping with arrows between the mapped elements when finite sets are mapped (see, for example, Fig. 5.2). For mappings of totally ordered, even non-finite sets, a coordinate system is usually used.

But there is another important type of graphical representation, which no longer cares about how individual elements are mapped, but about which compositions of mappings yield the same result: these are the *diagrams*. Capital letters denote sets, labeled arrows stand for mappings between these sets. A diagram is called *commutative* if all compositions of mappings with the same domain and the same range describe the same mapping. That is, the diagram in Fig. 5.3 is commutative if and only if $g \circ f = q \circ p$, i.e. if for every $x \in A$ one has $g(f(x)) = q(p(x))$.

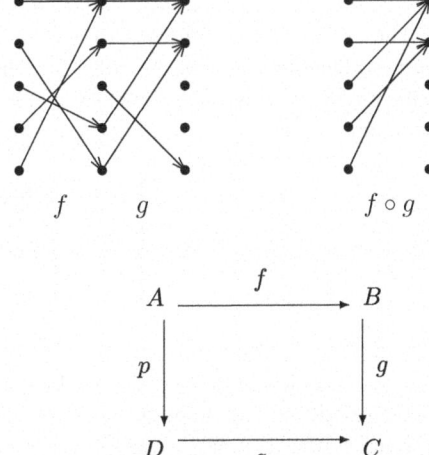

Fig. 5.2 From the composition of the relation graphs of functions we can derive the relation graph of their composition

Fig. 5.3 Commutative diagram

Restrictions and Extensions

Let $f : A \to B$ be a mapping. For each subset A' of A we define a mapping $(f|_{A'}, A', B)$ by

$$f|_{A'} : \begin{cases} A' \to B \\ x \mapsto f(x) \end{cases}$$

It is called *the restriction of the domain A of* (f, A, B) *to* A'.

If V is a "superset" of A (i.e., $A \subseteq V$) and (g, V, B) is a mapping with $g|_A = f$, then (g, V, B) is called *an extension of the domain A of* (f, A, B) *to* V.

There are many ways to construct an extension of the domain A of (f, A, B) to $V \supseteq A$. It is convenient to choose an element $b \in B$ and map all elements from $V \setminus A$ onto this b:

$$g : \begin{cases} V \to B \\ x \mapsto \begin{cases} f(x), & \text{if } x \in A, \\ b & \text{otherwise.} \end{cases} \end{cases}$$

Occasionally, the range B of a mapping (f, A, B) is restricted (to $B' \subseteq B$ with ran $f \subseteq B'$) or extended (to W with $B \subseteq W$). The mapping rule f is not affected by this. We speak of a *restriction* (f, A, B') or an *extension* (f, A, W) *of the range B of* (f, A, B). Restriction and extension are graphically represented as *diagrams*, see Example 5.5.

The terms just introduced are helpful when we transform a partial mapping (f, A, B) into a total mapping that adequately describes the partial mapping. Two variants have proven particularly useful:

One is $(f, \text{dom } f, B)$, the *restriction of* (f, A, B) *to* dom $f \subseteq A$.

The other is the *canonical extension of* (f, A, B) *from* dom $f \subseteq A$ *to* A. We map all elements of A not in dom $f \subseteq A$ to a new element $\star$ which has to be added to the range B. Thus we have an extension of dom f and of the range B. The result is the total mapping $(\overline{f}, A, B \dot\cup \{\star\})$ with the mapping rule

$$\overline{f} : \begin{cases} A \to B \dot\cup \{\star\} \\ x \mapsto \begin{cases} f(x), & \text{if } x \in \text{dom } f, \\ \star & \text{otherwise.} \end{cases} \end{cases}$$

If instead of $\star$ an element $b \in B$ would be chosen, then the partial mapping (f, A, B) could not be reconstructed from $(\overline{f}, A, B \dot\cup \{\star\})$.

We revisit the mappings from Example 5.2. The mapping F_2 is an extension of both the partial mapping F_1 and the mapping F_5, but only F_5 is a restriction of F_2. The mapping rules are $f_1 = f_2|_{\{a,b\}}$ and $f_5 = f_2|_{\{a,c\}}$.

In the following example we describe restrictions and extensions graphically by diagrams.

5.2 Composition and Diagrams

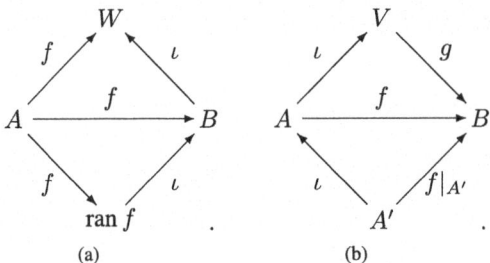

Fig. 5.4 Restriction and extension of the range. Let ran $f \subseteq B \subseteq W$. (**a**) The mapping $(f, A, \operatorname{ran} f)$ is **the** surjective restriction of the range B of (f, A, B) to ran f. The mapping (f, A, W) is **an** extension of the range B of (f, A, B) to W. (**b**) The mapping (g, V, B) is **an** extension of the domain A of (f, A, B) to V, while $(f|_{A'}, A', B)$ is **the** restriction of the domain A of (f, A, B) to A'.

Example 5.5 (Restriction and Extension) The commutative diagrams (a) and (b) represent restrictions and extensions of a mapping using injections (see Example 5.3 (7)) (Fig. 5.4).

Exercise 5.6 (Composition of Partial Mappings) Show that for the composition $g \circ f$ of partial mappings $f : A \multimap B$ and $g : C \multimap D$ one gets:

(a) $\operatorname{dom}(g \circ f) = f^{\leftarrow}[\operatorname{ran} f \cap \operatorname{dom} g] = f^{\leftarrow}[\operatorname{dom} g] \subseteq \operatorname{dom} f$.
(b) $\operatorname{ran}(g \circ f) = g[\operatorname{ran} f \cap \operatorname{dom} g] = g[\operatorname{ran} f] \subseteq \operatorname{ran} g$.

In Fig. 5.5, for example, the three dotted areas from left to right denote:

$$f^{\leftarrow}[\operatorname{dom} g], \quad \operatorname{ran} f \cap \operatorname{dom} g, \quad g[\operatorname{ran} f].$$

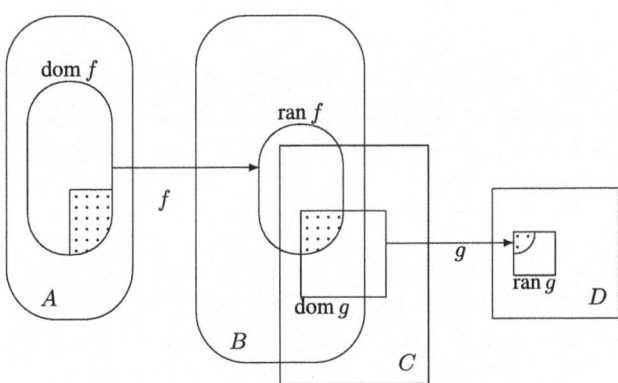

Fig. 5.5 Composition of partial mappings

5.3 Jectivities and Inversion

> Injective, surjective, bijective, injection, one-to-one, 1-1-mapping, surjection, mapping onto, bijection, (left/right)inverse, inverse mapping, inverse, uniqueness of the inverse, invertibility of the composition, partial left/right inverse, set of total mappings, partial transformation, permutation

We have chosen the actually unusual word "jectivities" as an abbreviation for the triple "injectivity, surjectivity, bijectivity".

A mapping (f, A, B) is called

- **injective,** if f is left unique, i.e. if

$$(\forall a_1, a_2 \in A) : f(a_1) = f(a_2) \Rightarrow a_1 = a_2,$$

- **surjective,** if f is **r**ight-total,[1] i.e. if

$$(\forall b \in B \; \exists a \in A) : f(a) = b,$$

- **bijective,** if f is injective and surjective.

The terms injective and surjective are also used for partial mappings, but a bijective mapping is always assumed to be total. Injective mappings are also called **injections, one-to-one**, **1-1-mappings**, or **mapping from** A **into** B. For surjective mappings the terms **surjection** or **mapping from** A **onto** B are common. Bijective mappings are also called **bijections** or **one-to-one correspondence**. In this case, A and B are called isomorphic, written $A \cong B$.

Instead of using the defining implication for injectivity, the equivalent logical converse is often used in proofs:

$$(\forall a_1, a_2 \in A) : \quad a_1 \neq a_2 \Rightarrow f(a_1) \neq f(a_2).$$

Obviously, a total mapping $f : A \to B$ is

      surjective if and only if $f[A] = B$,

      injective if and only if $f : A \to f[A]$ is bijective.

Example 5.7 (Mappings from Example 5.3) We check the mappings from Example 5.3 and find that the identity is always bijective. The integer functions are surjective, but not injective, the distance function is surjective, but not injective,

---

[1] The boldface **r** should help to remember that su**r**jective is **r**ight-total. In German **in**jective is analogously related to **l**inkseindeutig.

5.3 Jectivities and Inversion

the injections defined there are always injective, and the projections defined there are always surjective.

Lemma 5.8 (Jectivities) *For the mapping $f : A \to B$ one gets*

1. *f is injective* $\iff$ $(\forall b \in B) :$ $|f^{\leftarrow}(b)| \leq 1$.
2. *f is surjective* $\implies$ $(\forall b \in B) :$ $|f^{\leftarrow}(b)| \geq 1$.
3. *f is bijective* $\iff$ $(\forall b \in B) :$ $|f^{\leftarrow}(b)| = 1$.

Proof To 1. "$\Rightarrow$". Since f is injective $f(a_1) = f(a_2) = b$ implies $a_1 = a_2$ for any $b \in B$. Thus $|f^{\leftarrow}(b)| \leq 1$.

"$\Leftarrow$". For $a_1, a_2 \in A$ with $f(a_1) = f(a_2) =: b$ i.e., $a_1, a_2 \in f^{\leftarrow}(b)$, the assumption gives $a_1 = a_2$.

To 2. f is surjective exactly if for each $b \in B$ a preimage $a \in A$ with $f(a) = b$ exists. This means the set $f^{\leftarrow}(b)$ contains at least one element for each $b \in B$.

3. follows with the definition of "bijective" from 1. and 2., since $\leq$ is antisymmetric. □

For the compositions of surjective or injective mappings we get:

Theorem 5.9 (Composition) *Let $f : A \to B$ and $g : B \to C$ be mappings, then*

1. *f and g injective $\Rightarrow g \circ f$ injective $\Rightarrow f$ injective, but if $g \circ f$ is injective, g does not have to be injective.*
2. *f and g surjective $\Rightarrow g \circ f$ surjective $\Rightarrow g$ surjective, but if $g \circ f$ is surjective, f does not have to be surjective.*
3. *f and g bijective $\Rightarrow g \circ f$ bijective $\Rightarrow f$ injective and g surjective, but if $g \circ f$ is bijective, neither g has to be injective nor f has to be surjective.*

Note: injectivity is inherited by $g \circ f$ from the latter factor (hinterer Faktor in German) and surjectivity is inherited from the former factor.

Proof The example in Fig. 5.6 shows that g does not necessarily have to be injective and f does not necessarily have to be surjective if $g \circ f$ is injective and surjective.

Fig. 5.6 Compositions and injectivity/surjectivity

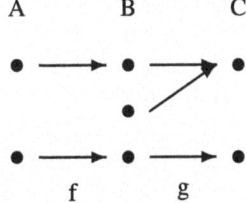

This shows the invalid implications, now we turn to the valid ones.

To 1. For $a_1, a_2 \in A$ with $(g \circ f)(a_1) = (g \circ f)(a_2)$ it follows that $f(a_1) = f(a_2)$, since g is injective, and $a_1 = a_2$, since f is injective, i.e. $g \circ f$ is injective.

Now assume that $g \circ f$ is injective. Then $f(a_1) = f(a_2)$ implies that $(g \circ f)(a_1) = (g \circ f)(a_2)$ and thus $a_1 = a_2$, i.e. f is injective.

To 2. Let $c \in C$. As g is surjective, there exists $b \in B$ with $g(b) = c$ and as f is also surjective, we find an $a \in A$ with $f(a) = b$. Consequently, $(g \circ f)(a) = g(b) = c \in (g \circ f)[A]$, i.e. $g \circ f$ is surjective.

If $g \circ f$ is surjective, for each $c \in C$ we first find a preimage $a \in A$ of c under $g \circ f$. Thus we have $f(a) \in B$ with $g(f(a)) = c$, i.e. g is surjective.

To 3. The claim follows from what has been proven under 1. and 2. □

If two mappings $f : A \to B$ and $g : B \to A$ are such that $g \circ f = \mathrm{id}_A$, then g is called **left inverse to** f and f is called **right inverse to** g. If g is both left and right inverse to f, i.e. if

$$g \circ f = \mathrm{id}_A \quad \text{and} \quad f \circ g = \mathrm{id}_B,$$

then f and g are called **inverses of each other**. Then g is called the **inverse mapping** or **inverse of** f denoted as $g = f^{-1}$, because we have the following

Lemma 5.10 (Uniqueness of the Inverse) *If a mapping $f : A \to B$ has both a right inverse and a left inverse, then these are equal. In particular, every mapping has at most one inverse.*

Proof If $g_1 : B \to A$ and $g_2 : B \to A$ are inverses to f, then:
$$g_1 = g_1 \circ \mathrm{id}_B = g_1 \circ (f \circ g_2) = (g_1 \circ f) \circ g_2 = \mathrm{id}_A \circ g_2 = g_2.$$
□

Lemma 5.11 (Characterization of Bijectivity) *For every mapping $f : A \to B$ the following are equivalent:*

(i) The mapping f is bijective.
(ii) The converse relation to f, $f^\leftarrow$, is a mapping.
(iii) There exists a mapping $f^{-1} : B \to A$ with $f^{-1} \circ f = \mathrm{id}_A$ and $f \circ f^{-1} = \mathrm{id}_B$.

Proof "(i) $\Rightarrow$ (ii)". Let $f : A \to B$ be bijective. By definition, $f^\leftarrow \subseteq B \times A$ is a binary relation. Since f is surjective, thus right-total, $f^\leftarrow$ is left total and since f is injective, thus left unique, $f^\leftarrow$ is right-unique.

"(ii) $\Rightarrow$ (iii)". Since $f^\leftarrow$ is left total and right unique, $f^{-1} := f^\leftarrow$ is a (total) mapping from B to A. Consequently $(\forall a \in A) : (f^{-1} \circ f)(a) \in f^\leftarrow(f(a)) = \{x \in A \mid f(x) = f(a)\} = \{a\}$. This means $f^{-1} \circ f = \mathrm{id}_A$. Similarly $(\forall b \in B) : (f \circ f^{-1})(b) = \{f(x) \mid x \in f^\leftarrow(b)\} = \{f(x) \mid f(x) = b\} = \{b\}$.

"(iii) $\Rightarrow$ (i)". If $a_1, a_2 \in A$ with $f(a_1) = f(a_2)$, then $a_1 = (f^{-1} \circ f)(a_1) = (f^{-1} \circ f)(a_2) = a_2$, thus f is injective. For every b in B one has $f^{-1}(b) \in A$ with $(f \circ f^{-1})(b) = b$. Thus, f is also surjective, and all together bijective. □

5.3 Jectivities and Inversion

Theorem 5.12 (Invertibility) *For every mapping $f : A \to B$ we get the following*

1. f is injective $\Leftrightarrow$ $(\exists g : B \to A)$ with $g \circ f = id_A$.
2. f is surjective $\Leftrightarrow$ $(\exists g : B \to A)$ with $f \circ g = id_B$.
3. f is bijective $\Leftrightarrow$ $(\exists g : B \to A)$ with $g \circ f = id_A$ and $f \circ g = id_B$.

*Note: **in**jective = **l**eft (**l**inks) invertible, **sur**ective = **r**ight invertible.*

Proof Since the identity is always bijective, the implications "$\Leftarrow$" in 1., 2. and 3. follow from Theorem 5.9.

For $f : A \to B$ and some $a_0 \in A$ consider the mapping g defined by

$$g : \begin{cases} B \to A \\ b \mapsto \begin{cases} a \in f^{\leftarrow}(b), & \text{if } f^{\leftarrow}(b) \neq \emptyset, \\ a_0 & \text{otherwise.} \end{cases} \end{cases}$$

To 1. As f is injective we have $|f^{\leftarrow}(b)| = 1$ So, for each $a \in A$ we have $g(f(a)) = (g \circ f)(a) = a = id_A(a)$, i.e. $g \circ f = id_A$.

To 2. If f is surjective, we can choose $g(b) = a \in f^{\leftarrow}(b)$. However, there may be several possibilities here as $|f^{\leftarrow}(b)| \geq 1$. So we obtain the mapping $g : B \to A$ with $(f \circ g)(b) = b = id_B(b)$ for each $b \in B$, i.e. $f \circ g = id_B$.

To 3. See Lemma 5.8. □

Theorem 5.13 (Invertibility of Composition) *If for two mappings f and g the inverse mappings f^{-1} and g^{-1} as well as the composition $g \circ f$ exist, then $(g \circ f)^{-1}$ also exists and $(g \circ f)^{-1} = f^{-1} \circ g^{-1}$.*

Proof One has $(f^{-1} \circ g^{-1}) \circ (g \circ f) = f^{-1} \circ g^{-1} \circ g \circ f = f^{-1} \circ id_B \circ f = id_A$ and correspondingly $(g \circ f) \circ (f^{-1} \circ g^{-1}) = g \circ f \circ f^{-1} \circ g^{-1} = g \circ id_B \circ g^{-1} = id_A$. □

If for partial mappings $f : A \multimap B$ and $g : B \multimap A$ one has $g \circ f \subseteq \triangle_A$, i.e. their composition is a "partial identical mapping", then g is called **partial left inverse to** f and f is called **partial right inverse to** g. In contrast to inverses (cf. Lemma 5.10), partial inverses are usually not uniquely determined. Partial left inverses can also be called **partial inverses**.

The following theorem shows why this makes sense.

Theorem 5.14 (Partial Inverse) *Every mapping has a partial right inverse. A mapping has a partial left inverse exactly if it is injective. In this case, the partial left inverse is also a partial right inverse.*

Proof We obtain a partial right inverse $g : B \multimap A$ to $f : A \multimap B$ by choosing an element $g(b) \in f^{\leftarrow}(b)$ for each $b \in B$, for which $f^{\leftarrow}(b) \neq \emptyset$. Then

$(f \circ g)(b) = f(g(b)) = b$. For $y \in B$ with $f^{\leftarrow}(y) = \emptyset$, $(f \circ g)(y)$ is not defined. Overall we have $f \circ g \subseteq \Delta_B$.

If f is injective, then $|f^{\leftarrow}(b)| \leq 1$ (cf. Lemma 5.8). With $g := f^{\leftarrow}$ for each $a \in A$ we obtain $(g \circ f)(a) \in f^{\leftarrow}(f(a)) \subseteq \{a\}$, i.e. $g \circ f \subseteq \Delta_A$.

If, conversely, the partial mapping f has a left inverse h, then for elements $a_1, a_2 \in A$ with $f(a_1) = f(a_2)$ we get $a_1 = (h \circ f)(a_1) = (h \circ f)(a_2) = a_2$, i.e. f is injective. □

Sets of Mappings

The *set of all total mappings* from the set A to the set B is denoted by B^A or also by $\mathrm{Map}(A, B)$. Accordingly, we can interpret M^n as the set of mappings of the n-element set $\{1, 2, \ldots, n\}$ into M. Each such mapping can be written as $(x_1, x_2, \ldots, x_n)$. So this is the mapping with the mapping rule $i \mapsto x_i$, $i \in \{1, 2, \ldots, n\}$.

Attention with the empty set Since total mappings are left total, it follows that $A \neq \emptyset$ implies $\mathrm{Map}(A, \emptyset) = \emptyset$. On the other hand, $\mathrm{Map}(\emptyset, B)$ contains the element $(\emptyset, \emptyset, B)$, denoted by θ. Apparently this is a total mapping. It usually is called the *empty mapping* θ. However, strictly speaking different sets B lead to different empty mappings.

A mapping from A to itself is also called *transformation of A*. For A^A one also writes $\mathcal{T}(A)$ and for the *set of partial transformations of* A correspondingly $\mathcal{PT}(A)$.

A bijective mapping from A to itself is called *permutation of A*. The *set of permutations of* A is denoted by $\mathrm{Sym}(A)$. If A has only finitely many elements, say $|A| = n$, then instead of $\mathrm{Sym}(A)$ one simply writes $\mathcal{S}_n$. We will discuss permutations and their composition in Chap. 7 under the keyword symmetric group.

Exercise 5.15 (Mappings and Sets) Let $f : S \to T$ be a mapping.

(a) Show that for arbitrary subsets $A, B \subseteq S$ one has $f(A \cap B) \subseteq f(A) \cap f(B)$.
(b) Show that f is injective if and only if: $(\forall A, B \subseteq S) : \quad f(A \cap B) = f(A) \cap f(B)$.
(c) Give an example where $f(A \cap B) \subset f(A) \cap f(B)$.
(d) Prove $f(A \cup B) = f(A) \cup f(B)$.

Exercise 5.16 (Transformations) Consider the two mappings $f(x) := 2x$ and $g(x) := \lfloor \frac{x}{2} \rfloor$ from $\mathbb{Z}$ to itself.

(a) Give mapping rules for $f \circ g$ and $g \circ f$.
(b) Are f, g, $g \circ f$, $f \circ g$ injective, surjective, bijective? (Proof or counterexample!)
(c) Give an extension of f to $\mathbb{Q}$.

Exercise 5.17 (Partial Transformations) Let $\mathcal{PT}(M)$ be the set of partial transformations of $M := \{1, 2, 3\}$. For a representation of their elements see (b). Set $A := \{f \in \mathcal{PT}(M) \mid f|_{\{1,2\}} = id_{\{1,2\}}\}$.

(a) How many elements does $\mathcal{PT}(M)$ contain?
(b) List the elements of the sets $g_i \circ A := \{g_i \circ f \mid f \in A\}$ for

$$g_1 := \begin{pmatrix} 1 & 2 & 3 \\ 3 & 1 & 2 \end{pmatrix} \quad g_2 := \begin{pmatrix} 1 & 2 & 3 \\ 2 & 2 & 2 \end{pmatrix} \quad g_3 := \begin{pmatrix} 1 & 2 & 3 \\ * & 3 & 1 \end{pmatrix}.$$

(c) By $\Phi_i : f \mapsto f \circ g_i$, $(i = 1, 2, 3)$, three transformations Φ_1, Φ_2, Φ_3 of M are defined.—Which of these mappings is injective, surjective, bijective?

Exercise 5.18 (Mappings and Inverse Relation) Let X and Y be sets, A and A' subsets of X, B and B' subsets of Y, and $f : X \to Y$ a mapping.—Show the following, and also, that equality does not hold in (a) and (c).
(a) $A \subseteq f^{\leftarrow}(f(A))$. (b) $f^{\leftarrow}(B \cup B') = f^{\leftarrow}(B) \cup f^{\leftarrow}(B')$.
(c) $f(f^{\leftarrow}(B)) \subseteq B$. (d) $f^{\leftarrow}(B \cap B') = f^{\leftarrow}(B) \cap f^{\leftarrow}(B')$.

Exercise 5.19 (Sets of Mappings) List S_4 as well as $\mathcal{T}(A)$, $\mathcal{PT}(A)$ and Sym(A) for $A \in \{\{1\}, \{1, 2\}, \{1, 2, 3\}\}$.

5.4 Homomorphism Theorem

Induced equivalence relation, canonical surjection/projection, Homomorphism Theorem

Lemma 5.20 (Induced Equivalence Relation) *For each mapping $f : A \to B$, by*

$$x \varrho_f y :\Leftrightarrow f(x) = f(y)$$

an equivalence relation on A is defined. For a partial mapping f one sets

$$x \varrho_f y :\Leftrightarrow f(x) = f(y) \text{ or } x, y \notin \text{dom } f.$$

Moreover, for every equivalence relation $\varrho \subseteq A \times A$ on a set A the mapping

$$\pi_\varrho : \begin{cases} A \to A/\varrho \\ a \mapsto [a]_\varrho, \end{cases}$$

is surjective.

Proof The relation ϱ_f is obviously reflexive, symmetric and transitive, since it is defined by an equality. Each ϱ_f-class contains exactly those elements of A that have the same (or, if f is partial, no) image under f.

Conversely, π_ϱ is left total, since the equivalence relation ϱ provides a class division on the set A, and right-unique, since different classes are disjoint (main theorem about equivalence relations, Theorem 4.20). Surjectivity of π_ϱ follows since for every equivalence class $[a]_\varrho$ one has $\pi_\varrho(a) = [a]_\varrho$. □

The relation ϱ_f, also called ker f, is called the **induced equivalence relation** of f on A.

The mapping π_ϱ is called the **canonical surjection**, or **canonical projection** from A to the set $A/_\varrho$.

Theorem 5.21 (Homomorphism Theorem for Sets) *Let $f : A \to B$ be a mapping and $\varrho \subseteq A \times A$ an equivalence relation on A with $\varrho \subseteq \varrho_f$. Then the mapping*

$$f' : \begin{cases} A/_\varrho \to B \\ [a]_\varrho \mapsto f(a) \end{cases}$$

is unique such that the diagram

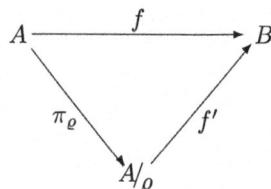

is commutative.

Furthermore, (a) f is surjective $\Rightarrow f'$ is surjective;

(b) $\varrho = \varrho_f \Rightarrow f'$ is injective.

In particular: $A/_{\varrho_f} \cong f[A]$.

All statements are valid also if f is a non-total mapping. Then of course f' is also a non-total mapping.

Proof Lemma 5.20 shows that π_ϱ is a surjective mapping. We show next that f' is indeed a mapping. We show that the relation f' is right-unique: For all $a_1, a_2 \in A$ one has

$$[a_1]_\varrho = [a_2]_\varrho \Leftrightarrow (a_1, a_2) \in \varrho \subseteq \varrho_f$$
$$\Rightarrow f(a_1) = f(a_2) \Leftrightarrow f'([a_1]_\varrho) = f'([a_2]_\varrho),$$

5.4 Homomorphism Theorem

i.e., the definition of the image of an equivalence class under f' is independent of the choice of the representative of this class. Moreover, the relation f' is left total: Since equivalence classes are always non-empty, $f'([a]_\varrho)$ is defined for every class $[a]_\varrho$.

The diagram is commutative, that is for every $a \in A$ one has $(f' \circ \pi_\varrho)(a) = f'([a]_\varrho) = f(a)$.

The mapping f' is uniquely determined: Suppose also f'' makes the diagram commutative, i.e. $f' \circ \pi_\varrho = f = f'' \circ \pi_\varrho$. Then for each $[a]_\varrho$ in $A/_\varrho$ one has $f'([a]_\varrho) = (f' \circ \pi_\varrho)(a) = (f'' \circ \pi_\varrho)(a) = f''([a]_\varrho)$ is, i.e. $f' = f''$.

To (a). Let f be surjective and $b \in B$, then there exists $a \in A$ with $f(a) = b$. Thus $b = (f' \circ \pi_\varrho)(a) = f'([a]_\varrho)$.

To (b). Let $\varrho = \varrho_f$ and let $[a_1]_\varrho, [a_2]_\varrho \in A/_\varrho$ with $f'([a_1]_\varrho) = f'([a_2]_\varrho)$. Then $f(a_1) = f(a_2)$, i.e. $(a_1, a_2) \in \varrho_f = \varrho$ and therefore $[a_1]_\varrho = [a_2]_\varrho$.

If f is a non-total mapping, we use the corresponding variant of Lemma 5.20. Then f' will also be a non-total mapping. The proof does not change. $\square$

This theorem will reappear in several variations, for graphs, groups, rings, modules. It is basic for the work with all sorts of mappings. Here we have the form for sets without any additional structure. Therefore it is also called the **Mapping theorem**.

Exercise 5.22 (Playing Cards) Form the Cartesian product $F \times Z$ of the sets $F := \{\clubsuit, \spadesuit, \heartsuit, \diamondsuit\}$ and $Z := \{7, 8, 9, 10, B, D, K, As\}$. Let $\pi_1 : F \times Z \to F$ and $\pi_2 : F \times Z \to Z$ be the first and second projection, respectively. With $(, As)$ and $(\heartsuit,)$ we denote the following mappings

$$(, As) : \begin{cases} F \to F \times Z \\ f \mapsto (f, As) \end{cases}$$

$$(\heartsuit,) : \begin{cases} Z \to F \times Z \\ z \mapsto (\heartsuit, z). \end{cases}$$

(a) Draw a diagram of the mappings $\pi_1, \pi_2, (, As), (\heartsuit,)$. Is it commutative? For which elements $(f, z) \in F \times Z$ one has $(, As) \circ \pi_1(f, z) = (\heartsuit,) \circ \pi_2(f, z)$.
(b) Give the sets $\pi_1^\leftarrow(\clubsuit)$, $\pi_2^\leftarrow(B)$, as well as $((, As) \circ \pi_1)^\leftarrow(f, z)$ for every element $(f, z) \in F \times Z$.
(c) Show that through $x \sim y :\iff \pi_1(x) = \pi_1(y)$ an equivalence relation $\sim$ is defined on the set $F \times Z$ and give the equivalence classes.

Exercise 5.23 (Modulo 5) Consider the following mappings:

$$(mod\ 5) : \begin{cases} \mathbb{Z} \to \mathbb{Z}_5 := \{0, 1, 2, 3, 4\} \\ z \mapsto n \in \mathbb{Z}_5, \text{ if } 5 \mid (z - n), \end{cases}$$

$$\oplus : \begin{cases} \mathbb{Z} \times \mathbb{Z} \to \mathbb{Z} \\ (x, y) \mapsto x + y \,(\text{ordinary addition in } \mathbb{Z}). \end{cases}$$

(a) Determine $(mod\,5)(-7)$, $(mod\,5)(234)$, $\oplus((29, 13))$, as well as $\oplus(\{(1, 1), (13, -11), (0, 2)\})$.
(b) Determine $(mod\,5)^{\leftarrow}(3)$, $(mod\,5)^{\leftarrow}(0)$, as well as $\oplus^{\leftarrow}(0)$.
(c) Show that $\oplus \circ (mod\,5) = (mod\,5) \circ \oplus$. Visualize this fact with a commutative diagram.
(d) What could an "addition" on $\mathbb{Z}_5$ look like?—Why do we need the equation from (c) for this?

Exercise 5.24 (Exemplification) Set $B := \{0, 1, 2, 3, 4, 5, 6, 7, 8, 9, 10\}$. For $x \in \mathbb{Z}$ set $m(x) := \overline{x}$, where $\overline{x} \in \mathbb{N}$ is such that $x - \overline{x}$ is divisible by 9 and $0 \leq \overline{x} \leq 8$. For example $m(33) = 6$ and $m(-2) = 7$. Then m is a mapping from $\mathbb{Z}$ to B.

(a) Give the equivalence classes with respect to ϱ_m and the canonical surjection π_{ϱ_m} associated with ϱ_m. Draw the corresponding commutative diagram.
(b) Describe the mapping m', as given in the Homomorphism Theorem.
(c) Is m' injective, surjective, bijective?

5.5 Cardinality of Sets

Countable, countably infinite, uncountable, first/second Cantor's diagonal method, continuum hypothesis, cardinal number, cardinality, Cantor/Schröder/Bernstein theorem, Tarski's fixed point theorem

We remember that the cardinality of a finite set is the number of its elements (see Sect. 2.3). How do we compare cardinalities when we don't feel like counting, or can't? With the help of mappings, we can compare the cardinalities of any sets, finite or infinite.

A set M is called **countable** if there is an injective mapping $\zeta : M \to \mathbb{N}$. In this case, we can imagine that the elements of M are numbered with natural numbers. For finite sets, this is always possible, i.e., every finite set is countable.

A countable, non-finite set is called **countably infinite**, and a non-countable set is called **uncountably infinite**. Whether such sets exist at all, we do not yet know.

Theorem 5.25 (Countability) *The sets $\mathbb{N}$, $\mathbb{Z}$ and $\mathbb{Q}$ are countable.*

5.5 Cardinality of Sets

Fig. 5.7 Cantor's diagonal method

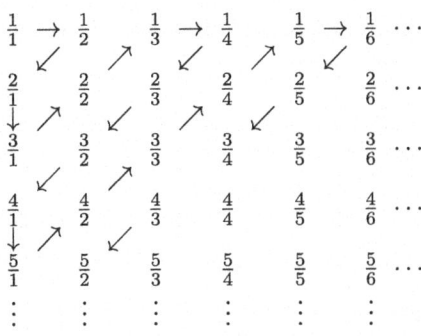

Proof As id : $\mathbb{N} \to \mathbb{N}$ is injective, $\mathbb{N}$ is countable. The mapping

$$\zeta : \begin{cases} \mathbb{Z} \to \mathbb{N}, \\ z \mapsto \begin{cases} 2z - 1, & \text{if } z > 0, \\ 0, & \text{if } z = 0, \\ -2z & \text{otherwise}. \end{cases} \end{cases}$$

is injective, so $\mathbb{Z}$ is also countable.

For the rational numbers, we use **Cantor's**[2] ***first diagonal method*** For this, all positive elements from $\mathbb{Q}$ are represented as fractions and written, as indicated in Fig. 5.7.

Since each row and each column contains infinitely many elements, we cannot count row by row or column by column! The counting is therefore done as indicated by the arrows: we start in the upper left corner with $\frac{1}{1}$, go from there one step to the right to $\frac{1}{2}$, then diagonally backwards down to $\frac{2}{1}$, continue one step down to $\frac{3}{1}$ etc. For each new element it is clarified whether it has already been counted (the representation of the elements of $\mathbb{Q}$ as a fraction is not unique). So, in this counting, for example $\frac{2}{2}(=\frac{1}{1})$ is omitted, $\frac{4}{2}(=\frac{2}{1})$ is omitted, etc. The elements of the created sequence we map to the odd natural numbers 1, 3, 5, …. Correspondingly, we consider the elements of the above sequence as the negative elements of $\mathbb{Q}$ and map them to the even natural numbers. Finally set $0 \mapsto 0$. The so defined mapping $\mathbb{Q} \to \mathbb{N}$ is injective, i. e. $\mathbb{Q}$ is also countable. □

After these famous results of Cantor, we find a much more general result which contains these results, since $\mathbb{N}$, $\mathbb{Z}$ and $\mathbb{Q}$ consist of algebraic numbers. However, its proof needs some more algebra and it is not " constructive". This result is due to Cantor and Dedekind.

---

[2] Georg Cantor, German mathematician, 1845–1918. Cantor himself was so astonished by the equivalence of apparently different sized sets that he expressed this in a letter to Richard Dedekind on June 20, 1877, not in his mother tongue: "Je le vois, mais je ne le crois pas."

Theorem 5.26 (Countability of Algebraic Numbers) *The set $\mathbb{A}$ of algebraic numbers is countable.*

Proof For $n \in \mathbb{N}$ let $\mathbb{A}_n \subseteq \mathbb{A}$ be the set of complex solutions of all algebraic equations $a_n x^n + a_{n-1} x^{n-1} + \ldots + a_1 x + a_0 = 0$ with coefficients $a_i \in \mathbb{Z}$ and $|a_i| \leq n$. We use the well known fact (here without proof), that an algebraic equation of degree n has at most n complex solutions. Thus, every $\mathbb{A}_n$ is countable, even finite. Now $\mathbb{A} = \bigcup_{n \in \mathbb{N}} \mathbb{A}_n$. Thus $\mathbb{A}$ is countable. □

Notation The cardinality of countably infinite sets is denoted by $\aleph_0$ (*aleph zero*), the first letter of the Hebrew alphabet indexed with 0.

This notation suggests that further cardinal numbers, like $\aleph_1, \aleph_2, \ldots$, might be meaningful. We first show that there are uncountable sets. Observe that this theorem implies that the set of transcendental numbers is uncountable.

Theorem 5.27 (Uncountability of $\mathbb{R}$) *The set $\mathbb{R}$ is uncountable.*

Proof With the help of the ***Cantor's second diagonal method***, we show in an indirect proof that the set $[0, 1[:= \{x \in \mathbb{R} \mid 0 \leq x < 1\}$ is uncountable. This clearly implies that $\mathbb{R}$ is uncountable.

We proceed as follows. For any arbitrary partial injective mapping of the set $[0, 1[$ into the set $\mathbb{N}$, we find an element $x \in [0, 1[$ for which the mapping is not defined. In other words, however we write the elements of $[0, 1[$ in a row, there will always be an element of $[0, 1[$, which is not in this row. That is, we show that there are only non-total partial injective mappings from $[0, 1[$ into $\mathbb{N}$.

So let $\zeta : [0, 1[\to \mathbb{N}$ be a partial injective mapping. That is, the elements x_i of $\operatorname{dom} f \subseteq [0, 1[$ can be numbered with the elements of $\mathbb{N}$. That is, they can be written in a row. We write these x_i as decimal numbers. We denote their digits behind the zero as follows:
$$x_1 =: 0.a_{11} a_{12} a_{13} \ldots,$$
$$x_2 =: 0.a_{21} a_{22} a_{23} \ldots,$$
$$x_3 =: 0.a_{31} a_{32} a_{33} \ldots,$$
$$\ldots$$

Now we choose digits $b_1, b_2, b_3, \ldots \in \mathbb{N}$ between 0 and 9 so that $b_1 \neq a_{11}$, $b_2 \neq a_{22}$, $b_3 \neq a_{33}$, …. Then $x := 0.b_1 b_2 b_3 \ldots$ is an element from $[0, 1[$ that does not appear in the above row. Thus $\zeta(x)$ is not defined, i.e., ζ is not a total mapping. □

Theorem 5.28 (Uncountability of Transcendental Numbers) *The set of transcendental numbers is uncountable.*

Proof Clear since algebraic numbers are countable and $\mathbb{R}$ is uncountable. □

5.5 Cardinality of Sets

Comparison of Cardinal Numbers

We now also compare the cardinalities of uncountable sets. The first problem that arises is the famous *continuum hypothesis*, proposed by Cantor at the beginning of the twentieth century. It states

$$\text{card}(\mathbb{R}) = \aleph_1.$$

This means: there is no cardinal number that is greater than $\aleph_0 = \text{card}(\mathbb{N})$ and smaller than $\text{card}(\mathbb{R})$.

It turns out that this hypothesis cannot be decided. More precisely: the continuum hypothesis is undecidable in the axiomatic system of set theory by Zermelo and Fraenkel. This means that this axiomatic system is consistent with the additional axiom $\text{card}(\mathbb{R}) = \aleph_1$, but also with the additional axiom $\text{card}(\mathbb{R}) \neq \aleph_1$, always under the assumption that the axiomatic system of Zermelo and Fraenkel is consistent. Only in 1963 P. Cohen[3] completed this proof, begun in 1940 by K. Gödel.[4]

First, we clarify the concept of *cardinal number*. All natural numbers are cardinal numbers, as they indicate the cardinalities of finite sets. Another cardinal number is $\aleph_0$, the cardinality of the set $\mathbb{N}$. In the proof of Theorem 5.25, we defined a bijective mapping $\zeta : \mathbb{Z} \to \mathbb{N}$ to show that $\mathbb{Z}$ is also countable.

As in the case of finite sets, we define for arbitrary sets A and B:

$$|A| = |B| \quad :\Leftrightarrow \quad (\exists f : A \to B) : f \quad \text{bijective}.$$

We say A and B are of *the same cardinality*, if $|A| = |B|$.

Thus, sets of the same cardinality are grouped into equivalence classes and a cardinal number is assigned to each such class. This way, cardinal numbers are defined as classes of sets of the same cardinality.

We define:

$$|A| \leq |B| \quad :\Leftrightarrow \quad (\exists f : A \to B) : f \quad \text{injective}.$$

In Theorem 5.27, we proved that there is no injective mapping $\zeta : \mathbb{R} \to \mathbb{N}$. However, there are obviously injective mappings from $\mathbb{N}$ to $\mathbb{R}$.

Thus $|\mathbb{N}| \leq |\mathbb{R}|$ and $|\mathbb{N}| \neq |\mathbb{R}|$—this we are willing to believe—but also $|\mathbb{N}| = |\mathbb{Z}| = |\mathbb{Q}|$—this we proved but it does not correspond to our imagination. In particular,

$$M \subsetneq N \text{ does not imply } |M| < |N|.$$

---

[3] Paul Cohen, US-American mathematician, 1934–2007.
[4] Kurt Friedrich Gödel, Austrian-American mathematician, 1906–1978.

Is $\leq$ an *order relation between cardinal numbers*? The totality of cardinal numbers is a proper class, i.e. not a set—which we cannot prove here. Therefore we cannot speak in the strict sense of an order relation. But the properties *reflexive, antisymmetric* and *transitive* can be easily transferred to classes. The relation $\leq$ is reflexive, because for every set A, $\mathrm{id}_A : A \to A$ is an injective mapping. It is transitive according to Theorem 5.9, because if $f : A \to B$ and $g : B \to C$ are injective mappings, then $g \circ f : A \to C$ is also an injective mapping.

It remains to show that $\leq$ is antisymmetric, i.e. if there is an injective mapping $f : A \to B$ and an injective mapping $g : B \to A$, then there exists a bijective mapping $h : A \to B$. This statement is known as the *Theorem of Cantor, Schröder and Bernstein*,[5] but the first correct proof comes from Dedekind. This suggests that proofs will not be easily understandable. The version given here is "ingenious ", it is not quite easy to understand. If you do not want to get into it now, it is enough to continue reading at Lemma 5.31.

Theorem 5.29 (Cantor, Schröder and Bernstein) *Let A and B be sets. There exists a bijective mapping $h : A \to B$ if and only if there is an injective mapping $f : A \to B$ and an injective mapping $g : B \to A$.*

Proof " $\Rightarrow$ ". If there is a bijective mapping $h : A \to B$, then $|A| \leq |B|$, because h is injective, and $|B| \leq |A|$, because the inverse $h^{-1} : B \to A$ of h is also injective.

" $\Leftarrow$ ". Conversely, let there be injective mappings $f_1 : A \to B$ and $f_2 : B \to A$. We use these to construct a bijective mapping $h : A \to B$:

First, we define a mapping

$$\Phi : \wp(A) \to \wp(A), \quad X \mapsto A \setminus f_2[B \setminus f_1[X]].$$

We get $X \subseteq Y \Rightarrow f_i[X] \subseteq f_i[Y]$ for both subsets $X, Y \subseteq A$ and $i = 1$ as well as for subsets $X, Y \subseteq B$ and $i = 2$. Due to the double complement formation, the following also applies to Φ:

$$(\forall X, Y \in \wp(A)) : X \subseteq Y \Rightarrow \Phi(X) \subseteq \Phi(Y). \qquad (\star)$$

We now show that Φ has a *fixed point*, i.e., that there exists a $W \in \wp(A)$ with $\Phi(W) = W$. We set

$$W := \bigcap_{Z \in \mathcal{W}} Z \text{ where } \mathcal{W} := \{ Z \in \wp(A) \mid \Phi(Z) \subseteq Z \}.$$

For each $Z \in \mathcal{W}$, $W \subseteq Z$, so $\Phi(W) \subseteq \Phi(Z) \subseteq Z$, due to $(\star)$ and by definition of $\mathcal{W}$. Thus, $\Phi(W) \subseteq \bigcap_{Z \in \mathcal{W}} Z = W$ follows. This yields the following implications

---

[5] Friedrich Wilhelm Karl Ernst Schröder, German mathematician, 1841–1902; Felix Bernstein, German mathematician, 1878–1956.

5.5 Cardinality of Sets

$$\Phi(\Phi(W)) \subseteq \Phi(W) \Rightarrow \Phi(W) \in \mathcal{W} \Rightarrow W = \bigcap_{Z \in \mathcal{W}} Z \subseteq \Phi(W).$$

So in total $W = \Phi(W)$, i.e. we have found the fixed point W of Φ. If we insert W into the definition of Φ, we get $A \setminus W = f_2[B \setminus f_1[W]]$, as shown in the figure. Thus we can compose the mapping h from f_1 and $f_2^{\leftarrow}$ as follows.

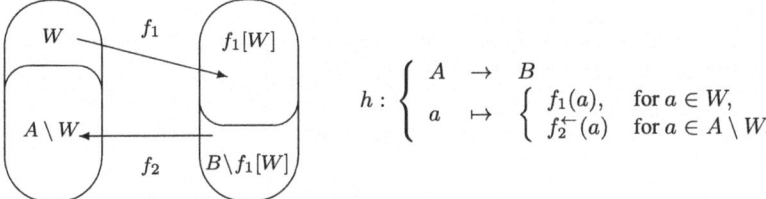

$$h : \begin{cases} A & \to & B \\ a & \mapsto & \begin{cases} f_1(a), & \text{for } a \in W, \\ f_2^{\leftarrow}(a) & \text{for } a \in A \setminus W. \end{cases} \end{cases}$$

It is now easy to see that h is injective and surjective. □

The given proof implicitly contains a special version of *Tarski's fixed point theorem*.[6] Without any significantly new ideas, this theorem can be proven for complete lattices using the same scheme. We therefore formulate it immediately in its more general form. From the construction of W, it is evident that it is the smallest fixed point of Φ.

Corollary 5.30 (Tarski's Fixed Point Theorem) *Let $(V, \leq)$ be a complete lattice and let $\Phi : V \to V$ be an isotone mapping, for which*

$$(\forall x, y \in V) : x \leq y \Rightarrow \Phi(x) \leq \Phi(y).$$

Then $w := \inf\{ z \in V \mid \Phi(z) \leq z \}$ is a fixed point of Φ, i.e. $\Phi(w) = w$.

Now we continue the comparison of cardinal numbers. We have defined a relation $\leq$ between cardinal numbers using injective mappings, which is reflexive, antisymmetric and transitive. As usual, we define the converse relation $\geq$ by $|A| \geq |B| :\Leftrightarrow |B| \leq |A|$.

Lemma 5.31 (Comparison of Cardinalities) *For sets A and B, we have*

1. $|A| \leq |B| \Leftrightarrow (\exists f : A \to B) : f$ *is injective,*

2. $|A| \geq |B| \Leftrightarrow (\exists f : A \to B) : f$ *is surjective or $B = \emptyset$.*

---

[6] Alfred Tarski, Polish mathematician, 1901–1983.

Proof

1. is the definition of $\leq$.
2. We show

"$\Rightarrow$". $|A| \geq |B|$ if and only if $|B| \leq |A|$. So there is an injective mapping $g : B \to A$. We choose any element $b \in B$ and set $f(x) := g^{\leftarrow}(x)$, if $x \in \operatorname{ran} g$ and $f(x) := b$ otherwise. This defines a surjective mapping $f : A \to B$ as g is left total.

"$\Leftarrow$". If there is a surjective mapping $f : A \to B$, then we choose for each element $b \in B$ one of the elements from $f^{\leftarrow}(b)$, say $x_b \in f^{\leftarrow}(b)$, and set $g(b) := x_b$. This defines a mapping $g : B \to A$ that is injective, since f is right unique. It follows that $|B| \leq |A|$, i.e. $|A| \geq |B|$.

□

Theorem 5.32 (Injective = Surjective for Finite Sets) *For any set M, the following statements are equivalent:*

(i) *M is a finite set.*
(ii) *Every surjective mapping from M to itself is injective.*
(iii) *Every injective mapping from M to itself is surjective.*

Proof "(i) $\Rightarrow$ (ii)". Take a finite set M and a surjective mapping $f : M \to M$. For $x \in M$ set $A := M \setminus f^{\leftarrow}(x)$. Just like f, the mapping $f|_A : A \to M \setminus \{x\}$ is also surjective. With Lemma 5.31 we get $|A| \geq |M \setminus \{x\}| = |M| - 1$. Thus $1 \geq |M| - |A| = |f^{\leftarrow}(x)|$ for each $x \in M$ and with Lemma 5.8 it follows that f is injective.

"(ii) $\Rightarrow$ (iii)". Assume (ii) and let $f : M \to M$ be injective. Choose $z_0 \in M$. According to the assumption, the set $f^{\leftarrow}(x)$ contains at most one element for each $x \in M$. With this we define a mapping

$$g : \begin{cases} M \to M \\ x \mapsto \begin{cases} y \in f^{\leftarrow}(x), & \text{if } f^{\leftarrow}(x) \neq \emptyset, \\ z_0 & \text{otherwise.} \end{cases} \end{cases}$$

As for all $y \in M$ one has $y = g(f(y)) \in g[M]$, g is surjective and in particular $g(f(z_0)) = z_0$. With (ii) it follows, that g is injective, i.e. in particular the set $g^{\leftarrow}(z_0)$ contains no further elements besides $f(z_0)$. This means by definition of g, that $f^{\leftarrow}(x) \neq \emptyset$ for each $x \in M$. According to Lemma 5.8, f is surjective.

"(iii) $\Rightarrow$ (i)" We show this by contraposition, i.e. we show "$\neg(i) \Rightarrow \neg(iii)$". For this let M be a non-finite set, thus $|\mathbb{N}| \leq |M|$. According to Lemma 5.31 there is an injective mapping $f : \mathbb{N} \to M$, i.e. M contains pairwise different elements $m_0 := f(0), m_1 := f(1), m_2 := f(2), \ldots$. We set $A := f(\mathbb{N})$ and define

5.5 Cardinality of Sets

$$g : \begin{cases} M \to M \\ x \mapsto \begin{cases} m_{2j}, & \text{if } x \in A \text{ with } x = m_j, \\ x & \text{otherwise}, \end{cases} \end{cases}$$

i.e. each element m_j from A is mapped to a different element m_{2j} from A and, if M contains further elements, these are mapped to themselves. The mapping g is not surjective, because none of the elements $m_{2j+1} \in M$, $j \in \mathbb{N}$, lies in the image of g. On the other hand, g is injective, because if $g(x) = g(y)$ is one of the elements from A, then it is a m_{2j}, $j \in \mathbb{N}$, i.e. $x = m_j = y$. And if $g(x) = g(y) \notin A$, then $x = g(x) = g(y) = y$. Thus $\neg(iii)$ is proven. □

Cardinality of the Power Set

In Theorem 2.16 we have given the cardinality of the power set of an n-element set as 2^n. Also for infinite sets, the power set is "larger" than the set itself, i.e. in particular there is no greatest cardinal number:

Theorem 5.33 (Cantor) *For every set M we have $\text{card}(M) < \text{card}(\wp(M))$. In particular, the set $\wp(\mathbb{N})$ is uncountable.*

Proof Let M be a set. In a way that resembles the proof of the existence of uncountable sets, we show that there is no surjective mapping $f : M \to \wp(M)$. For this, let $f : M \to \wp(M)$, $x \mapsto f(x) =: A_x$ be any mapping.
The set

$$Z := \{ x \in M \mid x \notin A_x \}$$

is certainly an element of $\wp(M)$. Suppose $Z \in \text{Im } f$, i.e., there exists $z \in M$ with $Z = f(z) = A_z$.
This gives a contradiction since
either $z \in Z$, then from the definition of Z, it follows that $z \notin A_z = Z$,
or $z \notin Z$, then from the definition of Z, it follows that $z \in A_z = Z$.
Therefore, Z is not in the image of f, i.e., f is not surjective. □

Exercise 5.34 (Equivalence Classes) Is "has the same cardinality as" an equivalence relation on the set of all finite sets?—Do the finite sets form a set at all?—If yes: how many equivalence classes do we get?

Exercise 5.35 (On the Cantor/Schröder/Bernstein Theorem) In the proof of the Cantor/Schröder/Bernstein Theorem, it is shown that the mapping

$$\Phi : \begin{cases} \wp(\mathbb{N}) \to \wp(\mathbb{N}) \\ X \mapsto \mathbb{N} \setminus f_2(\mathbb{N} \setminus f_1(X)) \end{cases}$$

has a fixed point. Find it for

$$f_1 : \begin{cases} \mathbb{N} \to \mathbb{N}, \\ x \mapsto x+5 \end{cases}$$

and for

$$f_2 : \begin{cases} \mathbb{N} \to \mathbb{N}, \\ x \mapsto x+3 \,. \end{cases}$$

Hint: The (with respect to inclusion) largest X with $X \subseteq \Phi(X)$ is a fixed point of Φ. Proceed as follows:

1. Determine $\Phi(X)$ for various sets X, for example by filling out a table like this:

	$X =$	$\{2, 3, 4\}$	$\{0, 1\}$	$\{0, 1\} \cup \{8, 9\}$	...
	$f_1(X) =$				
	$\mathbb{N} \setminus f_1(X) =$				
	$f_2(\mathbb{N} \setminus f_1(X)) =$				
$\Phi(X) =$	$\mathbb{N} \setminus f_2(\mathbb{N} \setminus f_1(X)) =$				
	$X \subseteq \Phi(X)$? (yes/no)				

2. If a fixed point W is found, then use it to determine the bijection

$$h : \begin{cases} \mathbb{N} \to \mathbb{N}, \\ x \mapsto \begin{cases} f_1(x), & \text{if } x \in W, \\ f_2^{-1}(x) & \text{otherwise.} \end{cases} \end{cases}$$

Represent f_1, f_2 and h graphically.

If no fixed point is found, show that $X \subseteq Y$ implies $f(X) \subseteq f(Y)$ for every mapping $f : A \to B$ and all subsets X, Y of A.

Moreover show that for each $a \in \mathbb{N}$ the mapping

$$g_a : \begin{cases} \mathbb{N} \to \mathbb{N}, \\ x \mapsto x+a \end{cases}$$

is injective.

Represent g_5 graphically.

Graphs 6

Graph theory is a relatively young subfield of mathematics, whose roots reach into physics, chemistry, and playful problem settings. Applications can also be found in engineering, linguistics, social sciences, and increasingly in numerous problems associated with computer science: data structures, computer networks, computer architecture, circuit theory, etc. For the famous completeness problems of algorithm theory (e.g., NP-completeness), examples can usually be given from graph theory. Conversely, many algorithms refer to questions from graph theory or underlying questions from corresponding problem areas; see also the book on algorithms by Schöning [66]. Additionally, problems from automata theory can often be described by graphs, as well as questions from the theory of formal languages (e.g., graph grammars).

We define a graph as a mathematical object that consists of *vertices* and *edges*, which are related such that vertices are at the beginning or end of edges. Such a structure always represents a binary relation. Therefore, everything we know about binary relations can be applied to graphs. However, with this new viewpoint, other questions become important. More than before for relations in general, it is advisable to draw suitable sketches and pictures for everything discussed in this chapter.

On the Literature There is extensive literature on graph theory, also from various aspects of applications. The book by Jungnickel [43] is focused on applications in computer science. Harary's [32] book describes the development of the theory and provides a good overview of its diversity within mathematics. Very comprehensive and yet concise is the book by Chartrand and Lesniak [16]. Long-awaited, and therefore particularly mentioned here, are books on the theory and practice of the graphical representation of graphs on the computer: G. Di Battista, P. Eades, R. Tamassia, I.G. Tollis [22], and M. Kaufmann, D. Wagner [44].

6.1 Basics

Directed/undirected graph, simple graph, multigraph, multiple edge, vertices, nodes, edges, incidence mapping, incident, adjacent, isolated vertex, loop, circuit

Graphs are objects that we have already encountered frequently in this book, e.g., relation graphs or Hasse diagrams, see Sect. 4.3. Now we start with a formal approach to graphs. A graph is a set of points, which are called nodes or vertices, and a set of possibly directed edges that connect these points. As the name suggests, graphs can best be visualized graphically. Consider Fig. 6.1.

Even though the picture gives us a good idea of a graph, we do not want the graph to really depend on the picture. We want to define a structure that, for example, is independent of whether one of the points is moved somewhere else.

So, we define a *graph* G as a triple $G = (V, E, p)$, where V is a non-empty set, called *vertex set* (or *node set*), E is a set, called *edge set*, and p is a mapping, called *incidence mapping*. We call G

directed graph, $\quad p : E \to V \times V$, and
undirected graph, $\quad p : E \to \{ Z \subseteq V \mid 1 \leq |Z| \leq 2 \}$.

The incidence mapping can be understood as the mapping that maps the edges to their endpoints. Here vertices and edges both are sets. The difference is that the edge set is the domain of the respective incidence mapping, while the vertex set is the range of this mapping. We could just as well assign to each vertex of the graph by means of a (differently defined) "incidence mapping" those edges that have "to do with" the vertex.

A graph $G = (V, E, p)$ is called

simple graph, $\quad$ if p is injective,
multigraph $\quad$ otherwise,

i.e., a multigraph contains *multiple* or *parallel edges*, a simple graph does not. Again, Fig. 6.1 can serve as an illustration.

Let $G = (V, E, p)$ be a graph. If $e \in E$ with $p(e) = (x, y)$ or with $\{x, y\} \subseteq p(e)$, then the edge e is called *incident* with the vertices x and y and the vertices x and y are called *adjacent*. Two edges are called *incident* if there is a vertex that is incident with both. A vertex that is not incident with any edge is called an *isolated vertex*. An edge that is incident with only one vertex is called a *loop*. In the literature, often a graph is only called simple if (in addition to our requirements) it has no loops.

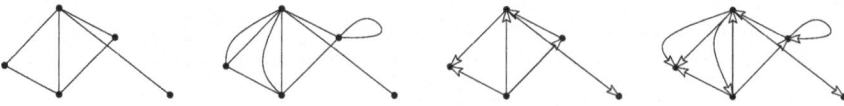

Fig. 6.1 From left to right: an undirected simple graph, an undirected multigraph, a directed simple graph, and a directed multigraph

In the directed case, furthermore one forbids ***antiparallel edges***, i.e., $e, f \in E$ with $p(e) = (x, y)$ and $p(e) = (y, x)$.

If $G = (V, E, p)$ is a simple graph, we can identify each edge e of G with its image $p(e)$ under the incidence mapping.

If G is a directed simple graph, then E is a subset of $V \times V$ and we write (x, y) or xy for an edge e with $p(e) = (x, y)$. This makes clear that the set of edges of a simple directed graph, even with loops, is a binary relation on $V \times V$.

If G is undirected, then E is a subset of $\{Z \subseteq V \mid |Z| \leq 2\}$ and we write $\{x, y\}$ for an edge e with $p(e) = \{x, y\}$. Every undirected graph $G = (V, E, p)$ obviously becomes a directed graph by replacing each edge e with $p(e) = \{x, y\}$ for $x \neq y$ with two edges e_1 and e_2 with $p(e_1) = (x, y)$ and $p(e_2) = (y, x)$. When making this replacement, it becomes clear that the set of edges of a simple undirected graph, even with loops, is a binary symmetric relation on $V \times V$.

In these two cases, we write $G = (V, E)$, i.e. we omit p to express that we have made the corresponding identifications.

For multigraphs, $p(E)$ becomes a multiset.

6.2 Directed Graphs

Projections, initial vertex, source, final vertex, target, input/output degree, predecessor, successor, edge sequence, cycle, path, circuit, semiedge sequence, semicycle, semipath, semicircuit, length, distance, distance function, (strongly/weakly) connected, (strongly/weakly) connected component

Let $G = (V, E, p)$ be a directed graph. Since each edge is assigned an (ordered) pair of vertices, it is natural to ask about the beginning and the end of an edge. The following mappings serve this purpose. The first and the second projection of the edge set E onto the vertex set V of G are defined as follows:

$$p_1 : \begin{cases} E \to V \\ e \mapsto x \end{cases} \quad \text{if} \quad p(e) = (x, y),$$

$$p_2 : \begin{cases} E \to V \\ e \mapsto y \end{cases} \quad \text{if} \quad p(e) = (x, y).$$

Using the 1st and 2nd projections π_1 and π_2 from the Cartesian product (see Example 5.3 (8)) we have:

$$p_i = \pi_i \circ p \quad \text{for } i = 1, 2.$$

We call $p_1(e)$ the ***initial vertex (source)*** and $p_2(e)$ the ***final vertex (target)*** of the edge $e \in E$.

For each vertex $x \in V$ we define

the *indegree* of x by $\delta^{\leftarrow}(x) := |\{e \in E \mid p_2(e) = x\}|$,
the *outdegree* of x by $\delta^{\rightarrow}(x) := |\{e \in E \mid p_1(e) = x\}|$,
the *predecessor set* of x by $N^{\leftarrow}(x) := \{y \in V \mid (y, x) \in p(E)\}$,
the *successor set* of x by $N^{\rightarrow}(x) := \{y \in V \mid (x, y) \in p(E)\}$.

Paths and Edge Sequences

Let $P_n := (e_1, e_2, \ldots, e_n)$ be a sequence of edges of G. We call P_n an *edge sequence*, if $n = 1$ or

$$(\forall i \in \{1, 2, \ldots, n-1\}) : \quad p_2(e_i) = p_1(e_{i+1}).$$

An edge sequence P_n with $p_1(e_1) = p_2(e_n)$ is called a *cycle*. An edge sequence P_n is called a *path*, if for all i, j with $1 \leq i < j \leq n$ one has

$$p_1(e_i) \neq p_1(e_j) \text{ and } p_2(e_i) \neq p_2(e_j).$$

A path therefore has no repetitions of edges. A path P_n with $p_1(e_1) = p_2(e_n)$ is called a *circuit*.

If we disregard the direction of the edges and replace in the above definitions each p_1 and p_2 with p_i and p_j with $i, j \in \{1, 2\}$, we obtain the definitions of a *semiedge sequence*, a *semicycle*, a *semipath* and a *semicircuit*. These are edge sequences in which consecutive edges can have different directions.

Distance and Connectivity

If $P_n = (e_1, e_2, \ldots, e_n)$ is a path or an edge sequence with $p_1(e_1) = x$ and $p_2(e_n) = y$, we also speak of an x, y-*path* or x, y-*edge sequence* or of a path or edge sequence *from x to y*. The number n of edges of P_n is called the *length* $\ell(P_n)$. The length of a shortest x, y-edge sequence is called *distance from x to y* and is denoted $d(x, y)$. This is equal to the length of a shortest x, y-path, since we have:

Lemma 6.1 (Paths and Circuits) *Every x, y-edge sequence contains an x, y-path and every cycle contains a circuit.*

Proof We prove the lemma for edge sequences P by mathematical induction according on the length of P. After that, we show the statement for cycles.

(IA) If P is an x, y-edge sequence with $\ell(P) = 1$, then P consists of only one edge and is therefore already a path.

6.2 Directed Graphs

(IH) Take $0 < n \in \mathbb{N}$, and suppose that for arbitrary vertices u and v in a graph G, every u, v-edge sequence of length less than n contains a u, v-path.

(IS) Now let P be an x, y-edge sequence of length $\ell(P) = n$, say $P = (e_1, e_2, \ldots, e_n)$.—We distinguish three cases:

Case 1. $p_2(e_1) = y$. Then $Q := (e_1)$ is an x, y-path.

Case 2. P contains an edge e_i with $1 < i$ starting at x, i.e. $p_1(e_i) = x$.
Then $P_1 := (e_i, e_{i+1}, \ldots, e_n)$ is an x, y-edge sequence with $\ell(P_1) = n - i + 1 < n$. According to (IH) it contains an x, y-path Q, which is also contained in P.

Case 3. $p_2(e_1) \neq y$ and $p_1(e_i) \neq x$ for all $1 < i < n$. With $z := p_2(e_1) \neq y$ we have that $P_2 := (e_2, e_3, \ldots, e_n)$ is a z, y-edge sequence with $1 \leq \ell(P_2) = n - 1 < n$. According to (IH) P_2 contains a z, y-path $Q_2 := (r_1, r_2, \ldots, r_m)$, which is also contained in P. We extend Q_2 by the edge e_1 and obtain an x, y-edge sequence $Q := (e_1, r_1, r_2, \ldots, r_m)$. This is an x, y-path, because by assumption $p_1(r_i) \neq x = p_1(k_1)$ for every edge r_i, since these edges are all contained in P_2.

If $Z = (e_1, e_2, \ldots, e_n)$ is a cycle, we choose an edge of Z that is not a loop, say e_1. (If there is no such, then each edge is a loop, i.e. a circuit in Z.) Then

$$p_1(e_2) = p_2(e_1) \neq p_1(e_1) = p_2(e_n),$$

so $P := (e_2, e_3, \ldots, e_n)$ is a $p_2(e_1), p_1(e_1)$-edge sequence. This contains a $p_2(e_1), p_1(e_1)$-path Q which, extended by the edge e_1, is a circuit contained in Z. □

For every graph $G = (V, E)$ we define the following mapping

$$d : \begin{cases} V \times V \to \mathbb{N} \cup \infty \\ (x, y) \mapsto \begin{cases} 0, & \text{if } x = y, \\ d(x, y), & \text{length of a shortest } x, y\text{ - path in } G, \\ \infty & \text{otherwise.} \end{cases} \end{cases}$$

This is called a ***distance function on*** G. Note that d is not a metric on V because in the directed case $d(x, y) \neq d(y, x)$ is possible.

We call a directed graph ***strongly connected***, if for any two vertices $x \neq y$ of the graph an x, y-path exists. A graph is called ***weakly connected*** if there exist semipaths between any two vertices. If a graph is not (strongly or weakly) connected, its maximal (strongly or weakly) connected parts are called ***(strongly or weakly) connected components***.

Exercise 6.2 (Riverman) A farmer (F) wants to bring a wolf (W), a goat (G), and a cabbage (C) across a river. He has a boat that can only fit one of these three objects in addition to him. During the action, neither the goat with the cabbage nor the wolf with the goat can be left alone, otherwise the cabbage or the goat would be eaten.

Draw a graph whose vertices are all states, e.g., (FWGC,–) denotes that all are on the left bank, and whose edges are all actions.

Give all solutions to the problem as edge sequences of this graph.

6.3 Undirected Graphs

Degree, neighbor, cycle, path, circuit, connected, complement graph, $K_n^{(s)}$, $\overline{K_n}^{(s)}$, P_n, C_n, bipartite, (spanning) tree/forest, independent edges, matching, maximum matching, perfect matching

Since every undirected graph can also be considered as a directed graph, we can use all definitions from Sect. 6.2 also for undirected graphs. Some terms become simpler, as the distinction of directions is omitted. So let $G = (V, E, p)$ be an undirected (multi)graph. For each vertex $x \in V$ we define

the ***degree of*** x by $\delta(x) := |\{e \in E \mid x \in p(e)\}| + |\{e \in E \mid x = p(e)\}|$,
the ***neighbors of*** x by $N(x) := \{y \in V \mid \{x, y\} \in p(E)\}$.

Since each (undirected) edge of a graph $G = (V, E, p)$ is incident with exactly two different vertices or counted twice with one vertex of G, it contributes 2 to the sum of all vertex degrees of G. Therefore, this sum is twice as large as the number of edges of G, i.e.

$$\sum_{x \in V} \delta(x) = 2|E|.$$

In particular, this implies the following

Lemma 6.3 (Number of Vertices with Odd Degree) *In undirected (multi) graphs, the number of vertices of odd degree is even.*

Paths, Moves, Distance and Connectivity

In undirected (multi)graphs it proves convenient, to define edge sequences as sequences of vertices (instead of edges):

A sequence $P_n := (x_0, x_1, x_2, \ldots, x_n)$ of vertices of G is called x_0, x_n-***(edge) sequence*** in G, if

$$(\forall i \in \{0, 1, \ldots, n-1\}): \quad \{x_i, x_{i+1}\} \in p(V).$$

6.3 Undirected Graphs

If $x_0 = x_n$, then P_n is called a *cycle*. The edge sequence P_n with pairwise different vertices (except possibly x_0 and x_n) is called a x_0, x_n-*path* in G. A path P_n is called a *circuit* if $x_0 = x_n$. A circuit with one vertex is thus a loop at this vertex and a circuit with two vertices consists of a double edge between these vertices. The terms strongly or weakly connected can now no longer be distinguished: we call an undirected graph *connected*, if any two different vertices are connected by a path.

Unlike a directed graph, an undirected graph is naturally equipped with a metric:

Lemma 6.4 (Metric) *The distance function d on a connected undirected graph $G = (V, E)$ is a metric on the set of vertices V.*

Examples 6.5 (Special Graphs) Among the simple, undirected graphs there are some for which special symbols and names are common.

(1) The *complement graph* $\overline{G}$ of a graph G without loops has the same set of vertices as G and contains exactly the edges that G does not have, and furthermore no loops. If this principle is also extended to loops, i.e. existing loops are omitted and vertices without loops receive a loop, this is called the *loop complement*.
(2) The *complete graph* K_n has n vertices and an edge between every two different vertices.
(3) The *totally disconnected graph* $\overline{K_n}$ also has n vertices, but no edge.
(4) If loops are to be added to K_n or $\overline{K_n}$, this can be expressed by a bracketed exponent: $K_n^{(s)}$ with $0 \leq s \leq n \in \mathbb{N}$ denotes a K_n, to which a loop is added at each of s vertices. Similarly for $\overline{K_n}$.
(5) The vertex set of an (m, n)-*bipartite graph* is the disjoint union of an m-element set A and an n-element set B. Edges exist only between vertices from A and vertices from B. A bipartite graph is called a *complete bipartite graph*, if all edges between vertices from A and B exist, it is denoted by $K_{m,n}$.
(6) A *path* of length n is denoted by P_n, i.e. P_n has n edges and $n + 1$ vertices. Thus, for example, $K_{1,2} = P_2$.
(7) A *circuit* with n edges is denoted by C_n, i.e., C_n has n edges and n vertices. So, for example, $K_{2,2} = C_4$.
(8) A graph without circles (in the directed case without semicircles) is called a *forest*, a connected forest is called a *tree*. For example, every path is a tree. If $G = (V, E)$ is a connected graph, we call a tree $T = (V, E')$ with $E' \subseteq E$ a *spanning tree* of G. If G is not connected, we form the union of the spanning trees of the connected components of G and obtain a so-called *spanning forest* of G. The Kruskal algorithm (Schöning [66]) determines spanning trees in given graphs, search and sorting algorithms often take place on trees (see Sections 4.2, 2.2 and 2.4 in Schöning [66]).

Matchings

Two edges of a graph G are commonly called ***independent edges*** if they are not incident in G. A set of pairwise independent edges is called a ***matching*** in G.

In the (directed) graph at the beginning of Sect. 6.4, Fig. 6.2, the edges $\{e_2, e_4\}$ or $\{e_2, e_3, e_8\}$ form matchings, both of which are also maximal with respect to inclusion. But only $\{e_2, e_3, e_8\}$ is a ***maximum matching***, i.e., it has maximal cardinality. It is even a ***perfect matching***, i.e., every vertex of the graph is incident with an edge of the matching.

Obviously, a graph that has a perfect matching has an even number of vertices.

A graph is called ***k-regular*** if each of its vertices has degree k. It is called ***regular*** if it is k-regular for some $k \in \mathbb{N}$. It is called ***strongly regular*** if the number of common neighbors of any two vertices u, v only depends on u, v being adjacent or not.

Exercise 6.6 (Strongly Regular Graphs) The graph $S_r = (V, E)$ is defined for $r \in \mathbb{N}$, $r \geq 3$ as follows:

$$V := \{Z \subseteq \{1, 2, \ldots, r\} \mid |Z| = 2\}, \ E := \{\{Z_i, Z_j\} \mid Z_i \cap Z_j \neq \emptyset\}.$$

(a) Draw S_r for $r = 3, 4, 5$.
(b) Show: The graph S_r is $(2r - 4)$-regular.
(c) Show: The graph S_r is strongly regular. (Any two adjacent vertices have exactly $r - 2$, any two non-adjacent vertices exactly 4 common neighbors.)

Exercise 6.7 (Vertex Degrees)

(a) In any undirected (multi-)graph, the number of edges is half as large as the sum of the degrees of the vertices.
(b) In any undirected (multi)graph, the number of vertices of odd degree is even.
(c) How could (a) and (b) be generalized to directed (multi)graphs?

Exercise 6.8 (Trees) Show that the three statements are equivalent:

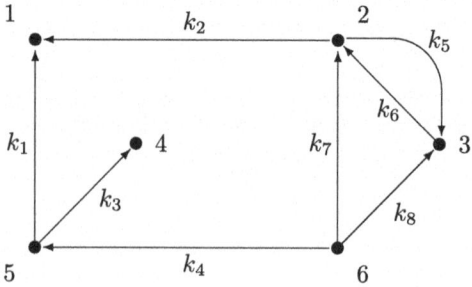

Fig. 6.2 Graphical representation of a graph

(i) $T = (V, E)$ is a tree.
(ii) $T = (V, E)$ is connected and any graph $G = (V, E')$, which is obtained by removing an edge, is not connected. In this case T is called **minimally connected**.
(iii) $T = (V, E)$ contains no cycles and any graph $H = (V, E'')$, which is obtained by adding a new edge, contains a cycle. In this case T is called **maximally cycle free**.

Exercise 6.9 (Bipartite Graphs)

(a) Draw all bipartite graphs with 2, 3, 4, 5, 6 vertices.
(b) How many bipartite graphs with n vertices are there?
(c) An undirected graph $G = (V, E)$ is bipartite if and only if there is a mapping $f : V \to \{\text{red, green}\}$ such that any two adjacent vertices are of different colors.
(d) Every tree is bipartite.
(e) A graph is bipartite if and only if it contains no cycles of odd length.

Exercise 6.10 (Shortest Paths) The table to the right gives the distances (in units of 100 miles) between the airports of the cities London, Mexico City, New York, Paris, Beijing, and Tokyo. Represent the connections between the cities by a graph whose edges are labeled with the distances. Remove edges from the graph and find a flight network of minimal total length, in which all six cities are still connected.

	L	MC	NY	Pa	Pe	T
L	0	56	35	2	51	60
MC	56	0	21	57	78	70
NY	35	21	0	36	68	68
Pa	2	57	36	0	51	61
Pe	51	78	68	51	0	13
T	60	70	68	61	13	0

6.4 Representations of Graphs

Adjacency list, matrix, adjacency matrix, symmetric matrix, incidence matrix, characterization of forests

There are many possibilities to represent a graph. One of them is the obvious generalization of the graph of a binary relation, known from Chap. 4. These graphical representations are very suitable to get an idea of the graph. They can be produced for larger graphs by a computer. We consider the graph with the vertex set $V := \{1, 2, 3, 4, 5, 6\}$ and the relation

$$E := \{(2, 1), (2, 3), (3, 2), (5, 1), (5, 4), (6, 2), (6, 3), (6, 5)\}.$$

To store a graph in a computer, one rather would input the relation directly. This would also take less storage space.

A common way is to represent a graph by ***adjacency lists***. The adjacency list $A(x)$ of the vertex x of a graph G consists in the undirected case of all neighbors of x (in any order) and in the directed case of all successors of x, i.e., all vertices y of G with $(x, y) \in p(E)$. For the above example we get

$$A(1) = (), \quad A(4) = (),$$
$$A(2) = (1, 3), \quad A(5) = (4, 1),$$
$$A(3) = (2), \quad A(6) = (3, 5, 2).$$

Occasionally, it may also be useful to list the predecessors of a vertex instead of the successors.

The representation of a graph by adjacency lists, implemented as linear lists, requires little storage space. Also, for example, the deletion and addition of vertices and edges is easily possible.

In the following, we will get to know further representation possibilities, which depending on what one wants to do with a graph, may be more or less favorable.

Matrices

The concept of a *matrix* is extensively studied in Linear Algebra. We introduce it here since it is also very useful for graphs. It will be taken up in Chap. 9.

An $m \times n$-***matrix over a set*** M is a rectangular scheme with m rows, n columns and entries from the set M. The entry of a matrix A in the i-th row and j-th column is denoted by a_{ij} or A_{ij}, the $m \times n$ matrix with entries a_{ij} is also written as (a_{ij}) or more precisely as $(a_{ij})_{1 \leq i \leq m, 1 \leq j \leq n}$.

The $m \times n$ matrix A, over $M = \{0, 1, 2\}$ with $m = 3$, $n = 2$ looks like this

$$A = (a_{ij}) = \begin{pmatrix} 1 & 1 \\ 0 & 2 \\ 2 & 0 \end{pmatrix}.$$

The ***adjacency matrix*** $A(G)$ of a (multi)graph G with n vertices is an $n \times n$ matrix $(a_{ij})_{1 \leq i, j \leq n}$, whose entry a_{ij} is the number of edges e of the graph with $p_1(e) = x_i$ and $p_2(e) = x_j$.

The undirected case is again reduced to the directed one, by considering each undirected edge $\{x, y\}$ as two directed edges (x, y) and (y, x).

6.4 Representations of Graphs

Obviously, the adjacency matrix of a graph is a $|V| \times |V|$ matrix over the set $M = \mathbb{N}$ (or over $M = \{0, 1\}$, if G is a simple graph.) The graph at the beginning of this section then has the adjacency matrix

$$A(G) = \begin{pmatrix} 0 & 0 & 0 & 0 & 0 & 0 \\ 1 & 0 & 1 & 0 & 0 & 0 \\ 0 & 1 & 0 & 0 & 0 & 0 \\ 0 & 0 & 0 & 0 & 0 & 0 \\ 1 & 0 & 0 & 1 & 0 & 0 \\ 0 & 1 & 1 & 0 & 1 & 0 \end{pmatrix}.$$

Lemma 6.11 (Symmetric Adjacency Matrix) *The adjacency matrix of an undirected graph is symmetric, i.e.* $(\forall i, j): a_{ij} = a_{ji}$.

Theorem 6.12 (Matrices and Graphs) *There is a bijective mapping of the set of all $n \times n$ matrices with entries from $\mathbb{N}$ to the set of all graphs with vertex set $V = \{x_1, \ldots, x_n\}$.*

Proof Take the matrix $A = (a_{ij})_{1 \leq i,j \leq n}$, set $V := \{x_1, \ldots, x_n\}$, $E := \{(x_i^s, x_j^s) \mid 1 \leq s \leq a_{ij} \ s \in \mathbb{N}\}$ and $p(x_i^s, x_j^s) := (x_i, x_j)$ for all i, j, s. We get a (multi)graph $G(A) := (V, E, p)$, whose adjacency matrix is A.

Conversely, given a (multi-)graph G with n vertices, the adjacency matrix $A(G)$ is an $n \times n$ matrix with entries from $\mathbb{N}$, for which $G(A(G)) = G$. □

The next theorem follows directly from the definition of $A(G)$:

Theorem 6.13 (Column and Row Sums) *In any directed graph G with adjacency matrix $A(G)$, for each vertex $x_i \in V$:*

$$\delta^{\leftarrow}(x_i) = \sum_{j=1}^{n} a_{ji} \quad \text{(sum of the elements of the i-th column),}$$

$$\delta^{\rightarrow}(x_i) = \sum_{j=1}^{n} a_{ij} \quad \text{(sum of the elements of the i-th row).}$$

Accordingly, in undirected graphs:

$$\delta(x_i) = \sum_{j=1}^{n} a_{ij} + a_{ii} = \sum_{j=1}^{n} a_{ji} + a_{ii}.$$

For a loopless graph $G = (V, E, p)$ with

$$V = \{x_1, x_2, \ldots, x_n\} \quad \text{and} \quad E = \{e_1, e_2, \ldots, e_m\}, \tag{6.1}$$

the *incidence matrix* is a $n \times m$ matrix $B(G) = (b_{ij})_{1 \le i \le n, 1 \le j \le m}$

with $b_{ij} := \begin{cases} 1, & \text{if } \quad x_i = p_1(e_j) \\ -1, & \text{if } \quad x_i = p_2(e_j) \\ 0 & \text{otherwise} \end{cases}$ in the directed case

and $b_{ij} := \begin{cases} 1, & \text{if } \quad x_i \in p(e_j) \\ 0 & \text{otherwise} \end{cases}$ in the undirected case.

So we have defined matrices with $|V|$ rows, $|E|$ columns and entries from $M = \{-1, 0, 1\}$.

Attention Here, undirected graphs cannot be considered as special directed graphs. An undirected edge e_j between vertices x_i and x_k provides entries $b_{ij} = b_{kj} = 1$ in the incidence matrix of an undirected graph. In the incidence matrix of a directed graph it would be considered as two edges, namely e_{j_1} from x_i to x_k and e_{j_2} from x_k to x_i. Thus we get the entries $b_{ij_1} = b_{ej_2} = 1$ and $b_{ij_2} = b_{ej_1} = -1$.

The above graph has the incidence matrix

$$B(G) = \begin{pmatrix} -1 & -1 & 0 & 0 & 0 & 0 & 0 & 0 \\ 0 & 1 & 0 & 0 & 1 & -1 & -1 & 0 \\ 0 & 0 & 0 & 0 & -1 & 1 & 0 & -1 \\ 0 & 0 & -1 & 0 & 0 & 0 & 0 & 0 \\ 1 & 0 & 1 & -1 & 0 & 0 & 0 & 0 \\ 0 & 0 & 0 & 1 & 0 & 0 & 1 & 1 \end{pmatrix}.$$

Characterization of Forests

Anyone who knows the concept of linear independence from Linear Algebra will enjoy the following characterization of forests.

Lemma 6.14 (Initial or Terminal Vertex) *Every forest with at least one edge has a vertex x with*

$$\delta^\leftarrow(x) + \delta^\rightarrow(x) = 1.$$

Proof Such a vertex can be found as one of the terminal vertices of a maximal semipath in the graph. □

Theorem 6.15 (Characterization of Forests) *A graph with at least one edge is a forest if and only if the columns of its incidence matrix are linearly independent.*

Proof Let $G = (V, E, p)$ be a directed graph with m edges and the incidence matrix $B(G)$. We prove both implications by contraposition.

" $\Rightarrow$ ". Suppose that the columns $s_1, s_2, \ldots, s_m$ of $B(G)$ are linearly dependent. Number them such that in the linear combination

$$\lambda_1 s_1 + \lambda_2 s_2 + \ldots + \lambda_r s_r + \ldots + \lambda_m s_m = 0$$

exactly the first r coefficients $\lambda_1, \lambda_2, \ldots, \lambda_r$ are non-zero. Due to the linear dependence, we can always achieve $r \geq 1$. Let $e_1, e_2, \ldots, e_r$ be the edges of G corresponding to $s_1, s_2, \ldots s_r$. The graph $G' := (V', E', p|_{V' \times V'})$ with $E' := \{e_1, e_2, \ldots e_r\}$ and $V' := p_1(E') \cup p_2(E')$ has an incidence matrix, which has at least two entries not equal to 0 in each row. If G' were a forest, then according to Lemma 6.14 there must be a vertex x whose corresponding row of the incidence matrix of G' has exactly one entry not equal to 0. It follows that G' contains a semicircuit. Since G' is contained in G, G is also not a forest.

" $\Leftarrow$ ". Let $C := (e_1, e_2, \ldots, e_r)$ be a semicircuit in G. The columns of the incidence matrix $B(G)$ of G corresponding to the edges $e_1, e_2, \ldots, e_r$ we denote by $s_1, s_2, \ldots, s_r$. If we traverse the semicircuit in the direction of e_1 (i.e., we start at $p_1(e_1)$ and continue via $p_2(e_1)$), then, if C is not a circuit, we traverse some edges in the opposite direction. For each such edge i we set $\lambda_i := -1$, for all other edges e_j (including e_1) we set $\lambda_j := 1$.

This implies:

$$\lambda_1 s_1 + \lambda_2 s_2 + \cdots + \lambda_r s_r = 0.$$

Indeed, the columns s_i, $1 \leq i \leq r$, have only entries not equal to 0 in the rows that correspond to vertices on the semicircuit. After multiplication with the λ_i, $1 \leq i \leq r$, each such row has in the columns s_i, $1 \leq i \leq r$, exactly one 1 and exactly one -1. This shows, that the columns of $B(G)$ are linearly dependent. □

6.5 Operations with Matrices

Product and sum of matrices, Schur-Hadamard multiplication, power of a matrix, distance matrix, reachability matrix, transitive closure, Boolean matrix, hypercube

Product and Sum of Matrices

The classical justification of the following definitions of multiplication and addition of matrices comes from Linear Algebra. But the surprising result in Theorem 6.16 is another good reason to use these definitions.

The *product* AB of an $m \times q$-matrix $A = (a_{ij})$ with a $q \times n$-matrix $B = (b_{ij})$ is an $m \times n$-matrix $C = (c_{ij})$ with entries

$$c_{ij} := \sum_{e=1}^{q} a_{ik} b_{kj}.$$

To form the product AB of two matrices A and B the number of columns of A must be equal to the number of rows of B. In addition, the entries of the matrices must be such that they can be added and multiplied.

Using a 2×3 and a 3×2 matrix with entries from $\mathbb{Z}$ this looks like :

$$\begin{pmatrix} a_{11} & a_{12} & a_{13} \\ a_{21} & a_{22} & a_{23} \end{pmatrix} \begin{pmatrix} b_{11} & b_{12} \\ b_{21} & b_{22} \\ b_{31} & b_{32} \end{pmatrix} =$$

$$\begin{pmatrix} a_{11}b_{11} + a_{12}b_{21} + a_{13}b_{31} & a_{11}b_{12} + a_{12}b_{22} + a_{13}b_{32} \\ a_{21}b_{11} + a_{22}b_{21} + a_{23}b_{31} & a_{21}b_{12} + a_{22}b_{22} + a_{23}b_{32} \end{pmatrix}.$$

For completeness, we mention here that the *sum of two matrices* is defined "elementwise", requiring that the two matrices have the same number of rows and columns and that their entries can be added. Using two 2×3 matrices with entries from $\mathbb{Z}$ as an example, this looks like this:

$$\begin{pmatrix} a_{11} & a_{12} & a_{13} \\ a_{21} & a_{22} & a_{23} \end{pmatrix} + \begin{pmatrix} b_{11} & b_{12} & b_{13} \\ b_{21} & b_{22} & b_{23} \end{pmatrix} =$$

$$\begin{pmatrix} a_{11}+b_{11} & a_{12}+b_{12} & a_{13}+b_{13} \\ a_{21}+b_{21} & a_{22}+b_{22} & a_{23}+b_{23} \end{pmatrix}.$$

There is also the so-called *Schur-Hadamard multiplication*[1] of matrices, where the entries of two matrices of the same size are multiplied elementwise ("bad students multiplication"). However, this multiplication does not have the properties that we need here.

Powers of the Adjacency Matrix

If $A = (a_{ij})_{1 \leq i,j \leq n}$ is a matrix, we denote with

$$A^r =: (a_{ij}^{(r)})_{1 \leq i,j \leq n}$$

---

[1] Issai Schur, Belarusian mathematician, 1875–1941; Jacques Salomon Hadamard, French mathematician, 1865–1963.

6.5 Operations with Matrices

the *r-th power of* A. We obtain this by r times multiplying A by itself. This is possible because A is a $n \times n$ matrix. **Attention:** In general, $a_{ij}^{(r)} \neq a_{ij}^r$. We must first calculate A^r, to be able to see $a_{ij}^{(r)}$! We set $A^1 := A$ and thus $a_{ij}^{(1)} = a_{ij}$.

Theorem 6.16 (Powers of the Adjacency Matrix) *If $A^r = (a_{ij}^{(r)})_{1 \leq i,j \leq n}$ is the r-th power of the adjacency matrix of $G = (V, E, p)$, then $a_{ij}^{(r)}$ is the number of x_i, x_j-edge sequences of length r in G.*

Proof We use mathematical induction on r:

(IA) The assertion is true for $r = 1$ by definition of the adjacency matrix $A^1 = A$.
(IH) Let $r \geq 1$ and $a_{ij}^{(r)}$ be for all $1 \leq i, j \leq n$ the number of x_i, x_j-edge paths of length r in G.
(IS) We have $A^{r+1} = A^r A$, therefore by definition of matrix multiplication

$$(\forall 1 \leq i, j \leq n): a_{ij}^{(r+1)} = \sum_{e=1}^{n} a_{ik}^{(r)} a_{kj}.$$

In $a_{ij}^{(r+1)}$, for each $k \leq n$, for each x_i, x_k-edge path of length r, all possible extensions by one more edge from x_k to x_j are counted. This gives all x_i, x_j-edge paths of length $r + 1$, as claimed.

□

Theorem 6.17 (Cycle-Free) *A directed graph with n vertices contains no cycles if and only if for all $1 \leq i \leq n$ and all $1 \leq r \leq n$ we have $a_{ii}^{(r)} = 0$.*

Proof Every cycle that contains x_i is in particular an x_i, x_i-edge path. Conversely, every x_i, x_i-edge path contains a cycle on which x_i lies, according to Lemma 6.1. The rest follows from Theorem 6.16. □

Distance Matrix

We take the values of the distance function from Sect. 6.3 of a graph G as the entries of an $n \times n$-matrix. This way we obtain the *distance matrix* of G:

$$D(G) = (d_{ij})_{1 \leq i,j \leq n} := (d(x_i, x_j))_{1 \leq i,j \leq n}.$$

For some applications the "surplus" of information that the distance matrix provides, compared to the adjacency matrix, is quite useful.

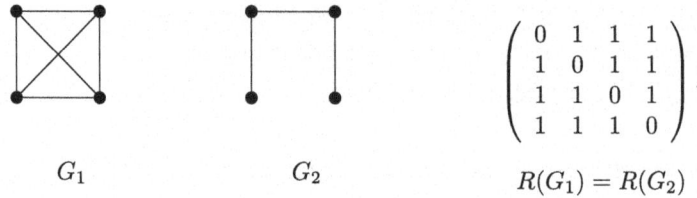

Fig. 6.3 Graphs with identical transitive closure

Theorem 6.18 (Distances) *For $i \neq j$, $d(x_i, x_j)$ is the smallest $r \in \mathbb{N}$ with $r < n$ and $a_{ij}^{(r)} \neq 0$. If such an r does not exist, then $d(x_i, x_j) = \infty$.*

Proof The assertion follows directly from Theorem 6.16 and the definition of $d(x_i, x_j)$. □

Reachability Matrix and Transitive Closure

Sometimes it is not interesting how long an x, y-edge path in G is, and also not how many there are, but only whether it is possible at all to get from x to y in G.

The *reachability matrix* $R(G) := (r_{ij})_{1 \leq i,j \leq n}$ of $G = (V, V, p)$ is defined by

$$r_{ij} := \begin{cases} 1, & \text{if there is an } x_i, x_j\text{-path,} \\ 0 & \text{otherwise.} \end{cases}$$

In the diagonal of the reachability matrix, there is a 1 if and only if the corresponding vertex lies on a cycle.

If we consider the adjacency matrix A of a simple directed graph as a matrix of the binary relation $E \subseteq V \times V$), then A^2 is the matrix of the relation $E^2 := E \circ E = \{(x, y) \in V \times V \mid (\exists z \in V) : (x, z), (z, y) \in E\}$. Analogously, A^r is the matrix of the relation $E^r := E^{r-1} \circ E$ for each $1 < r \in \mathbb{N}$.

We can consider a simple undirected graph as a symmetric relation by including the pairs (x, y) and (y, x) in the relation for each edge $\{x, y\}$.

The matrix $R(G)$ is the matrix of the smallest transitive relation with respect to inclusion that contains E. This relation is called *the transitive closure E^+ of* E.

Of course, different binary relations can have the same transitive closure. Figure 6.3 shows two undirected graphs with the same reachability matrix.

Methods for constructing the transitive closure from an initially non-transitive relation are quite important. In particular, the transitive closure of the directed graph obtained from the Hasse diagram of an order by orienting all edges upwards

6.5 Operations with Matrices

is exactly the graph of the underlying order relation. An algorithm (*Warshall*[2] *Algorithm*) for determining the transitive closure can be found in Schöning [66].

Boolean Matrices

A ***Boolean***[3] ***matrix*** is a matrix whose entries are either 0 or 1. The reachability matrix is an example of a Boolean matrix. Another example is the adjacency matrix of a simple undirected Graphs. Boolean operations of Boolean matrices analogous to the addition and multiplication of ordinary matrices are defined using the operations $\vee$ and $\wedge$. These are defined on the set $\{0, 1\}$ by the following tables:

$\vee$	0	1
0	0	1
1	1	1

$\wedge$	0	1
0	0	0
1	0	1

For Boolean $n \times n$ matrices A and B, we now define $S := A \vee B$ and $P := A \wedge B$ as follows by specifying their elements:

$$s_{ij} := a_{ij} \vee b_{ij} \quad \text{and} \quad p_{ij} := \bigvee_{e=1}^{n}(a_{ik} \wedge b_{kj}) \quad \text{for} \quad 1 \leq i, j \leq n.$$

We call $A \vee B$ the ***Boolean sum*** and $A \wedge B$ the ***Boolean product*** of A and B. The matrix operations are performed according to the usual addition and multiplication of matrices. We write $A^{<r>}$ for the r-th Boolean power of A and obtain:

Theorem 6.19 (Boolean Powers of the Adjacency Matrix) *Suppose that $A^{<r>} =: (a_{ij}^{<r>})_{1 \leq i,j \leq n}$ for $r > 1$ is the r-th Boolean power of the adjacency matrix of $G = (V, E, p)$. Then $a_{ij}^{<r>} = 1$ exactly if there is at least one x_i, x_j-edge path of length r in G.*

Theorem 6.20 (Powers of $A(G)$ yield $R(G)$) *The reachability matrix $R(G)$ of a graph G is obtained from the Boolean powers of its adjacency matrix A as follows:*

$$R(G) = A \vee A^{<2>} \vee A^{<3>} \vee \ldots \vee A^{<n>} = \bigvee_{e=1}^{n} A^{<e>}.$$

A suitable algorithm can be found in Schöning [66].

Exercise 6.21 (Graphs and Matrices) Consider the graph in Fig. 6.4.

---

[2] Stephen Warshall, US-American mathematician 1935–2006.
[3] George Boole, English (self-taught) mathematician, logician and philosopher, 1815–1864.

Fig. 6.4 The graph $G = (V, E)$

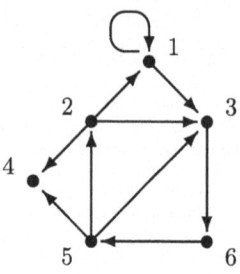

Set up its adjacency matrix $A(G)$, determine its first to sixth power and list all 1, 3-edge sequences and all 5, 3-edge sequences of length ≤ 6 in this graph. Number the edges in lexicographic order (i.e. $e_1 = (1, 1)$, $e_2 = (1, 3)$, $e_3 = (2, 1)$ etc.) and provide the incidence matrix $B(G)$.

Determine the reachability matrix $R(G)$ and decide and justify whether G is strongly connected. Why is it sufficient to consider edge sequences of length ≤ 6?

Exercise 6.22 (Transitive Closure) (cf. Exercise 4.30) Consider the binary relation ϱ on the set $M := \{1, 2, 3, 4, 5, 6\}$,

$$\varrho := \{(1, 2), (1, 4), (2, 3), (4, 5), (5, 6), (6, 2), (3, 5)\}.$$

(a) Draw the graphs of ϱ, ϱ^2, ϱ^3 and ϱ^4 and label them with G_1, G_2, G_3 and G_4.
(b) Provide a shortest path of maximum length in G_1.
(c) Show that for the adjacency matrices one has $A(G_1) \cdot A(G_1) = A(G_2)$.
(d) How can the edges of G_i, $i = 1, 2, 3, 4$, be interpreted?
(e) The relation $\varrho \cup \varrho^2 \cup \varrho^3 \cup \varrho^4$ is transitive.—Justify this using the corresponding graphs.

Exercise 6.23 (Hypergraphs) A triple $H = (V, E, p)$ is called *hypergraph*, if $p : E \to \wp(V) \setminus \{\emptyset\}$. Here, again, V is a non-empty set, called *vertex set* (or *node set*), E is a non-empty set, called *edge set*, and p is a mapping, called *incidence mapping*.

Note: Hypergraphs should not be confused with hypercubes (cf. Exercise 6.24)!

(a) Give V, E and p for the hypergraph $H = (V, E, p)$ in Fig. 6.5.
(b) Find a definition for paths in hypergraphs that meaningfully generalizes the corresponding definition for graphs.
(c) In analogy to the corresponding matrices for graphs, set up the matrices $A(H)$, $B(H)$, $R(H)$ and $D(H)$.
(d) Give formal definitions for all four matrices for hypergraphs.

6.6 Structure Preserving Mappings

Fig. 6.5 A hypergraph

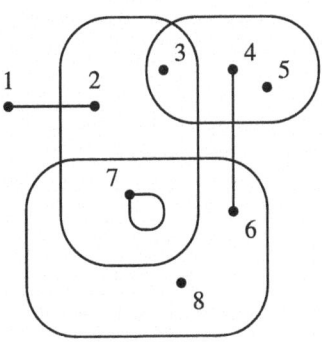

Exercise 6.24 (Hypercube) The *hypercube* Q_n is defined as the following graph. The vertex set $V(Q_n)$ consists of all 0, 1—sequences of length n. An (undirected) edge is put between two such sequences, if they differ at exactly one position.

(a) Draw Q_1, Q_2, Q_3 and Q_4. Represent Q_{n+1} by extending the sequences, that form the vertices of Q_n, by a 0 and by a 1.
(b) Give a bijective mapping of the vertex set of Q_3 to the power set $\wp(\{a, b, c\})$ of a three-element set.—Try it also with bijective $f : Q_n \to \wp(\{a_1, \ldots, a_n\})$.
(c) Give the 1st projection of the vertex set of Q_3 onto the set $\{0, 1\}$ and use this principle to find a surjective mapping of the vertex set of Q_3 onto that of Q_2.— Try it also with surjective $p : V(Q_n) \to V(Q_k)$ for $1 \le k \le n$.
(d) Can this procedure be reversed to describe an injective mapping $\iota : V(Q_2) \to V(Q_3)$ or $\iota : V(Q_k) \to V(Q_n)$ for $1 \le k \le n$?
(e) What regularities does the graph Q_n show?—Degree of a vertex, number of shortest paths between two vertices, etc.

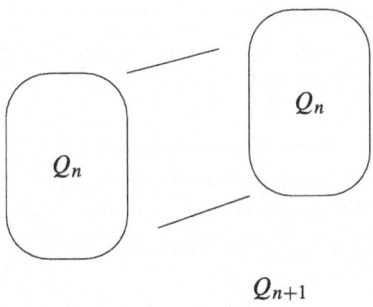

6.6 Structure Preserving Mappings

(Strong) graph (ega)morphism, edge preserving mapping, isomorphism, endo(ega)morphism, automorphism, congruence, factor graph, canonical surjection, Homomorphism Theorem

To investigate mathematical structures such as sets or graphs, "suitable" mappings are particularly useful. The question of "suitable" or "unsuitable" aims at whether an image resembles its preimage "to a sufficient extent," i.e., whether the mapping preserves the structure in question. Once a type of mapping has been chosen for a given structure, these are usually called *morphisms, homomorphisms*, or *structure-preserving mappings*. Sometimes, they also receive a special name: the homomorphisms of vector spaces, for example, are called *linear mappings*. Many results in this direction can be found in Knauer/Knauer, "Algebraic Graph Theory: Morphisms, Monoids, and Matrices" [50].

Although the results of this section can be transferred to multigraphs and graphs with loops, we formulate them only for simple, loopless graphs.

Let $G_1 := (V_1, E_1)$ and $G_2 := (V_2, E_2)$ be graphs. A mapping $f : V_1 \to V_2$, also written as $f : G_1 \to G_2$, is called **(graph) morphism** or also *(graph) homomorphism*, if it is **edge preserving**, i.e. if

$$(\forall x, y \in V_1): \quad (x, y) \in E_1 \Rightarrow (f(x), f(y)) \in E_2$$

in the directed case and

$$(\forall x, y \in V_1): \quad \{x, y\} \in E_1 \Rightarrow \{f(x), f(y)\} \in E_2$$

in the undirected case.

In particular, f induces a mapping $E_1 \to E_2$. A graph morphism is called **strong graph morphism**, if in the preceding definitions $\Rightarrow$ is replaced by $\Leftrightarrow$.

A mapping $f : V_1 \to V_2$ is called **(graph) egamorphism**, if it preserves edges or identifies vertices, i.e. if

$$(\forall x, y \in V_1): \quad (x, y) \in E_1 \text{ and } f(x) \neq f(y) \Rightarrow (f(x), f(y)) \in E_2$$

in the directed and analogously in the undirected case. For data compression egamorphisms are important. They are used under various names. See for example 1.10 in Schöning [66].

If $G_1 = G_2$, then f is called an **endo(ega)morphism**.

A bijective graph morphism, whose inverse mapping is also a graph morphism, is called a **(graph) isomorphism**. To express that $f : G_1 \to G_2$ is an isomorphism, one also writes $f : G_1 \cong G_2$. An isomorphism thus induces a bijection between E_1 and E_2. An isomorphism of a graph onto itself is called **automorphism**.

See Fig. 6.6 for a strong graph homomorphism. The inverse mapping of a bijective graph morphism does not have to be a graph morphism, as shown in Fig. 6.7. Graph egamorphisms do not have to be graph homomorphisms, see Fig. 6.8.

6.6 Structure Preserving Mappings

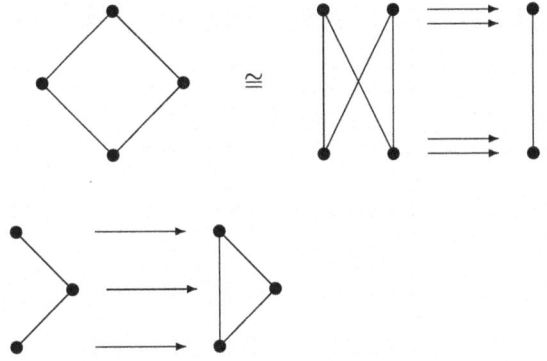

Fig. 6.6 Two graphs isomorphic to C_4, which are mapped onto K_2 by a strong graph morphism

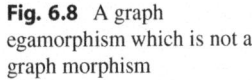

Fig. 6.7 A bijective graph morphism, whose inverse mapping is not edge preserving, i.e. it is not an isomorphism. In particular, it is not strong

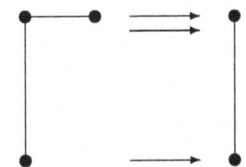

Fig. 6.8 A graph egamorphism which is not a graph morphism

Graph Congruence, Quotient Graph, Homomorphism Theorem

The equivalence relation $\varrho_f = \{(x, y) \mid f(x) = f(y)\}$, induced by a graph morphism $f : G_1 \to G_2$ on the vertex set V_1 of G_1, cf. Lemma 5.20, serves as a prototype of structure preserving equivalence relations on graphs. We note:

Lemma 6.25 (Induced Equivalence Relation) *Let $f : G_1 \to G_2$ be a graph morphism of the graph G_1 into the loopless graph G_2. Then the equivalence classes of the relation ϱ_f do not contain adjacent vertices of G_1.*

As usual, structure preserving equivalence relations are called *congruences*.

We define: An equivalence relation ϱ on the vertex set V of the graph $G = (V, E)$ is called **(graph) congruence**, if

$$\varrho \subseteq (V \times V) \setminus E,$$

i.e., if the individual ϱ-classes do not contain adjacent vertices.

If now ϱ is a congruence on $G = (V, E)$, we set

$$E_\varrho := \{([x]_\varrho, [y]_\varrho) \mid (\exists a \in [x]_\varrho, \exists b \in [y]_\varrho) : (a, b) \in E \}.$$

Thereby we define the **quotient graph of G with respect to ϱ** by

$$G/_\varrho := (V/_\varrho, E_\varrho).$$

With this, we obtain an analogue to Lemma 5.20:

Lemma 6.26 (Canonical Surjection) *Let $G = (V, E)$ be a graph and $\varrho \subseteq V \times V$ a congruence on G. By*

$$\pi_\varrho : \begin{cases} V \to V/_\varrho \\ x \mapsto [x]_\varrho \end{cases}$$

*a surjective graph morphism from G to $G/_\varrho$ is defined, the **canonical surjection of** ϱ.*

Proof Due to Lemma 5.20, we only need to show that π_ϱ preserves edges. Indeed, according to the definition of E_ϱ:
$(\forall x, y \in E) : \quad (x, y) \in V \quad \Rightarrow \quad (\pi_\varrho(x), \pi_\varrho(y)) = ([x]_\varrho, [y]_\varrho) \in E_\varrho.$ □

Theorem 6.27 (Homomorphism Theorem for Graphs) *Let $f : G \to H$ be a graph morphism and ϱ a congruence on G with $\varrho \subseteq \varrho_f$. Then the graph homomorphism*

$$f' : \begin{cases} G/_\varrho \to H \\ [x]_\varrho \mapsto f(x) \end{cases}$$

is unique such that the diagram

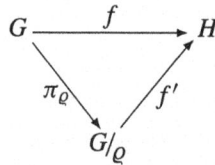

is commutative.
Furthermore,

(a) f is surjective $\Rightarrow$ f' is surjective;

(b) $\quad \varrho = \varrho_f \Rightarrow f'$ is injective.

In particular: $G/_{\varrho_f} \cong f[G]$ as sets of vertices, not as graphs.

6.7 Euler and Hamilton Problems

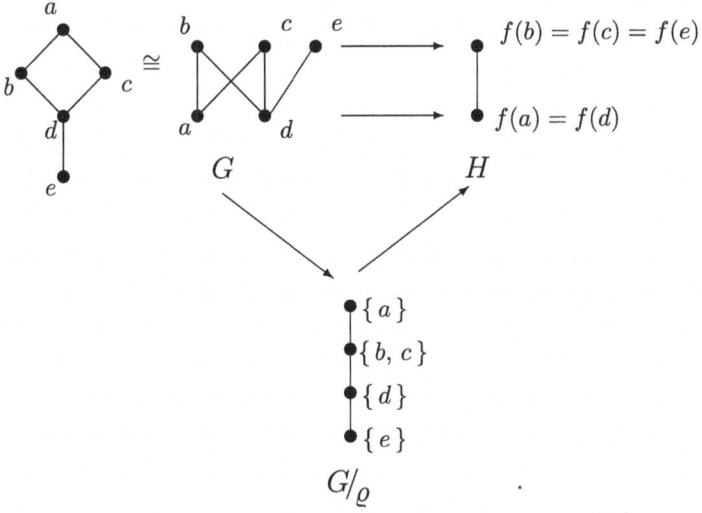

Fig. 6.9 Example for the Homomorphism Theorem

Proof Much of the work has already been done in the proof of the Homomorphism Theorem for sets (Theorem 5.21). It remains only to show that the above-defined mapping f' preserves edges. For π_ϱ we have just shown this in Lemma 6.26.

By definition of the factor graph, for each edge $([x]_\varrho, [y]_\varrho) \in E_\varrho$ there are vertices $a \in [x]_\varrho$ and $b \in [y]_\varrho$ in G, such that $(a, b) \in E(G)$. Now

$(a, x), (b, y) \in \varrho \subseteq \varrho_f \Rightarrow f(a) = f(x), f(b) = f(y)$ and thus
$(f'([x]_\varrho), f'([y]_\varrho)) = (f(x), f(y)) = (f(a), f(b)) \in E(H)$, since $(a, b) \in E(G)$ and f is a graph morphism. □

Figure 6.9 exemplifies the details of the Homomorphism Theorem.

Exercise 6.28 (Composition of Graph Morphisms) If f and g are two graph morphisms, strong graph morphisms, graph egamorphisms of a graph G into itself, then $f \circ g$ has the same property.

Exercise 6.29 (Edge Bijection But No Isomorphism) Find a graph morphism that induces a bijection on the edges, but is not an isomorphism.

6.7 Euler and Hamilton Problems

Euler graph, Euler trail, Euler cycle, Hamilton path, Hamilton circuit, Hamilton graph, Travelling Salesman Problem, decision tree, closure of a graph, Theorems of Bondy and Chvátal, of Dirac, of Ore, edge graph, box product

Fig. 6.10 Graphs with Euler Trails and Cycles

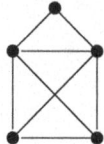
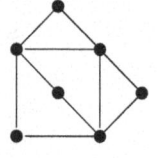

The House of Santa Claus   An Euler graph

The two problems presented here mark in some sense the beginning of Graph Theory. Many applications of Graph Theory can be traced back to them, although both probably originated as games. It is also noteworthy that the formulations of these problems are very similar, while the possibilities to solve them differ greatly from each other. The Euler problem is satisfactorily solved with Euler's theorem, while there is still no complete solution for the Hamilton problem.[4]

Euler Trail, Cycle, Graph

Let $G = (V, E, p)$ be an undirected multigraph. An x, y-edge path, which contains each edge of G exactly once, is called x, y-**Euler trail**. A cycle in G that is an Euler trail is called an **Euler cycle**. A graph that has an Euler cycle is called an **Euler graph**.

Example 6.30 (House of Santa Claus) A well-known example of a graph with an Euler trail that is not an Euler graph is the so-called "House of Santa Claus". Next to it, an Euler graph is depicted. The readers are encouraged to find the Euler trails in Fig. 6.10 themselves. We recall that in German the walking on an Euler trail of the "House of Santa Claus" is accompanied with the words "Das ist das Haus vom Ni ko laus", one syllable per edge.

Theorem 6.31 (Euler Graphs) *For every connected undirected multigraph G, the following statements are equivalent:*

(i) *G is an Euler graph.*
(ii) *Every vertex of G has even degree.*
(iii) *The edge set of G can be partitioned into circuits.*

Proof "(i) $\Rightarrow$ (ii)". Let C be an Euler cycle in G. This contains all edges of G (and thus also all vertices, some possibly several times). Each vertex x of G lies i_x times, with $1 \leq i_x$, on the cycle and therefore has the degree $\delta(x) = 2i_x$.

---

[4] Sir William Rowan Hamilton, Irish mathematician, 1805–1865.

"(ii) $\Rightarrow$ (iii)". We prove this for not necessarily connected graphs by induction on the number k of edges of G.

(IA) For $k = 0$, the statement is trivial.
(IH) In every graph with fewer than k edges, the implication is valid.
(IS) Let G be a graph with k edges, $k \geq 1$. If (ii) is fulfilled, even $k \geq 2$. Choose a path of maximal length in G and write it as a sequence of vertices, say $P = (x_0, x_1, \cdots, x_p)$ with $1 \leq p$, where any two consecutive vertices are adjacent. Since $\delta(x_p)$ is even, x_p is incident with another edge besides the one on the path. Let z be the second vertex incident with this edge. Since P is a maximal path, z is already on P, say $z = x_i$ where $0 \leq i < p - 1$. Thus, $C := (x_i, x_{i+1}, \cdots, x_p, z)$ is a circuit in G, given as a sequence of vertices. The graph G', which results if we remove all edges of this circuit from G, still has property (ii), but contains fewer than k edges. According to (IH), we can decompose its set of edges into circles and together with C we obtain a decomposition of the set of edges of G into circles.

"(iii) $\Rightarrow$ (i)". Let (iii') be the statement: "The set of edges of G can be decomposed into cycles."—Since every circuit is a cycle, $(iii) \Rightarrow (iii')$. We show $(iii') \Rightarrow (i)$ with induction on the number c of cycles in a given decomposition of the set of edges of G.

(IA) If $c = 1$, then this cycle is an Euler cycle of G.

(IH) In every graph, whose set of edges has been decomposed into fewer than c cycles, an Euler cycle exists.

(IS) Let $c \geq 2$ and $C_1, C_2, \cdots, C_c$ be a decomposition of the set of edges of G into cycles. Since G is connected, C_1 has at least one vertex in common with some cycle, say C_i. If we first traverse C_1 and then C_i from this vertex, we obtain a new cycle $C_1 C_i$ in G.

Now $C_2, C_3, \cdots, C_{i-1}, C_1 C_i, C_{i+1}, \cdots, C_c$ is a decomposition of the set of edges of G into $c - 1$ cycles, and so, according to (IH), G is an Euler graph. □

Corollary 6.32 (Euler Trails) *A connected undirected multigraph $G = (V, E, p)$ contains an Euler trail if and only if the number of its vertices with odd degree is equal to 0 or 2.*

Proof Let the number of vertices of G with odd degree be equal to u. For $u = 0$, according to Theorem 6.31, there is even an Euler cycle. According to Lemma 6.3, $u = 1$ is not possible. We assume that G has at least two vertices with odd degree, say x and y. If G' is the graph that results from G by adding an edge between x and y, then G contains an x, y-Euler trail $\Leftrightarrow$ G' contains an Euler cycle $\Leftrightarrow$ G' has only vertices of even degree $\Leftrightarrow$ G has exactly two vertices of odd degree. □

A suitable algorithm can be found in Schöning [66].

The Problem of Hamilton

A path or circuit that contains each vertex of a graph G exactly once is called **Hamilton path** or **Hamilton circuit**. A graph that contains a Hamilton circuit is called **Hamilton graph**.

The question of whether a graph contains a Hamilton circuit dates back to the game "Around the World" proposed by Sir William Hamilton in 1859. There the vertices of the graph of a *regular dodecahedron* corresponded to certain cities. A round trip through these cities was to be found. See Fig. 6.11.

Obviously, every complete graph K_n, $n \geq 4$, contains many Hamilton circuits.

However, if we assign lengths to the edges of a complete graph, another problem arises: Find a round trip of shortest length. This is the *Travelling Salesman Problem*, one of the famous *NP-complete problems*. This is a whole class of problems, for which no satisfactory solution has been found so far (and for which probably no solution exists).

The "House of Santa Claus" from Example 6.30 is a Hamilton graph, the graph depicted next to it, however, is not. A proof for this is often not easy even for such a small example. In Fig. 6.12 we indicate a (not very creative) method that systematically tries out all possibilities with the help of a *decision tree*. Here we find that there is no Hamilton circuit in the present graph.

Fig. 6.11 The dodecahedron graph

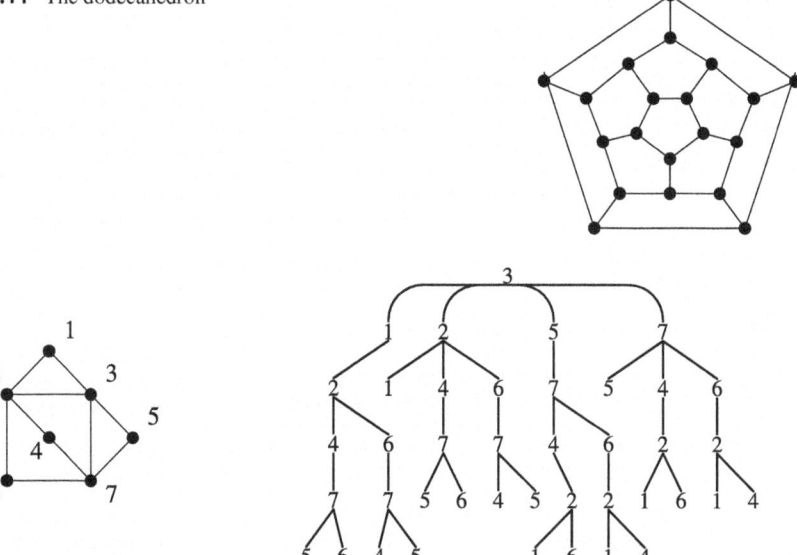

Fig. 6.12 A decision tree

6.7 Euler and Hamilton Problems

There are a number of sufficient conditions for Hamilton graphs, which were proven in the 1950s and 1960s of the twentieth century. They can now easily be deduced from the theorem of Bondy[5] and Chvátal[6] from 1976.

Let $G = (V, E)$ be an undirected simple graph with n vertices. If there are non-adjacent vertices u and v in G with

$$\delta(u) + \delta(v) \geq n,$$

we add an edge $\{u, v\}$ to V. We continue this until get a graph, in which $\delta(u) + \delta(v) < n$ for any two non-adjacent vertices u and v. This graph is denoted by $[G]$ and called the *closure of* G.

Theorem 6.33 (Bondy and Chvátal, 1976) *A simple undirected graph G is a Hamilton graph if and only if $[G]$ is a Hamilton graph.*

Proof It is clear that $[G]$ is a Hamilton graph if G already has this property.

We show conversely: If H is obtained from G by adding an edge according to the construction of $[G]$, and if H is a Hamilton graph, then G is already a Hamilton graph. Consequently, G is a Hamilton graph if $[G]$ is a Hamilton graph.

Let u and v be non-adjacent vertices of $G = (V, E)$ with $\delta(u) + \delta(v) \geq n$. And let H be derived from G by adding the edge $\{u, v\}$. If H is a Hamilton graph, then H contains a Hamilton circuit that includes the edge $\{u, v\}$, compare the graph further down on the left. Furthermore, G contains a u, v-path P, on which each vertex of G appears exactly once. We denote P as sequence of vertices $P := (x_1, x_2, \cdots, x_n)$ with $x_1 = u$ and $x_n = v$. Define the following sets of vertices: X the neighbors of u, excluding x_2, and Y the vertices that follow the neighbors of v on P, excluding x_{n-1}. Formally

$$X := \{x_i \mid ux_i \in E \text{ and } 3 \leq i \leq n-1\} \text{ and}$$
$$Y := \{x_i \mid vx_{i-1} \in E \text{ and } 3 \leq i \leq n-1\}.$$

Since $x_2 \notin X$, we have $|X| = \delta(u) - 1$ and since there is no vertex in Y corresponding to the edge vx_{n-1}, we have $|Y| = \delta(v) - 1$.

From the assumption about u and v, it follows that $|X| + |Y| = \delta(u) + \delta(v) - 2 \geq n - 2$. And as $X \cup Y$ has at most $n - 3$ elements due to $u, v, x_2 \notin X \cup Y$, it follows that $|X \cap Y| \neq \emptyset$.

---

[5] John Adrian Bondy, British-Canadian mathematician *1944.
[6] Vašek Chvátal, Czech-Canadian mathematician *1946.

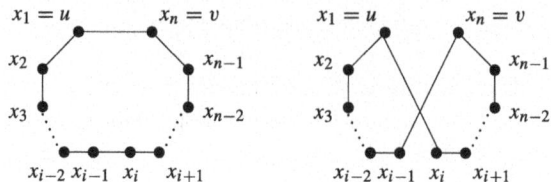

So there is an i with $3 \leq i \leq n-1$, for which both ux_i and vx_{i-1} are edges of G, compare the right graph. Then, however, $(x_1, x_2, \cdots, x_{i-1}, x_n, x_{n-1}, \cdots, x_i, x_1)$ is a Hamilton circuit in G. □

As immediate consequences, we obtain the theorems of Dirac[7] and of Ore,[8] which provide sufficient conditions for Hamilton graphs (using that the closure of G is complete in each case).

Corollary 6.34 (Dirac, 1952) *Let G be a graph with n vertices. If each vertex of G has at least the degree $\frac{n}{2}$, then G is a Hamilton graph.*

Corollary 6.35 (Ore, 1960) *Let G be a graph with n vertices. If for any two non-adjacent vertices u and v one has that*

$$\delta(u) + \delta(v) \geq n,$$

then G is a Hamilton graph.

Corollary 6.36 (Ore, 1961) *Let G be a graph with n vertices and m edges. If*

$$m \geq 2n + (n-2)(n-3),$$

then G is a Hamilton graph.

Proof (By contraposition) If G is not a Hamilton graph, then according to Corollary 6.35 there are vertices u and v of G with $\delta(u) + \delta(v) < n$. Since in every graph $2m = \sum_{x \in V} \delta(x)$ (cf. Lemma 6.3), we have here

$$2m = \delta(u) + \delta(v) + \sum_{x \in V \setminus \{u,v\}} \delta(x) < n + n + 2\binom{n-2}{2} = 2n + (n-2)(n-3).$$

□

Exercise 6.37 (Euler) Formulate and solve the *Königsberg Bridge Problem*.

---

[7] Paul Dirac, English physicist, 1902–1984.
[8] Oystein Ore, Norwegian mathematician, 1899–1966.

Fig. 6.13 The box product

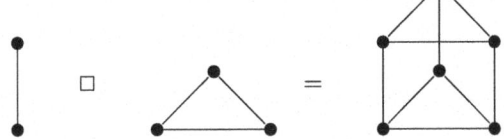

Exercise 6.38 (Hamilton) Find a Hamilton circuit in the graph of the dodecahedron depicted in Fig. 6.11.

Exercise 6.39 (Closure of a Graph) What is the minimal number of edges of an undirected, loopless graph G with 6 vertices, for which $[G]$ is K_6?

Exercise 6.40 (Line Graph) For an undirected graph $G = (V, E)$, the **line graph** or **edge graph** $L(G) = (V_{L(G)}, E_{L(G)})$ is defined by $V_{L(G)} := V$ and $E_{L(G)} := \{\{e_i, e_j\} \mid e_i \text{ and } e_j \text{ are incident edges of } G\}$.

(a) Show: If G is an Euler graph then $L(G)$ is an Euler graph.
(b) Prove or disprove the converse of (a).
(c) Find a necessary and sufficient condition for $L(G)$ to be an Euler graph.
(d) Show: If G is an Euler graph then $L(G)$ is a Hamilton graph.
(e) Prove or disprove the converse of (d).
(f) Draw $L(G)$ for $G = K_3, K_4, K_{2,2}, K_{2,3}$ and $K_{3,3}$.
(g) Show that for the graphs S_r (without loops) (cf. Exercise 6.3) one has $L(K_r) = S_r$.
(h) When is $L(K_{m,n})$ strongly regular (cf. Exercise 6.3)?

Exercise 6.41 (Box Product) The **box product** or sometimes also *Cartesian product* $G_1 \square G_2$ of two undirected graphs $G_1 = (V_1, E_1)$ and $G_2 = (V_2, E_2)$ is defined by

$V(G_1 \square G_2) := V_1 \times V_2$

$E(G_1 \square G_2) := \{\{(x_1, x_2), (y_1, y_2)\} \mid (x_1 = y_1 \text{ and } \{x_2, y_2\} \in E_2) \text{ or } (x_2 = y_2 \text{ and } \{x_1, y_1\} \in E_1)\}$.

Is $L(K_{m,n}) = K_m \square K_n$?

Hint: One draws $G_1 \square G_2$ like this: for each vertex u of G_1 take a copy G_2^u of G_2, for each edge $\{u, v\}$ of G_1 connect all corresponding corners of the copies G_2^u and G_2^v, see e.g. Fig. 6.13.

Note, that two more graph products, namely the cross product and the boxcross product will appear in Example 12.12(3).

Groupoid, Semigroup, Group 7

Sets with an inner composition form the starting point for the investigation of all algebraic structures. We provide a brief overview here and also introduce less common algebraic structures. They, in particular, provide a solid background for formalizations in computer science. In the last part of this chapter, groups are examined separately due to their ubiquitous importance. This part can essentially be worked on independently of the rest of this chapter.

All algebraic structures considered here can be viewed as examples of the specification of abstract data types. This also applies to the algebraic structures from the following two chapters.

On the Literature From the extensive literature, we only mention the following books by O. Borůvka [13], by R. H. Bruck [15], by J. M. Howie [37], and by Gerard Lallement [54]. For group theory, we recommend the book by H. Lüneburg [57]. For the "weak" structures, such as semigroups, acts, etc., Chapter 1 from the book by M. Kilp, U. Knauer and A. V. Mikhalev [45] is also suitable for reading and further study.

7.1 Inner Compositions

n-ary composition, unary, binary, ternary, 0-ary, distinguished element, partial, everywhere defined, inner composition, (not) closed, commutative (= abelian), algebraic structure, algebra, independence of representative, vector addition, scalar product, vector product, max, min, arithmetic/geometric mean, Boolean function, subalgebra, (left/right-)ideal, absorption property

For a set M and $n \in \mathbb{N}$, a mapping

$$\tau : \begin{cases} M^n \to M \\ (x_1, x_2, \cdots, x_n) \mapsto \tau(x_1, x_2, \cdots, x_n) \end{cases}$$

is called an (*n-ary*) *composition* on M. For 1-ary, 2-ary, and 3-ary, one says also **unary**, **binary** or **ternary**.

In the definition of an n-ary composition, we did not exclude the case $n = 0$.

Again, the Empty Set/Mapping According to the definition, a 0-ary composition on M is a mapping $\tau : M^0 \to M$.—But what is M^0?—If M^n is the set of mappings of the n-element set $\{1, 2, \ldots, n\}$ into M, then M^0 is consequently the set of mappings of the empty set $\emptyset$ into M. There is exactly one of these, namely the empty mapping θ, so $M^0 = \{\theta\}$ Consequently, a 0-ary composition is a mapping $\tau : \{\theta\} \to M$ with a single element domain $\{\theta\}$. So $\tau(M^0)$ is also single element, and that means the mapping τ selects exactly one element $\tau(\theta)$ from M:

$$\tau : \begin{cases} \{\theta\} \to M \\ \theta \mapsto \tau(\theta) \in M. \end{cases}$$

Usually, a 0-ary composition is identified with the element it selects. We then speak of a ***distinguished element***. This is usually a one or a zero, generalizations of the elements 1 and 0 in $\mathbb{R}$.

If the mapping τ is partial, we speak of a ***partial*** composition. If τ is not partial, this can be emphasized by the formulation τ is ***defined everywhere***. More precisely, we speak here of ***inner operations*** to distinguish them from the outer operations considered in Sect. 9.1. If we consider the composition τ only on a subset $A \subseteq M$ and there is an element

$$(x_1, x_2, \cdots, x_n) \in A^n \text{ for which } \tau(x_1, x_2, \cdots, x_n) \notin A,$$

we say A is ***not closed*** with respect to τ. An example is $\tau = - \subseteq \mathbb{N} \times \mathbb{N}$, because $\tau(1, 2) = 1 - 2 \notin \mathbb{N}$).

A set M, on which one or more operations are defined, we call an ***algebraic structure*** or just an ***algebra***. We write M together with the symbols for the operations in parentheses, e.g. $(\mathbb{N}, +, \cdot)$, $(\mathbb{Z}, +, \cdot, 0, 1)$, $(\mathbb{Q}, \cdot, ^{-1})$, $(\mathbb{R}, +, \cdot)$. Here $+$ and $\cdot$ are binary, $^{-1}$ is unary and 0 and 1 are 0-ary.

An algebraic structure $(M, \oplus)$ with binary composition is called ***commutative*** (***abelian***[1]), if $x \oplus y = y \oplus x$ for all $x, y \in M$.

---

[1] Niels Henrik Abel, Norwegian mathematician, 1802–1829.

7.1 Inner Compositions

Examples 7.1 (Numbers)

(1) The most well known examples are addition and multiplication of (natural, integer, rational, real, complex) numbers.
Let $M \in \{\mathbb{N}, \mathbb{Z}, \mathbb{Q}, \mathbb{R}, \mathbb{C}\}$, then

$$+ : \begin{cases} M \times M \to M \\ (x, y) \mapsto x + y \end{cases} \quad \text{and} \quad \cdot : \begin{cases} M \times M \to M \\ (x, y) \mapsto x \cdot y \end{cases}$$

are everywhere defined binary operations on M.

(2) *Division* is a partial binary composition on $\mathbb{R}$ and on $\mathbb{Q}$:

$$\text{div} : \begin{cases} \mathbb{R} \times \mathbb{R} \to \mathbb{R} \\ (x, y) \mapsto \text{div}(x, y) := \frac{x}{y}, \text{ if } y \neq 0. \end{cases}$$

(3) *Subtraction* is a binary composition on $\mathbb{Z}$:

$$- : \begin{cases} \mathbb{Z} \times \mathbb{Z} \to \mathbb{Z} \\ (x, y) \mapsto xy. \end{cases}$$

The set $\mathbb{N}$ is not closed with respect to this composition: $x - y$ is not in $\mathbb{N}$, if $x < y$.

(4) The *negative*

$$- : \begin{cases} \mathbb{Z} \to \mathbb{Z} \\ x \mapsto -x \end{cases}$$

is a unary composition on $\mathbb{Z}$.

(5) The *inverse*

$$^{-1} : \begin{cases} \mathbb{Q} \to \mathbb{Q} \\ x \mapsto x^{-1} := \frac{1}{x} \end{cases}$$

is only a partial unary composition on $\mathbb{Q}$, as it is not defined for $x = 0$.

(6) *Exponentiation* is a binary composition on $\mathbb{N}$:

$$\exp : \begin{cases} \mathbb{N} \times \mathbb{N} \to \mathbb{N} \\ (x, y) \mapsto exp(x, y) := x^y. \end{cases}$$

(7) The set $\mathbb{Z}_n$ of the **residue classes modulo** n naturally has an addition and a multiplication:

$$+ : \begin{cases} \mathbb{Z}_n \times \mathbb{Z}_n & \to \mathbb{Z}_n \\ ([x]_n, [y]_n) & \mapsto [x+y]_n \end{cases} \text{ and } \cdot : \begin{cases} \mathbb{Z}_n \times \mathbb{Z}_n & \to \mathbb{Z}_n \\ ([x]_n, [y]_n) & \mapsto [xy]_n, \end{cases}$$

For the last example, it is important to ensure that these definitions are *independent of representatives*, i.e., that the result of the composition of two residue classes does not depend on the choice of representatives. One also says that addition and multiplication are **well defined**. So, for example, $[1]_2+[3]_2$ must be the same class as $[3]_2+[3]_2$, namely $[0]_2$. In this case, we also write $1_2+1_2 = 0_2$ for $[1]_2+[3]_2 = [0]_2$ or even[2] $1+1 = 0$.

Lemma 7.2 *The following definitions of the operations on $\mathbb{Z}_n$*

$$+ : \begin{cases} \mathbb{Z}_n \times \mathbb{Z}_n & \to \mathbb{Z}_n \\ ([x]_n, [y]_n) & \mapsto [x+y]_n \end{cases} \text{ and } \cdot : \begin{cases} \mathbb{Z}_n \times \mathbb{Z}_n & \to \mathbb{Z}_n \\ ([x]_n, [y]_n) & \mapsto [xy]_n \end{cases}$$

are well defined, i.e. independent of representatives.

Proof We only show that this addition is well defined. The proof for the multiplication is similar, see also the proof of Theorem 8.15.

Consider $y \in \mathbb{Z}$ and $[x_1]_n = [x_2]_n$, then $n \mid (x_1 x_2)$. So there exists $z \in \mathbb{Z}$ with $nz = x_1 x_2 = x_1 + y - (x_2 + y)$. It follows that $n \mid ((x_1+y)-(x_2+y))$ and therefore $[x_1 + y]_n = [x_2 + y]_n$. So $[x_1]_n + [y]_n = [x_2]_n + [y]_n$. □

Examples 7.3 (Multi-ary Operations)

(1) The mapping max that maps each n-tuple of real numbers to the *maximum* of these numbers is an n-ary composition on $\mathbb{R}$.

$$\max : \begin{cases} \mathbb{R}^n & \to \mathbb{R} \\ (x_1, x_2, \ldots, x_n) & \mapsto \max\{x_1, x_2, \ldots, x_n\}. \end{cases}$$

Similarly, the *minimum* function min provides for each $n \geq 2$ an n-ary composition on $\mathbb{R}$.

(2) The *arithmetic mean* M_a of any n numbers, $n \in \mathbb{N}$, is an n-ary composition, which is defined everywhere. The *geometric mean* M_g, however, is not defined everywhere:

---

[2] This is a way to dispel the misconception that 1+1 always equals 2.

7.1 Inner Compositions

$$M_a : \begin{cases} \mathbb{R}^n \to \mathbb{R} \\ (x_1, x_2, \ldots, x_n) \mapsto \frac{1}{n}(x_1 + x_2 + \cdots + x_n), \end{cases}$$

$$M_g : \begin{cases} \mathbb{R}^n \to \mathbb{R} \\ (x_1, x_2, \ldots, x_n) \mapsto \sqrt[n]{x_1 x_2 \cdots x_n}. \end{cases}$$

(3) The three mappings that map any three points of $\mathbb{R}^2$ not lying on a straight line to the intersection of the perpendicular bisectors/of the angle bisectors/of the side bisectors of the triangle spanned by these points, are partial ternary operations on $\mathbb{R}^2$.

(4) From binary operations we can arbitrarily form n-ary operations. A very arbitrary, but formally correct example is this:

$$\begin{cases} \mathbb{R}^4 \to \mathbb{R} \\ (x_1, x_2, x_3, x_4) \mapsto x_1 x_4 + 5 x_3 x_2^3. \end{cases}$$

(5) In Mathematical Logic, we have considered truth tables. A table with n components (propositional variables) represents an n-ary composition on $\{0, 1\}$. Such a composition is also called a ***Boolean function***. Compare Example 1.7.

(6) Given three vertices u, v, w in a tree T, mapping them to their median $m(u, v, w)$, i.e., the unique vertex which lies on the shortest path between u, v, u, w, and v, w, yields a ternary operation.

Examples 7.4 (More Operations)

(1) Operations on a set M can be transferred component wise to any Cartesian power M^k. Here we take $k = 3$:

$$+ : \begin{cases} (M \times M \times M) \times (M \times M \times M) \to M \times M \times M \\ ((x_1, x_2, x_3), (y_1, y_2, y_3)) \mapsto (x_1 + y_1, x_2 + y_2, x_3 + y_3). \end{cases}$$

(2) ***Vector addition*** is a binary operation on $\mathbb{R}^n$ for $1 \leq n \in \mathbb{N}$:

$$+ : \begin{cases} \mathbb{R}^n \times \mathbb{R}^n \to \mathbb{R}^n \\ \left(\begin{pmatrix} v_1 \\ \vdots \\ v_n \end{pmatrix}, \begin{pmatrix} w_1 \\ \vdots \\ w_n \end{pmatrix} \right) \mapsto \begin{pmatrix} v_1 + w_1 \\ \vdots \\ v_n + w_n \end{pmatrix}. \end{cases}$$

We can consider this as the addition of two matrices each with one column and n rows, see Sect. 6.5. For $n = 1$ this is the usual addition on $\mathbb{R}$, see Example 7.1 (1).

(3) Composition of mappings is a binary operation on the set $M^M = \text{Map}(M, M)$ of all mappings of a set M into itself:

$$\circ : \begin{cases} M^M \times M^M \to M^M \\ \quad (f, g) \mapsto f \circ g. \end{cases}$$

(4) The *scalar product*, defined by

$$<,>: \begin{cases} \mathbb{R}^n \times \mathbb{R}^n \to \mathbb{R} \\ \left(\begin{pmatrix} v_1 \\ \vdots \\ v_n \end{pmatrix}, \begin{pmatrix} w_1 \\ \vdots \\ w_n \end{pmatrix} \right) \mapsto v_1 w_1 + v_2 w_2 + \ldots + v_n w_n, \end{cases}$$

is not an operation on $\mathbb{R}^n$ for $n \geq 2$.

Geometric interpretation: With the help of the scalar product we can define and calculate the *length of a vector* and the *angle between two vectors*. For $v \in \mathbb{R}^n$, the length of v is defined as $||v|| := \sqrt{<v, v>}$, and for $v, w \in \mathbb{R}^n$, $||v|| \cdot ||w|| \cdot \cos \sphericalangle(v, w) := <v, w>$, defines the *cosine* of the angle between v and w.

(5) The *vector product* is a binary operation on $\mathbb{R}^3$:

$$\times : \begin{cases} \mathbb{R}^3 \times \mathbb{R}^3 \to \mathbb{R}^3 \\ \left(\begin{pmatrix} v_1 \\ v_2 \\ v_3 \end{pmatrix}, \begin{pmatrix} w_1 \\ w_2 \\ w_3 \end{pmatrix} \right) \mapsto \begin{pmatrix} v_2 w_3 v_3 w_2 \\ v_3 w_1 v_1 w_3 \\ v_1 w_2 v_2 w_1 \end{pmatrix}. \end{cases}$$

This operation does not correspond to the product of matrices, see Sect. 6.5.

Geometric interpretation: The vector product of two vectors v and w in $\mathbb{R}^3$ can be interpreted as a vector $v \times w$, which is perpendicular to both v and w and whose length is equal to the area of the parallelogram spanned by v and w, where "perpendicular" and "area" can be defined using the scalar product.

Subalgebras

If $(M, \oplus)$ is an algebra, then $U \subseteq M$ is called a **subalgebra of** $(M, \oplus)$ (also: *algebraic substructure*), if U is closed with respect to the operation $\oplus$ (more precisely: with respect to the restriction of this operation to U). In this case, we write $U \leq (M, \oplus)$ or also $(U, \oplus) \leq (M, \oplus)$. For the following lemma, compare the theorem about subgroups cited in Sect. 1.8 (as an example of the structure of a mathematical theorem).

7.2 From Groupoid to Group

Lemma 7.5 (Subalgebra) *If $(M, \oplus)$ is an algebra with a binary operation $\oplus$, then $U \subseteq M$ is a subalgebra of $(M, \oplus)$, if and only if*

$$U \oplus U := \{ x \oplus y \mid x, y \in U \} \subseteq U.$$

If for a subalgebra U of $(M, \oplus)$ additionally

$M \oplus U \subseteq U$,   then U is called a *left ideal of* M,
$U \oplus M \subseteq U$,   then U is called a *right ideal of* M,
$M \oplus U \oplus M \subseteq U$,   then U is called an *ideal of*
M, written $U \trianglelefteq M$.

It is said that U has the ***absorption*** or ***swallowing property*** from the corresponding side.

Note that for rings a slightly different ideal definition applies, cf. the subsection "Ideal,...." in Sect. 8.2.

Examples 7.6 (Subalgebra)

(1) The algebra $(\mathbb{N} \setminus \{0\}, \cdot, {}^{-1})$ is not a subalgebra of $(\mathbb{R} \setminus \{0\}, \cdot, {}^{-1})$, because $\mathbb{N}$ is not closed with respect to ${}^{-1}$.
(2) The set $(\mathbb{Z}_4, +_4)$ with $[x]_4 +_4 [y]_4 := [x + y]_4$ of the equivalence classes $\equiv$ mod 4 of $\mathbb{Z}$ is not a subalgebra of $(\mathbb{Z}, +)$: $\mathbb{Z}_4$ is not a subset of $\mathbb{Z}$. Even if we choose, for example, $\{0, 1, 2, 3\}$ as a representative system for the equivalence classes, we do have a subset of $\mathbb{Z}$, but $+_4$ yields completely different results on these four elements than $+$ on $\mathbb{Z}$.
(3) The algebra $(2\mathbb{Z}, +)$ with $2\mathbb{Z} := \{2z \mid z \in \mathbb{Z}\}$ is a subalgebra, but not an ideal of $(\mathbb{Z}, +)$, because $2+1 \notin 2\mathbb{Z}$. But $(2\mathbb{Z}, \cdot)$ is an ideal of $(\mathbb{Z}, \cdot)$, because $2\mathbb{Z} \cdot \mathbb{Z} = \mathbb{Z} \cdot 2\mathbb{Z} \subseteq 2\mathbb{Z}$.
(4) For the set $2\mathbb{N} := \{0, 2, 4, \ldots\}$ of even natural numbers, we have $(2\mathbb{N}, \cdot) \trianglelefteq (\mathbb{N}, \cdot)$. For the set $2\mathbb{N}+1 := \{1, 3, 5, \ldots\}$ of odd natural numbers, $(2\mathbb{N}+1, \cdot, 1)$ is a subalgebra of the algebra $(\mathbb{N}, \cdot, 1)$, but not an ideal.

7.2 From Groupoid to Group

> Groupoid, associative, semigroup, (right/left) one (= neutral element), left zero semigroup, right zero semigroup, monoid, free word monoid, (right/left) zero (= absorbing element), associative law, uniqueness of one and zero, group, right/left cancellable, (uniquely) right/left solvable, quasigroup, loop, Moufang loop, composition table, diagonally symmetric, Latin square

Many of the operations considered in mathematics are binary. We have already encountered some examples of binary operations. A set G with a binary operation $\otimes$ is called ***groupoid*** $(G, \otimes)$, sometimes the name ***magma*** is also used. A groupoid

$(G, \otimes)$—more precisely: its operation—is called *associative*, if

$$(\forall x, y, z \in G) : x \otimes (y \otimes z) = (x \otimes y) \otimes z.$$

An associative groupoid is called **semigroup**.

An element e of a groupoid $(G, \otimes)$ is called **left neutral** with respect to $\otimes$, if

$$(\forall x \in G) : e \otimes x = x,$$

i.e., if applying it from the left changes nothing. Analogously, a **right neutral** element is defined. An element that is both right and left neutral, is called **neutral**. Since the operation of a semigroup is usually written as multiplication, it has become customary to refer to these elements as **right One, left One** or **One** in analogy to the situation in $(\mathbb{Z}, \cdot)$. A semigroup with one is called **monoid**.

Also the 0 in $(\mathbb{Z}, \cdot)$ is a prototype for an abstract concept: An element z of a groupoid $(G, \otimes)$ is called **right zero (right absorbing)**, if

$$(\forall x \in G) : x \otimes z = z,$$

i.e., if z "extinguishes" every element x by operating from the right. Analogously, the terms **left zero (left absorbing)** are defined, a two sided absorbing element is called **zero (absorbing)**.

Examples 7.7 (Zeros and Ones)

(1) In $(\mathbb{N}, \cdot)$ or $(\mathbb{Z}, \cdot)$ the element 0 is the zero and 1 is the One, the neutral element. In $(\mathbb{N}, +)$ or $(\mathbb{Z}, +)$ however, the element 0 is neutral, could thus be called "the One" of these algebraic structures, there is no zero element there. To avoid confusion, we prefer to speak in such a case of "the neutral element with respect to $+$". All 4 algebraic structures are monoids.

(2) The set A^* of all words over the alphabet A with the concatenation as operation, see the subsection "Formal Languages" in Sect. 1.1 is a monoid. It is also called *free word monoid*. Its neutral element is the empty word.

(3) Since the composition of mappings is always associative, compare Theorem 5.4, and since the identical mapping id_M is the neutral element, $(M^M, \circ)$ is a monoid for every set M. One calls $(M^M, \circ)$ **full transformation monoid**, submonoids are called **transformation monoids**. For every $x_0 \in M$ the constant mapping c_{x_0}, which maps every element of M to x_0, is a left zero of $(M^M, \circ)$. Right zeros do not exist in M^M, if M contains more than one element.

(4) The **zero vector** $(0, \ldots, 0) \in \mathbb{R}^n$ is a neutral element with respect to the vector addition in $\mathbb{R}^n$, cf. Example 7.4 (2).

7.2 From Groupoid to Group

(5) In a ***left zero semigroup*** every element is a left zero and every element is a right One. The following table shows the left zero semigroup $H = \{x, y, z\}$:

·	x	y	z
x	x	x	x
y	y	y	y
z	z	z	z

The analogue applies to analogously defined ***right zero semigroups***.

While, as we have seen, there can certainly be multiple unilaterally neutral or unilaterally absorbing elements in an algebraic structure, there is always at most one One (unit) and at most one zero.

Theorem 7.8 (Uniqueness of One and Zero) *Every algebraic structure $(M, \cdot)$ contains at most one One (unit) and at most one zero.*

Proof If e_1 and e_2 are Ones (units) of $(M, \cdot)$, then $e_1 = e_1 e_2 = e_2$. The proof of the second part proceeds analogously. □

On the way to the group, in Fig. 7.1 we indicate a number of other properties that operations can have. Combinations of such properties have led to further algebraic structures, the definitions and names of which are also included in the figure. The simplest structure with a binary operation is the groupoid, one of the richest is the group.

A Reading Aid Each arrow in this figure corresponds to an implication, e.g., every loop is a quasigroup. Thus, towards to the top, i.e. the groupoid, properties become weaker. A double arrow corresponds to an equivalence, with the two properties at the ends of the arc above it. So a groupoid is a quasigroup if and only if it is uniquely right and uniquely left solvable. From this we see in particular that every quasigroup is cancellable.

The name Moufang Loop goes back to Ruth Moufang, the discoverer of this structure.[3] As further reading aid, we provide some—mostly trivial—statements that can be read from the picture:

- Every loop is a quasigroup.
- A groupoid is uniquely solvable exactly if it is uniquely right solvable and uniquely left solvable.
- From the properties "uniquely right solvable and uniquely left solvable" the property "solvable" follows for every groupoid.

---
[3] German mathematician, 1905–1977.

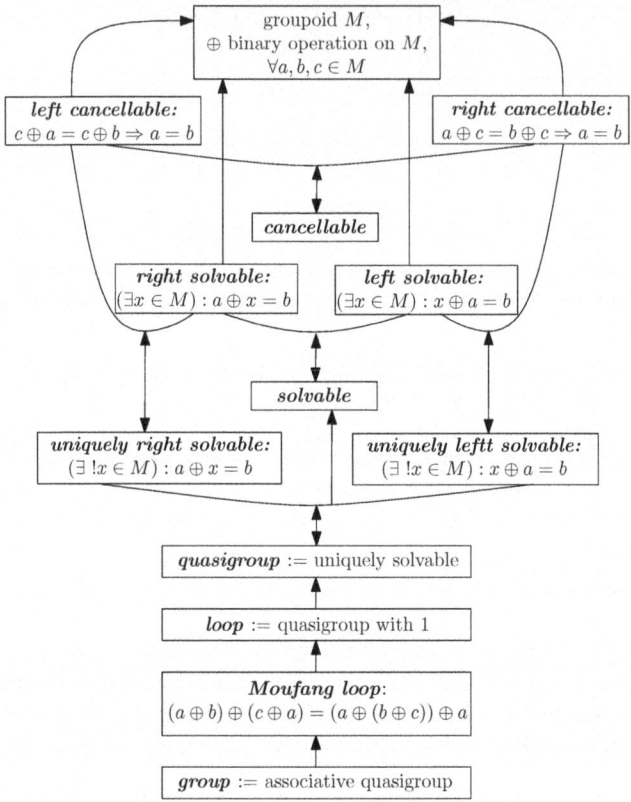

Fig. 7.1 Classes of groupoids

As an example, we prove:

Theorem 7.9 (Uniquely Right Solvable) *The property of being "left cancellable and right solvable" is equivalent to the property of being "uniquely right solvable".*

Proof " $\Rightarrow$ ". If $ax_1 = ax_2 = b$ in a right solvable groupoid $(M, \cdot)$, then $x_1 = x_2$ follows from left cancellability. So every equation $ax = b$ is uniquely solvable.

" $\Leftarrow$ ". For left cancellability, we have to show "$ca = cb \Rightarrow a = b$". If $ca = cb$, then a and b are solutions of the equation $cx = d$ (with $d := ca$). From the uniqueness of right solvability, it follows that $a = b$. □

Theorem 7.10 (Associative and Solvable = Group) *A semigroup is solvable if and only if it is uniquely solvable if and only if it is group.*

Proof We only prove the fist equivalence. A semigroup that is right and left solvable is already uniquely right and left solvable.

7.2 From Groupoid to Group

Let $(M, \cdot)$ be right and left solvable. We show that $(M, \cdot)$ is uniquely right solvable. For this, let $ax = ay = b$. Due to left solvability, there exists $p \in M$ with $x = px$ and $q \in M$ with $p = qa$, and due to right solvability, there exists $r \in M$ with $y = xr$. Thus, we have: $x = px = (qa)x = q(ax) = q(ay) = (qa)y = py = p(xr) = (px)r = xr = y$ and uniqueness is shown. The unique left solvability can be shown analogously. □

Now let's look at the figure again. With the previous theorem we have proven that the boxes "solvable", "quasigroup", "loop", "Moufang loop" all identify with the box "group" if $\oplus$ is associative. If $(M, \oplus)$ is even a monoid (cf. Exercise 7.15), then also the boxes "right solvable", "left solvable", "uniquely right solvable", "uniquely left solvable" identify with the box "group".

When we consider Fig. 7.1 as a graph, these identifications provide two graph egamorphisms from Figs. 7.1 to 7.2.

The transition to associativity then provides the graph in Fig. 7.2.

If M additionally has a unit, i.e., is a monoid, we obtain the graph in Fig. 7.3.

Examples 7.11

(1) For every graph G, the graph morphisms, the graph egamorphisms, the strong graph morphisms from G to itself with the composition of mappings as operation form monoids, cf. Exercise 6.28. The graph automorphisms with composition form a group.

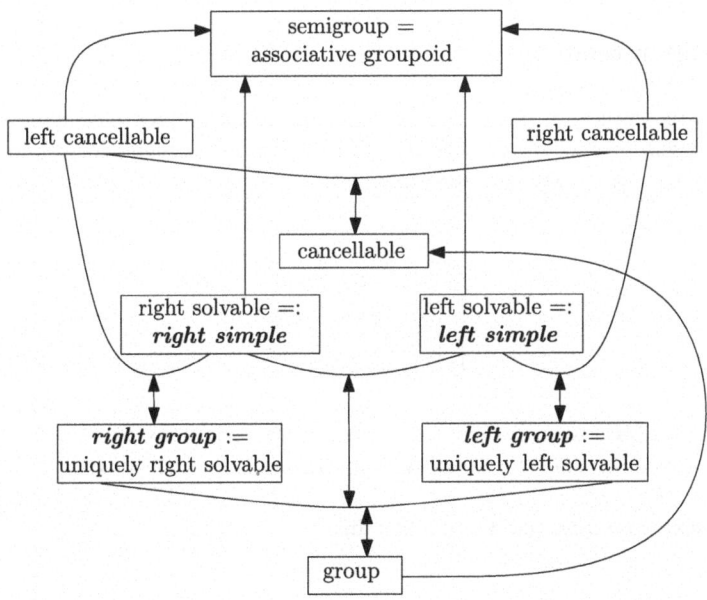

Fig. 7.2 Semigroup classes

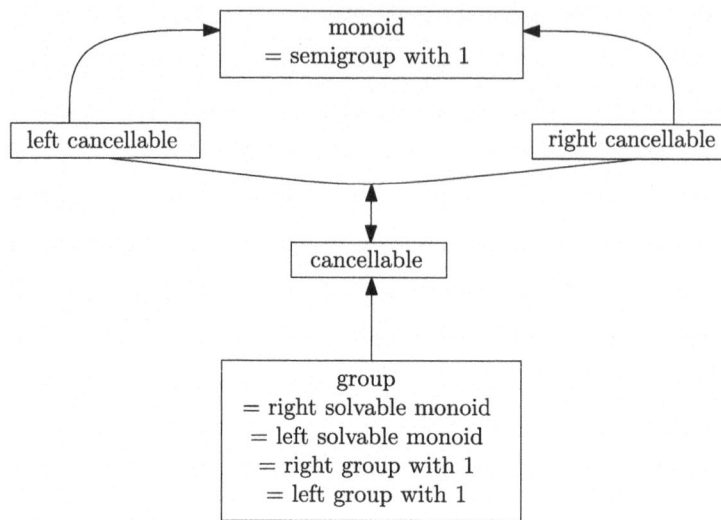

Fig. 7.3 Monoid Classes

(2) $(2\mathbb{N} + 1, +)$ is not a subgroup of the monoid $(\mathbb{N}, +)$.
(3) $(\mathbb{Z}, -)$ is a proper quasigroup, i.e. not associative, but uniquely solvable.
(4) $(\mathbb{N}, -)$ is not a subquasigroup of the quasigroup $(\mathbb{Z}, -)$.
(5) A left zero semigroup is a left group.

Composition Table

Look at the example of truth tables in logic. There we have represented compositions in *composition tables*. We proceed accordingly here and obtain, for example, for $M := \{a, b\}$ and $\tau : M^3 \to M$ the table:

a	a	a	$\tau(a, a, a)$
a	a	b	$\tau(a, a, b)$
a	b	a	$\tau(a, b, a)$
⋮	⋮	⋮	⋮

This table has $|M|^3$ rows or $|M|^n$ rows for an n-ary operations. So it only provides a practical representation for small sets. In the case $n = 2$, there is a method that works with $|M|$ rows. Let $M := \{x_1, x_2, \ldots, x_n\}$ and $\tau : M^2 \to M$, then we call this a *composition table* and write it like this:

7.2 From Groupoid to Group

$$\begin{array}{c|ccc} \tau & x_1 & x_2 & \cdots \\ \hline x_1 & \tau(x_1, x_1) & \tau(x_1, x_2) & \cdots \\ x_2 & \tau(x_2, x_1) & \tau(x_2, x_2) & \cdots \\ \vdots & \vdots & \vdots & \end{array}$$

We can read some properties of elements of an algebraic structure from the composition table.

An element a is

- a *right/left identity* exactly if the column/row of the element a in the composition table equals the front column/top row of the table (in the following table at the left, a is a left identity),
- a *right/left zero* exactly if the column/row of the element a in the composition table only contains this element (in the following table in the middle, a is a left zero).

A groupoid is

- *right/left solvable* exactly if each row/column of the composition table contains each element of the groupoid once; it is easy to see that in groupoids with finitely many elements then each element in each row and in each column appears exactly once, and this means that the groupoid is *uniquely right and left solvable* (see the following table on the right),
- *commutative*, if the composition table is ***diagonally symmetric***, that is, it does not change when flipped over the diagonal line from top left to bottom right (see the table on the right).

$$\begin{array}{c|ccc} & a & b & c \\ \hline a & a & b & c \\ \cdot & \cdot & \cdot & \cdot \\ \cdot & \cdot & \cdot & \cdot \end{array} \qquad \begin{array}{c|ccc} & a & b & c \\ \hline a & a & a & a \\ \cdot & \cdot & \cdot & \cdot \\ \cdot & \cdot & \cdot & \cdot \end{array} \qquad \begin{array}{c|ccc} & e & a & b \\ \hline e & e & a & b \\ a & a & b & e \\ b & b & e & a \end{array}$$

Attention Unfortunately, we cannot tell from the composition table whether the algebraic structure is associative or not. To check for associativity, we always have to calculate all triples with the two different bracketings. Since this is usually very laborious, one is pleased when the algebraic structure consists of mappings and their composition is the composition of the mappings. Then we already know that this composition is always associative (Theorem 5.4).

The composition table of a finite quasigroup is also called **Latin square**. This is a square table with n rows formed from n different elements with the property that each element appears exactly once in each row and in each column of the table.

Every **Sudoku** is a special 9 element quasigroup! A beautiful application can be found on page 253ff. in the book by K. Denecke [21].

Inverse Elements

Composition with a One does not change anything and composition with a zero creates a kind of end state. With the help of inverses we can describe how certain changes can be undone: the composition of an element with its inverse results in One.

If x is an element of a monoid $(G, \cdot, 1)$, then $y \in G$ is called a *right inverse* to x if $x \cdot y = 1$. Analogously, we define a *left inverse* to x. An *inverse* to x is both right inverse and left inverse to x. An inverse to x we denote by x^{-1}, as we already used for numbers and mappings.

Example 7.12 (Inverse) The inverse mapping f^{-1} of a bijective mapping f is inverse to f with respect to the composition of mappings. For one sided inverses, compare Theorem 5.14. The reciprocal $\frac{y}{x}$ of a fraction $\frac{x}{y} \in \mathbb{Q}\setminus\{0\}$ is inverse to this fraction with respect to the multiplication of rational numbers.

Exercise 7.13 (Ones and Zeros) Examine the examples in Example 7.1 (if necessary, choose $n := 2$) to see if one/two sided zeros or Ones exist and, if applicable, one or two sided inverses.

Exercise 7.14 (Three Element Groupoids) How many non isomorphic three element groupoids are there? Which are associative? Which are groups? Solve the corresponding problem for two element groupoids.

Exercise 7.15 (Right Solvable Monoid = Group) Show that an associative right or left solvable groupoid with 1 is already a group.

Exercise 7.16 (Mappings) Let $M := \{a, b, c\}$,

$$f_1 := \begin{pmatrix} a\,b\,c \\ a\,b\,b \end{pmatrix}, \quad f_2 := \begin{pmatrix} a\,b\,c \\ b\,b\,b \end{pmatrix}, \quad f_3 := \begin{pmatrix} a\,b\,c \\ a\,a\,a \end{pmatrix}, \quad f_4 := \begin{pmatrix} a\,b\,c \\ b\,a\,a \end{pmatrix}$$

(for notation compare Exercise 5.17 in Sect. 5.3).

(a) Set up the composition table for the composition ○ of these mappings.
(b) Is ○ an operation on $\mathcal{F} := \{ f_1, f_2, f_3, f_4 \}$?
    On which subsets of $\mathcal{F}$ is ○ an operation?

Exercise 7.17 (Quasigroup) Consider the odd circuit C_{2n+1}. If you map any two vertices to the vertex in the middle of the path of even length between them, you get a commutative, idempotent (non-associative) quasigroup.

7.3 Compatible Relations

Compatible relation, congruence, ordered algebraic structure/groupoid/semigroup/group, homomorphism, isomorphism, endomorphism, automorphism, algebraic factor structure, factor algebra, induced congruence, kernel congruence, canonical surjection/projection, Homomorphism Theorem

A binary relation ϱ on a set M is called *compatible* with a binary inner composition $\oplus$ on M, if for arbitrary elements $x_1, y_1, x_2, y_2 \in M$ one has

$$x_1 \varrho\, y_1 \text{ and } x_2 \varrho\, y_2 \;\Rightarrow\; (x_1 \oplus x_2)\, \varrho\, (y_1 \oplus y_2).$$

A compatible equivalence relation on $(M, \oplus)$ is called *congruence* on $(M, \oplus)$.

If $(M, \oplus)$ is an algebraic structure with a binary inner composition $\oplus$, and $(M, \leq)$ is a partially ordered set, then $(M, \oplus, \leq)$ is called *ordered algebraic structure* if $\leq$ is an order relation that is compatible with $\oplus$. So we get *ordered groupoids, ordered semigroups, ordered groups* etc. Since $\leq$ is not an equivalence relation, it is not a congruence.

Examples 7.18 (Compatibility) The following examples are not surprising, as they were the model for the concept of compatibility.

(1) $(\mathbb{N}, \cdot, \leq)$ is an ordered semigroup.
(2) $(\mathbb{Z}, +, \leq)$ is an ordered group.
(3) $(\mathbb{Z}, \cdot, \leq)$ is not an ordered monoid, as multiplication by -1 does not preserve the inequality $1 \leq 2$.
(4) The equivalence relation $\equiv$ mod n is a congruence on $(\mathbb{Z}, +)$ and on $(\mathbb{Z}, \cdot)$. The proof is already in Lemma 7.2.
(5) If $f : M \to M$ is a mapping of the algebraic structure $(M, \oplus)$ into itself, such that for all $x, y \in M$ $f(x \oplus y) = f(x) \oplus f(y)$, then the mapping rule f is a relation compatible with $\oplus$.

The last example leads to the following concept.

Structure Preserving Mappings

Let $(A, \oplus)$ and $(B, \otimes)$ be algebras. A mapping $f : A \to B$ is called *homomorphism* or *structure preserving* or *compatible mapping*), if for any elements $a_1, a_2 \in A$ one has

$$f(a_1 \oplus a_2) = f(a_1) \otimes f(a_2).$$

In particular, we get **groupoid, semigroup, group homomorphism** etc. If A and B are monoids and $f : A \to B$ is a semigroup homomorphism with $f(1_A) = 1_B$, then f is called a **monoid homomorphism**

Theorem 7.19 (Bijective Homomorphisms) *If $f : A \to B$ is a bijective homomorphism of the algebraic structures $(A, \oplus)$ and $(B, \otimes)$, then the inverse mapping f^{-1} is also a homomorphism.*

Proof For $b_1, b_2 \in B$ set $a_1 := f^{-1}(b_1)$ and $a_2 := f^{-1}(b_2)$. Then we get that $f(a_i) = f \circ f^{-1}(b_i) = b_i$, $i = 1, 2$, and

$$\begin{aligned}
f^{-1}(b_1 \otimes b_2) &= f^{-1}(f(a_1) \otimes f(a_2)) \\
= f^{-1}(f(a_1 \oplus a_2)) &= (f^{-1} \circ f)(a_1 \oplus a_2) \\
= a_1 \oplus a_2 &= (f^{-1} \circ f)(a_1) \oplus (f^{-1} \circ f)(a_2) \\
= f^{-1}(f(a_1)) \oplus f^{-1}(f(a_2)) &= f^{-1}(b_1) \oplus f^{-1}(b_2).
\end{aligned}$$

So f^{-1} is also a homomorphism. □

Looking again at Example 6.7, we see that an analogue of Theorem 7.19 is not valid for bijective graph morphisms.

We call a homomorphism $f : (A, \oplus) \to (B, \otimes)$

- **isomorphism**, if f is bijective, and write $(A, \oplus) \cong (B, \otimes)$,
- **endomorphism**, if $(A, \oplus) = (B, \otimes)$,
- **automorphism**, if f is a bijective endomorphism.

Let $(M, \leq)$ and $(N, \leq)$ be ordered sets, $x, y \in M$. A mapping $f : (M, \leq) \to (N, \leq)$ is called

- **isotone**, if $x \leq y$ implies $f(x) \leq f(y)$,
- **antitone**, if $x \leq y$ implies $f(x) \geq f(y)$.

Examples 7.20 (Structure Preserving)

(1) $(\mathbb{R}^+, \cdot) \to (\mathbb{R}, +)$, $x \mapsto \log x$, is a homomorphism, since for all $x, y \in \mathbb{R}^+ := \{r \in \mathbb{R} \mid r > 0\}$ one has $\log(xy) = \log x + \log y$. This is even an isomorphism.
(2) The absolute value function $|\ | : (\mathbb{Z}, +) \to (\mathbb{N}, +)$, $z \mapsto |z|$ is not a homomorphism, since for example $|1 + (-1)| = 0 \neq 2 = |1| + |-1|$.
(3) The squaring $^2 : (\mathbb{Z}, +) \to (\mathbb{N}, +)$, $z \mapsto z^2$ is not a homomorphism, since for example $(1 + 2)^2 = 9 \neq 5 = 1^2 + 2^2$.
(4) Isotone mappings preserve order, antitone mappings do not.

The following theorem shows how structure preserving mappings and substructures fit together.

7.3 Compatible Relations

Theorem 7.21 (Injection of a Subalgebra) *If $(M, \oplus)$ is an algebra, $U \subseteq M$ and $\iota : U \to M$, $x \mapsto x$ is the associated injection, then $(U, \oplus)$ is a subalgebra of $(M, \oplus)$, written $(U, \oplus) \leq (M, \oplus)$ if and only if ι is a homomorphism.*

Proof First, if $(U, \oplus)$ is a subalgebra of $(M, \oplus)$, then $x_1, x_2 \in U \Rightarrow x_1 \oplus x_2 \in U \Rightarrow \iota(x_1 \oplus x_2) = x_1 \oplus x_2 = \iota(x_1) \oplus \iota(x_2)$. Thus, ι is a homomorphism.

Now let $U \subseteq M$ and $\iota : U \to M$ be a homomorphism. We have to show that $\oplus$ is an inner composition on U. For $x_1, x_2 \in U$ we get $x_1 \oplus x_2 = \iota(x_1) \oplus \iota(x_2) = \iota(x_1 \oplus x_2)$, i.e. $x_1 \oplus x_2 \in U$. Thus, the restriction of $\oplus$ to U is a composition on U. □

Example 7.22 (Subalgebra) The set of permutations of a finite set M is a subalgebra and even a group, *the full permutation group* in the *full transformation monoid* M^M of M:

$$(\mathcal{S}_n, \circ, \mathrm{id}_M) \leq (M^M, \circ, id_M).$$

Kernel, Factor Algebra, Homomorphism Theorem

For sets with mappings and for graphs with graph morphisms, we have already proven Homomorphism Theorems. Now we can also establish the connection between homomorphic images and factor structures for algebras. Since the proofs already exist for sets with mappings, we only need to check compatibility properties here.

Lemma 7.23 (Homomorphism ⤳ Congruence) *Let $f : (A, \oplus) \to (B, \odot)$ be a homomorphism of algebras, $x, y \in A$. Then the equivalence relation induced by f (cf. Lemma 5.20)*

$$x \varrho_f y \quad :\Leftrightarrow \quad f(x) = f(y)$$

*is a congruence on the algebra $(A, \oplus)$. It is called the congruence **induced** by f, or **kernel congruence of** f, also written as $\ker f$.*[4]

Proof From $x \varrho_f y$ and $x' \varrho_f y'$ it follows that $f(x) = f(y)$ and $f(x') = f(y')$. Since f is a homomorphism, $f(x \oplus x') = f(x) \odot f(x') = f(y) \odot f(y') = f(y \oplus y')$ follows and thus $(x \oplus x') \varrho_f (y \oplus y')$. □

---

[4] This might cause confusion, because here $\ker f$ is not a subobject. Opposed to that normal divisors, ideals or submodules, which are induced by the kernel congruences for groups, rings or modules, are kernels of the corresponding mappings f, and are subobjects. They are also denoted by $\ker f$.

In Lemma 7.2 we have transferred addition and multiplication of $\mathbb{Z}$ to the residue classes modulo n. This procedure can be generalized:

Theorem 7.24 (Factor Algebra) *Let ϱ be a congruence on an algebra $(A, +)$. By*

$$\oplus : \begin{cases} A/\varrho \times A/\varrho \to A/\varrho \\ ([a]_\varrho, [b]_\varrho) \mapsto [a]_\varrho \oplus [b]_\varrho := [a+b]_\varrho \end{cases}$$

a binary inner composition is defined on the factor set A/ϱ. Then $(A/\varrho, \oplus)$ is called **algebraic factor structure** *or* **factor algebra**.

Proof To prove that $\oplus$ is indeed an inner composition on A/ϱ, we only need to show that its definition is independent of the representatives. That is, it does not matter which representatives of the equivalence classes we have chosen. Indeed, for $x \in [a]_\varrho$, $y \in [b]_\varrho$ one has $x \varrho a$ and $y \varrho b$. Since ϱ is a congruence, it follows that $(x+y) \varrho (a+b)$, i.e. $[x+y]_\varrho = [a+b]_\varrho$. □

Example 7.25 (Quasigroup) We examine the following 6 element quasigroup Q_6

	1 2 3 4 5 6
1	1 2 3 4 5 6
2	6 1 2 3 4 5
3	5 6 1 2 3 4
4	4 5 6 1 2 3
5	3 4 5 6 1 2
6	2 3 4 5 6 1

Let $\varrho_{1,4}$ denote the congruence on Q_6 generated by $\{(1,4)\}$. It yields the classes $[1] = \{1,4\}, [2] = \{2,5\}, [3] = \{3,6\}$. The factor set $Q_6/\varrho_{1,4}$ is the quasigroup Q_3 defined analogously to Q_6.

Analogously, we can define operations on the factor set A/ϱ for inner compositions of other arities on A. We need that ϱ is compatible with the respective composition on A. For example $[e]_\varrho$ is the neutral element (zero, right One, etc.) of $(A/\varrho, \oplus)$, if e has the corresponding property in $(A, \cdot)$. This leads to the following

Lemma 7.26 (Congruence $\rightsquigarrow$ Surjective Homomorphism) *Let $\varrho \subseteq A \times A$ be a congruence on the algebra A. Then*

$$\pi_\varrho : \begin{cases} (A, +) \to (A/\varrho, \oplus) \\ x \mapsto [x]_\varrho \end{cases}$$

is a surjective homomorphism. It is called the **canonical surjection of** ϱ, *or also the* **canonical projection onto** A/ϱ.

7.3 Compatible Relations

Theorem 7.27 (Homomorphism Theorem for Algebras) *Let $(A, +)$ and $(B, \cdot)$ be algebras, $f : (A, +) \to (B, \cdot)$ a homomorphism and ϱ a congruence on $(A, +)$ with $\varrho \subseteq \varrho_f$. Then the homomorphism*

$$f' : \begin{cases} (A/_\varrho, \oplus) \to (B, \cdot) \\ [a]_\varrho \mapsto f(a) \end{cases}$$

is unique such that the diagram

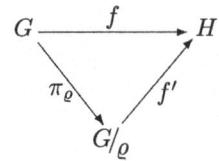

is commutative.
Furthermore

(a) f is surjective $\Rightarrow f'$ is surjective;

(b) $\varrho = \varrho_f \Rightarrow f'$ is injective.

In particular, $A/_{\varrho_f} \cong f[A]$.

Proof From Theorem 7.24, we know that the factor set $A/_\varrho$ with the composition defined there is a factor algebra. From the Homomorphism Theorem for sets (Theorem 5.21), we know that f' is the uniquely determined mapping that makes the diagram commutative, and that (a) and (b) apply. It remains to be shown that π_ϱ and f' are algebra homomorphisms. This is true for π_ϱ because of Lemma 7.26. Moreover, $f'([a]_\varrho) \cdot f'([b]_\varrho) = f(a) \cdot f(b) = f(a + b) = f'([a + b]_\varrho) = f'([a]_\varrho \oplus [b]_\varrho)$. This implies the last statement of the theorem: $A/_{\varrho_f} \cong f[A]$. □

Exercise 7.28 (Homomorphism Theorem for Algebras) Construct a semigroup homomorphism from the right zero semigroup $R_3 = \{r_1.r_2, r_3\}$ to the right zero semigroup $R_2 = \{r_1.r_2\}$. Determine the induced congruence and verify the Homomorphism Theorem.

Exercise 7.29 (Quasigroups and Congruences) Consider the quasigroup Q_6 from Example 7.25 and verify the statements made there.

7.4 Groups

Group, Klein four group, generating element, cyclic group, dihedral group, symmetric group, list notation, cycle notation, cycle, composition, length, fixed point, permutations and transpositions, subgroup, normal subgroup (= normal divisor), kernel, factor/quotient group, Homomorphism Theorem

As announced, we consider groups separately here. An algebraic structure is a group only if it is a right solvable or left solvable monoid, we already know that. Because of the great importance of the concept of a group—not only within mathematics— we start here again somewhat more elementary. This should also make it easier to get started at this point if a reader has not studied this chapter in detail up to this point. We give three equivalent definitions and show in Theorem 7.30 the missing steps of their equivalence.

The algebra $(M, \cdot)$ is a ***group***, if (I) or (II) or (III) apply:

(I)   (a) *Associativity:* $(\forall x, y, z \in M) : x(yz) = (xy)z$.
      (re) *Existence of a right identity:* $(\exists e \in M)(\forall x \in M) : xe = x$.
      (rinv) *Every element has a right inverse with respect to this right identity:* $(\forall x \in M)(\exists x^{-1} \in M) : xx^{-1} = e$.
(II)   (a) *Associativity:* $(\forall x, y, z \in M) : x(yz) = (xy)z$.
      (e) *Existence of an identity:* $(\exists e \in M)(\forall x \in M) : xe = ex = x$.
      (inv) *Every element has an inverse with respect to this identity:* $(\forall x \in M)(\exists x^{-1} \in M) : xx^{-1} = x^{-1}x = e$.
(III) $(M, \cdot)$ is an associative quasigroup.

Definition (I) is used if one wants to prove that an algebraic structure is a group, since relatively little has to be shown. Definitions (II) or (III) are used if one already knows that a group is present, one can then already use a few more properties, or if one wants to show that an algebraic structure is not a group, because then one has more properties, of which only one must not be fulfilled.

Thus, a group is a special algebraic structure $(M, \cdot, ^{-1}, e)$ with a binary, a unary and a 0-ary operation.

The following theorem provides in particular the proof that Definition (I) implies Definition (II), the converse direction is clear.

Theorem 7.30 (Groups) *If $(G, \cdot, ^{-1}, e)$ is a group according to Definition (I), then:*

1. *The right identity e is also a left identity, so e is the identity in G.*
2. *Every right inverse x^{-1} of every element $x \in G$ is also a left inverse of x, so it is the inverse of x.*
3. *The inverse of every element is uniquely determined.*
4. *For all $x, y \in G$ one has that $(x^{-1})^{-1} = x$ and $(xy)^{-1} = y^{-1}x^{-1}$.*

7.4 Groups

Proof 2. Let $x \in G$, x^{-1} be a right inverse of x and $(x^{-1})^{-1}$ be a right inverse of x^{-1}. We calculate: $x^{-1}x = x^{-1}xe = x^{-1}xx^{-1}(x^{-1})^{-1} = x^{-1}e(x^{-1})^{-1} = e$, so x^{-1} is also a left inverse of x.

1. With 2. for every $x \in G$ we get $ex = xx^{-1}x = xe = x$, so e is also a left identity in G.

3. If x^{-1} and x' are inverses of x in G, then: $x^{-1} = x^{-1}e = x^{-1}xx' = ex' = x'$, so the inverse of x is uniquely determined.

4. Due to 3. we have $(x^{-1})^{-1} = x$ and finally $xyy^{-1}x^{-1} = xex^{-1} = xx^{-1} = e$, so $(xy)^{-1} = y^{-1}x^{-1}$. □

Derived from the general definitions in Sect. 7.3 we now use the terms group homomorphism, isomorphism, endomorphism, automorphism. So: if $(G, \cdot)$ and $(H, +)$ are groups, then a mapping $f : (G, \cdot) \to (H, +)$ is a group homomorphism if $f(x \cdot y) = f(x) + f(y)$ for $x, y \in G$. In general, for algebraic structures it must be proven that $f(1_g) = 1_H$ and $f(x^{-1}) = (f(x))^{-1}$, since a mapping compatible with the operations must also be compatible with the 0-ary operation 1_G and the unary operation $^{-1}$. Here, both follow automatically from the group properties.

Examples 7.31 (Groups of Numbers)

(1) Known groups are the sets $\mathbb{C}$, $\mathbb{R}$, $\mathbb{Q}$ and $\mathbb{Z}$ with the respective addition as composition. But $\mathbb{N}$ is not a group, because there are no inverses. We write, for example, $(\mathbb{Z}, +, -, 0)$ or briefly $(\mathbb{Z}, +)$ for the *additive group* of $\mathbb{Z}$.
(2) With their multiplications, $\mathbb{C} \setminus \{0\}$, $\mathbb{R} \setminus \{0\}$ and $\mathbb{Q} \setminus \{0\}$ are groups, but not $\mathbb{Z} \setminus \{0\}$ or $\mathbb{N} \setminus \{0\}$. We write, for example, $(\mathbb{Q} \setminus \{0\}, \cdot, ^{-1}, 1)$ or briefly $(\mathbb{Q} \setminus \{0\}, \cdot)$ for the *multiplicative group* of $\mathbb{Q}$.

Examples 7.32 (Small Groups)

(1) The existence of a right identity and left inverses in a semigroup is not sufficient for this set to be a group. This is shown by the example of a ***left zero semigroup*** H, compare Example 7.7. If H contains at least two elements, H is not a group, because there is no left identity.
(2) For the smallest groups, we provide the composition tables. We denote the neutral element in each case with e. Simple case distinctions lead to the result that there is exactly one group with one, two or three elements, the last two being isomorphic to $(\mathbb{Z}_2, +)$ and $(\mathbb{Z}_3, +)$:

	e
e	e

	e	a
e	e	a
a	a	e

	e	a	b
e	e	a	b
a	a	b	e
b	b	e	a

(3) For four elements, the distinctions of cases already become more tedious.

Finally, we find four composition tables of groups.

In the first, each element is its own inverse and the product of any two different elements $\neq e$ is equal to the third $\neq e$. This group with 4 elements is called **Klein's**[5] **four group**. Its multiplication table follows below. In each of the three other multiplication tables, there is an element x (called the **generating element** of the group), such that all other elements are powers of this one: the first group is generated by both a (we have $a^1 = a$, $a^2 = c$, $a^3 = b$, $a^4 = e$) and b (we have $b^1 = b$, $b^2 = c$, $b^3 = a$, $b^4 = e$). A generating element of the second group is a and also c. The third group is generated by b and also by c. A group that contains a generating element is called a **cyclic group**.

	e	a	b	c
e	e	a	b	c
a	a	e	c	b
b	b	c	e	a
c	c	b	a	e

	e	a	b	c
e	e	a	b	c
a	a	c	e	b
b	b	e	c	a
c	c	b	a	e

	e	a	b	c
e	e	a	b	c
a	a	b	c	e
b	b	c	e	a
c	c	e	a	b

	e	a	b	c
e	e	a	b	c
a	a	e	c	b
b	b	c	a	e
c	c	b	e	a

Renaming the elements transforms each of the three multiplication tables into any other of them. We consider as mappings

$$f = \begin{pmatrix} e & a & b & c \\ e & a & c & b \end{pmatrix} \quad \text{and} \quad g = \begin{pmatrix} e & a & b & c \\ e & b & a & c \end{pmatrix},$$

which in the second row indicate the respective image of the element above. These mappings accomplish the renaming. They are group isomorphisms such that f transforms the first table into the second and g transforms the second table into the third. This means, these three groups are isomorphic to each other, they are $(\mathbb{Z}_4, +)$ "up to isomorphism", the **cyclic group of order 4**.

(4) Another family of finite groups are the **dihedral groups** D_n. The dihedral group D_n has $2n$ elements. They are the symmetries of a regular n-gon in $\mathbb{R}^2$, or in other words, the automorphisms (see Sect. 6.6) of the graph C_n (see Example 6.5 (7)). Geometrically speaking, they are the covering mappings of the regular n-gon.

---

[5] Felix Klein, German mathematician, 1849–1925.

Symmetric Groups

From Sect. 5.3 we already know Sym(M), the set of permutations of M.

Theorem 7.33 (Symmetric Group) *Let M be a set. The set Sym(M) of permutations of M forms a group with respect to the composition of mappings. The neutral element is the identity. The inverse to $f \in$ Sym(M) is the inverse mapping f^{-1}.*

The group Sym(M) is called the **symmetric group on** M. Symmetric groups and their subgroups are also called **permutation groups**. If M is finite, we do not bother about the names of its elements, we just set $M = \{1, 2, 3, \ldots, n\}$. In this case, Sym($M$) is also written as S_n and is called the **symmetric group of order** n.

Every permutation of a finite set M can also be represented by the so called **list notation**, which we have already used in Exercise 5.17.

$$f = \begin{pmatrix} 1 & 2 & 3 & \cdots & n \\ f(1) & f(2) & f(3) & \cdots & f(n) \end{pmatrix}.$$

More compact and often more convenient than this, although somewhat unusual, is the **cycle notation** of permutations. We explain it using an example. On the left we have the list notation, on the right the cycle notation:

$$f = \begin{pmatrix} 1 & 2 & 3 & 4 & 5 & 6 \\ 4 & 5 & 3 & 6 & 2 & 1 \end{pmatrix} = (146)(3)(52).$$

In a **cycle**, i.e., between two round brackets on the right, there are elements of $M = \{1, 2, \ldots, 6\}$. The rule is: each element is mapped to its direct neighbour on the right, the most right element is mapped to the most left in the bracket (i.e., cyclically). Thus, $(146) = (461) = (614)$, expressing that $f(1) = 4$, $f(4) = 6$, $f(6) = 1$. We can always find such a representation by starting with any element of M and then writing its image under f to the right of it until an element is mapped to the first of the cycle. After that, a new cycle begins. In this way, the cycles automatically become disjoint. So we have the following theorem.

Theorem 7.34 (Permutations and Cycles) *The representation of a permutation by disjoint cycles is unique up to the order of the cycles.*

The concatenation of arbitrary cycles, i.e., the **composition of cycles** is interpreted as a composition of mappings, if the cycles have elements in common, we must pay attention to the order of the cycles. We read them like other mappings from right to left, i.e., for $f := (13)(12)$ we get $f(2) = 3$, since 2 is mapped by the cycle at the right to 1 and 1 is subsequently mapped by the cycle at the left to 3. Furthermore, $f(1) = 2$ is derived from the cycle at the right, 2 does not appear further on the left.

And finally, $f(3) = 1$ is derived from the cycle at the left, 3 does not appear at the right. From this, it can be seen that $(13)(12) = (123)$.

The ***length of a cycle*** is the number of elements between the two brackets, cycles of length 1 are called ***fixed points*** of the permutation, they do not need to be listed in the cycle notation. Cycles of length 2 are called ***transpositions***. The latter are suitable as building blocks for permutations.

Theorem 7.35 (Permutations and Transpositions) *Every permutation can be represented (usually not uniquely) as a composition of transpositions (and cycles of length 1).*

Proof It suffices to show that we can write every cycle of length $n > 2$ as a composition of transpositions. This is done as follows (to be read from the back):

$$(a_1 a_2 \cdots a_n) = (a_1 a_2)(a_2 a_3) \cdots (a_{n2} a_{n1})(a_{n1} a_n).$$

This representation is obviously not unique, as

$$(16)(14) = (146) = (461) = (41)(46).$$

□

Normal Subgroup, Factor Group, Homomorphism Theorem

For groups, we obtain from Lemmas 7.23 and 7.26 that every group homomorphism f provides a congruence ϱ_f. And conversely, every congruence ϱ on a group provides a canonical surjection π_ϱ. Groups are algebraic structures in which congruences and homomorphisms can be described by special subgroups. This is not the case for graphs or semigroups. A subset U of G is called a ***subgroup*** of G if U itself is a group with the same composition as defined on G. A subgroup N of a group G is called a ***normal divisor*** or ***normal subgroup of*** G, written $N \trianglelefteq G$, if

$$(\forall a \in G): \quad aN = Na.$$

Here, $aN := \{an \mid n \in N\}$, is a so-called ***complex product***.

Let $(G, \cdot, e)$ and $(H, \cdot, e')$ be groups and $f : G \to H$ a group homomorphism, then $\ker f := \{x \in G \mid f(x) = f(e) = e'\}$ is called the ***kernel of*** f. The following theorem establishes the connection. In particular, it becomes clear that $N := \ker f$ induces a congruence ϱ_f. Previously (Lemma 7.23), this was called the kernel congruence of f, also written $\ker f$. So we have

$$x \varrho_{\ker f} y \Leftrightarrow x \ker f \, y \Leftrightarrow xy^{-1} \in \ker f = N.$$

7.4 Groups

Theorem 7.36 (Kernel, Normal Subgroup and Congruence) *Let $(G, \cdot, e)$ and $(H, \cdot, e')$ be groups and $f : G \to H$ a group homomorphism, let ϱ be a congruence on the group $(G, \cdot, e)$.*

1. *The set $N := [e]_\varrho$ is a normal divisor of G and $\ker f = [e]_{\varrho_f}$.*
2. *Let $N \leq G$ be a normal divisor of G, and take $x, y \in G$. By $x \varrho_N y :\Leftrightarrow xy^{-1} \in N$ we get a congruence on G, the **congruence induced by N**. For this congruence, N is just the equivalence class of the neutral element e of G. And $\pi_N := \pi_{\varrho_N} : G \to G/_{\varrho_N}$, the **canonical surjection induced by N**, is a group homomorphism with $\ker \pi_N = N$.*
3. *The assignment $N \mapsto \varrho_N$ from the set $\mathcal{N}(G)$ of all normal divisors of G to the set $\mathcal{C}(G)$ of all congruences on G, and $\varrho \mapsto [e]_\varrho$ in the opposite direction, each defines an order preserving bijection with respect to $\subseteq$.*

A simple proof of this theorem can be found in Körner [52].

If N is a normal divisor of G, then we write for the factor structure by the congruence ϱ_N induced by N instead of $G/_{\varrho_N}$ simply $G/_N$ and speak of the ***factor group*** or also the ***quotient group***.

Examples 7.37 (Normal Divisors)

(1) In a group G, G and e are normal divisors, such that $G/_G \cong \{e\}$ and $G/_{\{e\}} \cong G$.
(2) Obviously, every subgroup of an abelian group is a normal divisor. In the non commutative case, however, we also find subgroups that are not normal divisors.
(3) The symmetric group S_3 is not commutative. We have the subgroups

$$A := \{(1), (123), (132)\} \quad \text{and} \quad B := \{(1), (12)\}$$

of S_3. Only A is a normal divisor of S_3. Indeed, for B we get $(13)B = \{(13), (123)\} \neq \{(13), (132)\} = B(13)$. According to the above theorem,

$$x \varrho y \quad :\Leftrightarrow \quad xy^{-1} \in A, \quad (x, y \in S_3),$$

is a congruence. But

$$x \varrho y \quad :\Leftrightarrow \quad xy^{-1} \in B, \quad (x, y \in S_3),$$

is indeed an equivalence relation, but not a congruence on S_3.
(4) The sets $n\mathbb{Z} = \{nz \mid z \in \mathbb{Z}\}$, $n \in \mathbb{N}$, are exactly the subgroups of the (abelian) group $(\mathbb{Z}, +)$, so also their normal divisors. The relations $\equiv \mod n$, also written as $\cong \mod n$, are therefore exactly all congruences on $(\mathbb{Z}, +)$. We get $\mathbb{Z}_n = \mathbb{Z}/_{n\mathbb{Z}}$, since

$$x \equiv y \mod n \Leftrightarrow n|xy \Leftrightarrow xy \in n\mathbb{Z}.$$

This is the specialization of the relation ϱ_N used in Theorem 7.36 to $\varrho_{n\mathbb{Z}}$. It coincides with the relation $\equiv \bmod n$, cf. Example 7.1 (4). Thus, all factor groups of $\mathbb{Z}$ are given by the groups $(\mathbb{Z}_n, +_n), n \in \mathbb{N}$.

When we have found all normal divisors of a group, we therefore know all its congruences and thus also (up to isomorphism) all its homomorphic images:

Theorem 7.38 (Homomorphism Theorem for Groups) *Let G and H be groups, $f : G \to H$ a homomorphism and N a normal divisor in G with $N \subseteq \ker f$. Then*

$$f' : \begin{cases} G/N \to H \\ [a]_N \mapsto f(a) \end{cases}$$

is the uniquely determined homomorphism, such that

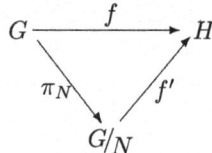

is commutative. Furthermore

(a) f is surjective $\Rightarrow f'$ is surjective;

(b) $N = \ker f \Rightarrow f'$ is injective.

In particular, $G/\ker f \cong f[G]$ as groups.

Proof The statements follow from Theorem 7.27. □

Exercise 7.39 (Permutations) Let $M := \{a, b, c\}$. Consider the set $\mathcal{S}_3$ of all permutations of M. Represent all elements of $\mathcal{S}_3$ in cycle notation and set up the multiplication table for the composition of permutations.

7.5 Product Structures

Product of algebras, projections, universal property of the product, solution of a universal problem

Let $(A, \oplus)$ and $(B, \odot)$ be two algebras, then we define a binary inner composition on the Cartesian product $A \times B$ by

7.5 Product Structures

$$\otimes : \begin{cases} (A \times B) \times (A \times B) \to A \times B \\ ((a_1, b_1), (a_2, b_2)) \mapsto (a_1, b_1) \otimes (a_2, b_2) := (a_1 \oplus a_2, b_1 \odot b_2) \end{cases}.$$

Then $(A \times B, \otimes)$ is also an algebra, which we call **the product of the algebras** A **and** B. Note that $\otimes$ is defined component wise.

If $A_1, A_2, \ldots, A_n$ are algebras and $n \geq 2$, we also define component wise a binary inner composition on the Cartesian product $A_1 \times A_2 \times \cdots \times A_n$. This algebra is called **the product of the algebras** $A_1, A_2, \ldots, A_n$. The same is even possible for infinitely many algebras.

Examples 7.40 (Products)

(1) $(\mathbb{R}, +) \times (\mathbb{R}, +)$ with $(x_1, y_1) + (x_2, y_2) = (x_1 + x_2, y_1 + y_2)$ is the product of the group $(\mathbb{R}, +)$ with itself. It is a group (compare the vector addition in Example 7.4 (2)).
(2) $(\mathbb{Z}_2, +) \times (\mathbb{Z}_2, +)$ is isomorphic to the Klein four group, compare Examples 7.32 (3).
(3) $(\mathbb{N}, \cdot) \times (\mathbb{N}, \cdot)$ with $(x_1, y_1) \cdot (x_2, y_2) = (x_1 \cdot x_2, y_1 \cdot y_2)$ is the product of the monoid $(\mathbb{N}, \cdot)$ with itself. It is a monoid.

In category theory (compare Sect. 12.12), property 2 of the following theorem is used for the definition of the product.

Theorem 7.41 (Universal Property) *If* $P := A_1 \times A_2$ *is the product of the algebras* A_1 *and* A_2*, then*

1. *For each* $i \in \{1, 2\}$*, the* i-*th projection* p_i *is a homomorphism:*

$$p_i : \begin{cases} P \to A_i \\ (a_1, a_2) \mapsto a_i \end{cases}$$

2. *For any arbitrary algebra* Q *with arbitrary homomorphisms,* $i = 1, 2$,

$$q_i : \begin{cases} Q \to A_i \\ z \mapsto q_i(z), \end{cases}$$

by

$$q : \begin{cases} Q \to P \\ z \mapsto (q_1(z), q_2(z)) \end{cases}$$

a unique homomorphism is defined, so that

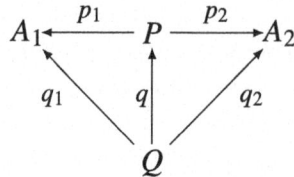

is commutative.

In 2., we speak of the *universal property of the product* and mean that the product $(A_1 \times A_2, p_1, p_2)$ not only provides projections onto the factors, but that it is also in a certain sense an extreme case: every other candidate (Q, q_1, q_2) can be "reduced" to the product by a uniquely determined homomorphism q. Therefore, we say that the product is the *solution to a universal problem*.

Exercise 7.42 (Product) Prove Theorem 7.41 and verify the statements for the quasigroup $Q_3 \times (\mathbb{Z}_3, \cdot)$ (cf. Example 7.25) with concrete morphisms q_1 and q_2.

From Semirings to Fields

8

In this chapter, we consider sets with at least two internal operations, which are connected by different distributive laws. Fields and rings are the most common representatives of this kind. It turns out that even those structures that initially appear rather artificial, have well-known and common models, or are useful in applications. Finally, we give a brief insight into the so-called Universal Algebra.

On the Literature Structures like ring and field are described in books on algebra, for example in Lorenz [55], Körner [52] or Lüneburg [57]. For more exotic structures, there are specialized books, the titles of which are each hint enough on the content: U. Hebisch and H.J. Weinert [34], G. Pilz [62], H. Wähling [71], K.A. Zhevlakov et al. [77]. An easily understandable book on universal algebra is Werner [74], also recommended are Lugowski [56] and Wechler [72].

8.1 From Semirings to Hypercomplex Systems

Right/left distributive, (non-associative)(semi-)ring, with one, (semi-) field, skew field, right/left(almost)-, right/left near -, max-plus semifield, Hamiltonian quaternions, Cayley numbers, octonions, (hyper)complex conjugate, quaternion group

If $(M, \oplus, \odot)$ is an algebraic structure with two binary internal operations, we say $\odot$ is

right distributive over $\oplus$, if $(b \oplus c) \odot a = (b \odot a) \oplus (c \odot a)$,
left distributive over $\oplus$, if $a \odot (b \oplus c) = (a \odot b) \oplus (a \odot c)$,
distributive over $\oplus$, if $\odot$ is right and left distributive over $\oplus$ for all $a, b, c \in M$.

Note If $(M, \odot)$ is commutative, then right distributive automatically follows from left distributive and vice versa.

Since $(M, \oplus)$ and $(M, \odot)$ are algebraic structures, we can assume algebraic properties for both independently of each other, combine these and relate them through various distributivities.

We proceed formally again. Table 8.1 shows the name of the algebraic structure $(R, \oplus, \odot)$, when $(R, \oplus)$ is one of the structures in the first column and $(R, \odot)$ is one of the structures in the first row, and if $\odot$ is distributive over $\oplus$. Here 0 stands for the neutral element with respect to $\oplus$.

We speak of a *commutative (semi)ring*, if the semigroup $(R, \odot)$ is commutative, and of a (semi)ring *with identity*, if in $(R, \odot)$ a neutral element exists.

The table shows that it would be more consistent to say field instead of skew field and add the word commutative in the case of a commutative multiplication. However, for historical reasons, this is less common.

Because of the importance of rings and fields, we will examine these in Sect. 8.2 and also in Chap. 10 in more detail, similar to groups in the previous chapter.

If $\odot$ is only distributive over $\oplus$ from the right or only from the left, then the prefixes *right/left* and even *right near/left near* are added, leading to monstrosities as non-associative right near semiring. So mostly, *"right ring"* is used.

Note There are also algebraic structures $(R, \oplus, \odot)$ for which $(R, \oplus)$ is a non-commutative group. We have suppressed the corresponding additional line in the preceding table, as special names for these structures are not common.

However, the following theorem applies (cf. Wähling [71], page 157). This shows in particular that fields always have commutative addition, although this is not required as part of the axiom system.

Theorem 8.2 (Fields Have Commutative Addition) *Let $(R, \oplus)$ and $(R \setminus \{0\}, \odot)$ be (not necessarily commutative) groups. If $\odot$ is left or right distributive over $\oplus$, then $(R, \oplus)$ is commutative.*

Table 8.1 Structures with two internal operations

	$(R, \odot)$			
$(R, \oplus)$	Groupoid	Semigroup	$(R \setminus \{0\}, \odot)$ group	$(R \setminus \{0\}, \odot)$ commutative group
Monoid	*Non associative semiring*	*Semiring*	*Semi field*	*Commutative semifield*
Commutative group	*Non associative ring*	*Ring*	*Skew Field*	*field*

8.1 From Semirings to Hypercomplex Systems

The following examples show that these seemingly arbitrary definitions have a real background.

Examples 8.3 (Numbers)

(1) With their usual addition and multiplication, $\mathbb{C}$, $\mathbb{R}$ and $\mathbb{Q}$ are fields.
(2) $(\mathbb{Z}, +, \cdot)$ is a ring, but not a field.
(3) $(\mathbb{N}, +, \cdot)$ is a (commutative) semiring, but not a ring.
(4) $(\mathbb{R} \cup \{-\infty\}, \max, +)$ is a commutative semifield, the so-called *max-plus semifield over* $\mathbb{R}$. Here $\{-\infty\}$ plays the role of zero and 0 the role of one. This, or more precisely the analogously defined *min-plus semifield over* $\mathbb{R}$, was already in the 1980s called *tropical semiring* by French mathematicians, honoring Imre Simon a founder of the theory.[1]

Example 8.4 (Right Semirings and Rings) If $(A, \diamond)$ is an algebra, then by

$$(f \oplus g)(a) := f(a) \diamond g(a) \quad \text{and} \quad (f \circ g)(a) := f(g(a)),$$

$f, g \in A^A$, $a \in A$, two binary operations on A^A are defined. In $(A^A, \oplus, \circ)$, $\circ$ is right distributive over $\oplus$. If $\diamond$ is commutative, so is $\oplus$.

(a) If $(A, \diamond)$ is a semigroup, then $(A^A, \oplus, \circ)$ is a right semiring.
(b) If $(A, \diamond)$ is a group, then $(A^A, \oplus, \circ)$ is a right ring, also called right near ring.
(c) Distributivity also applies from the left if we only consider the subset $\mathrm{Hom}(A, A)$ of structure-preserving mappings h from A^A, i.e., mappings for which

$$(\forall a, b \in A): \ h(a \diamond b) = h(a) \diamond h(b).$$

Thus, $(\mathrm{Hom}(A, A), \oplus, \circ)$ here provides examples of semirings and rings.

Hypercomplex Systems

In this section, we return to the complex numbers $\mathbb{C} \cong \mathbb{R}^2$ (cf. the corresponding part in Sect. 3.2). The method of defining addition and multiplication of elements from $\mathbb{R}^2$ based on $(\mathbb{R}, +, \cdot)$ can be repeated, now for elements from $\mathbb{C}^2$, and based on $(\mathbb{C}, +, \cdot)$, and so on. The resulting algebraic structures are called *hypercomplex systems* or also *division algebras*. It is interesting how the algebraic properties, especially of the respective multiplication, change, one might say, deteriorate.

---

[1] Hungarian-Brazilian computer scientist and mathematician, 1943–2009.

Theorem 8.5 (Complex Numbers) *The complex numbers* $\mathbb{C} \cong \mathbb{R}^2$ *with component-wise addition* $\oplus$:
$(x_1, x_2) \oplus (y_1, y_2) := (x_1 + y_1, x_2 +_2)$
and the following multiplication $\odot$:
$(x_1, x_2) \odot (y_1, y_2) := (x_1 y_1 - x_2 y_2, x_1 y_2 + x_2 y_1)$
for $(x_1, x_2), (y_1, y_2) \in \mathbb{C} = \mathbb{R}^2$, *form a field. Here for addition, subtraction and multiplication in $\mathbb{R}$ the usual symbols are used.*

Now we "continue in this way" and construct the *Hamiltonian quaternions*.[2]

Theorem 8.6 (Hamiltonian Quaternions) *The (Hamiltonian) quaternions* $\mathbb{H} := \mathbb{C}^2$ *with component-wise addition* $\oplus$:
$(x_1, x_2) \oplus (y_1, y_2) := (x_1 + y_1, x_2 + y_2)$
and the following multiplication $\odot$:
$(x_1, x_2) \odot (y_1, y_2) := (x_1 \cdot y_1 - x_2 \cdot \overline{y_2}, x_1 \cdot y_2 + x_2 \cdot \overline{y_1})$,
with $(x_1, x_2), (y_1, y_2) \in \mathbb{H}$, *form a non-commutative field, i.e., a proper skew field. Here $+$ and $\cdot$ denote the compositions in $\mathbb{C}$. For the complex number* $y_i = (y_{i1}, y_{i2}) \in \mathbb{C}$ *we put* $\overline{y_i} = \overline{(y_{i1}, y_{i2})} := (y_{i1}, -y_{i2})$, *the complex conjugate.*

We repeat this procedure once more, and construct the *Cayley octaves*,[3] also called *octonions*. Now we start from the Hamiltonian quaternions.

Theorem 8.7 (Cayley Octaves) *The (Cayley) octaves* $\mathbb{O} := \mathbb{H}^2$ *with component-wise addition* $\oplus$:
$(x_1, x_2) \oplus (y_1, y_2) := (x_1 + y_1, x_2 + y_2)$
and the following multiplication $\odot$:
$(x_1, x_2) \odot (y_1, y_2) := (x_1 \cdot y_1 - x_2 \cdot \overline{y_2}, x_1 \cdot y_2 + x_2 \cdot \overline{y_1})$,
with $(x_1, x_2), (y_1, y_2) \in \mathbb{O}$, *form a non-associative ring, so in particular not a skew field. Here $+$ and $\cdot$ denote the compositions in $\mathbb{H}$.*
For $y_i := (y_{i1}, y_{i2}) \in \mathbb{H}$ *we put* $\overline{y_i} = \overline{(y_{i1}, y_{i2})} := (\overline{y_{i1}}, -y_{i2})$ *the* **"hypercomplex"** *conjugate. Here $\overline{y_{i1}}$ is the complex conjugate.*

From Sect. 3.2 we know that $\mathbb{C} \cong \{a + bi \mid a, b \in \mathbb{R}, i^2 = -1\}$. Similarly, we can show
$$\mathbb{H} \cong \{a + bi + cj + dk \mid a, b, c, d \in \mathbb{R}, i^2 = j^2 = k^2 = -1, ij = k, jk = i, ki = j, ji = -k, kj = -i, ik = -j\}.$$
The 8 elements $\pm 1, \pm i, \pm j, \pm k$ with the given multiplication form the so-called **quaternion group**.

---

[2] Developed by W. Hamilton in 1843, who we already encountered concerning graphs, but already discovered in 1840 by Olinde Rodrigues, a French mathematician, banker and social reformer, 1795–1851.

[3] Published by Cayley in 1845, but already communicated in 1843 by John Thomas Graves, an Irish mathematician and jurist, 1806–1870.

A similar, even more complicated representation, now as eightfold sums, also exists for the elements from $\mathbb{O}$.

Exercise 8.8 (Right Semiring) Prove the statements from Example 8.4.

We have to show the following:

(α) $(A^A, \oplus)$ is a semigroup in (a), a group in (b). The properties of $\diamond$ of the algebra $(A, \diamond)$ are transferred to $\oplus$: if, for example, $e \in A$ is neutral, then the constant mapping $c_e : x \mapsto e, x \in A$, is the neutral element in A^A.
(β) $(A^A, \circ)$ is a semigroup.
(γ) $\circ$ is right distributive over $\oplus$, i.e. $(f \oplus g) \circ h = (f \circ h) \oplus (g \circ h)$ for all $f, g, h \in A^A$.
(δ) $\diamond$ commutative $\Rightarrow \oplus$ commutative
(ε) $\circ$ is left distributive over $\oplus$, for mappings from $Hom(A, A)$

Exercise 8.9 (Complex Numbers, Quaternions, and Octonions)

(a) Prove Theorem 8.5.
(b) Prove Theorem 8.6 and investigate the alternative representation of $\mathbb{H}$ through fourfold sums.
(c) Prove Theorem 8.7 and find the alternative representation of $\mathbb{O}$ through eightfold sums.

8.2 Rings and Fields

Ring, field, residue class ring/field, polynomials over a ring, degree, normalized, coefficient, polynomial ring, indeterminate, (formal) power series, matrix ring, unit matrix, ordered ring/field, ring homomorphism, preserving unity, sub(semi)ring/field, left-/right ring ideal, absorption property, kernel, induced congruence, factor/quotient ring, Homomorphism Theorem, zero divisor, prime field

We recall from the table from Sect. 8.1: An algebraic structure $(R, +, \cdot)$ is a *(commutative) ring* if $(R, +)$ is an abelian group, $(R, \cdot)$ is a (commutative) semigroup, and $\cdot$ is distributive over $+$. If $(R \setminus \{0\}, \cdot)$ is even an abelian group, then $(R, +, \cdot)$ is a *field*.

Fields thus are special rings.
The simplest example of a commutative ring which is not a field, is $(\mathbb{Z}, +, \cdot)$.
The rings $(\mathbb{Q}, +, \cdot), (\mathbb{R}, +, \cdot), (\mathbb{C}, +, \cdot)$ are fields. Compare also to Example 8.3.
Let R be a ring without zero divisors. This means that the product of non-zero elements is not zero, cf. Sect. 10.1. The smallest field into which R can be embedded is called the **quotient field** or **field of fractions** of R. One already known example is $\mathbb{Q}$, the quotient field of $\mathbb{Z}$. For a field K, the **field of rational functions** $K(x)$ is the quotient field of the *polynomial ring* $K[x]$, defined on the next page. Formally, this is of course possible. But one might wonder, why the hell we should divide polynomials. It will turn out that this is a very important device, especially over

finite fields, cf. Example 9.14. The so-called ***Laurent***[4] ***series*** are the quotient field of the ring of *power series* $\mathbb{R}[[x]]$, denoted by $\mathbb{R}((x))$, defined also on the next pages. In both cases this can be taken as definition of the respective construction. For some more elementary definitions compare the following subsection *Polynomial rings, formal power series*. Compare also Example 9.5 in Sect. 9.3 and Example 12.23 in Sect. 12.4. In both places additional features of these constructions are exhibited.

Intermediate Fields, Residue Class Rings, Matrix Rings

There exist many fields between $\mathbb{Q}$ and $\mathbb{C}$: We take a subset $S \subseteq \mathbb{C}$. The elements of $S \cup \mathbb{Q}$ are added, multiplied, subtracted, and divided finitely many times. The result is called ***adjunction of*** S or also ***extension field of*** $\mathbb{Q}$ ***by*** S, written $\mathbb{Q}(S)$.
Consider, for example $S = \{\sqrt{2}\} \subseteq \mathbb{R}$. The constructed field is denoted by $\mathbb{Q}(\sqrt{2})$. Then $\mathbb{Q}(\sqrt{2}) = \{a + b\sqrt{2} \mid a, b \in \mathbb{Q}\}$. This can immediately be verified using the usual rules of calculation in $\mathbb{R}$. In particular we ensure that the inverses are of the same form. Indeed, $(a+b\sqrt{2})^{-1} = -a/(2b^2-a^2) + (b/(2b^2-a^2))\sqrt{2}$ in $\mathbb{Q}(\sqrt{2})$.

In this field, we can thus solve polynomial equations of the form $x^2 - 2 = 0$. Similarly, $\mathbb{Q}(\sqrt[3]{5})$ is the set of all numbers $a + b\sqrt[3]{5} + c\sqrt[3]{25}$, with $a, b, c \in \mathbb{Q}$. The sets $\mathbb{Z}_n$ of residue classes modulo n with the addition and multiplication of residue classes from Example 7.1 (4) provide *finite rings* and *finite fields*. They are called ***residue class rings*** or as ***residue class fields***, cf. Theorem 8.15.

Consider the $n \times n$-***matrix ring*** with elements (entries) from a Ring $(R, +, \cdot)$ with matrix addition and matrix multiplication, compare Sect. 6.5. For $n > 1$ they form a non-commutative ring with one. The latter is the so-called ***identity matrix***, for which all elements on the diagonal are 1, all other elements are 0. The ***zero matrix*** has all entries 0 and is the zero element.

Polynomial Rings, Formal Power Series

Let $(R, +, \cdot)$ be a commutative ring and take $x \notin R$. A formal expression of the form

$$P := \sum_{i=0}^{n} a_i x^i := a_0 + a_1 x + a_2 x^2 + \cdots + a_n x^n$$

with $n \in \mathbb{N}$ and $a_0, \ldots, a_n \in R$ is called a ***polynomial in*** x ***over*** R. If $a_n \neq 0$, then n is called the ***degree of*** P. If $a_n = 1$ then the polynomial is called ***normalized***. The $a_0, \ldots, a_n$ are called ***coefficients*** of P. The set of all polynomials in x over R is denoted by $R[x]$. Sum $+$ and product $\cdot$ of polynomials are defined as follows

---

[4] Pierre Alphonse Laurent, 1813–1854, French mathematician.

8.2 Rings and Fields

$$\sum_{i=0}^{m} a_i x^i + \sum_{j=0}^{n} b_j x^j := \sum_{k=0}^{\max\{m,n\}} (a_k + b_k) x^k \quad \text{and}$$

$$\sum_{i=0}^{m} a_i x^i \cdot \sum_{j=0}^{n} b_j x^j := \sum_{\substack{k=0 \\ i+j=k}}^{m+n} (a_i b_j) x^k,$$

Here $a_{m+1} := \cdots := a_n := 0$ is set, if $m < n$, and $b_{n+1} := \cdots := b_m := 0$, if $n < m$.

$(R[x], +, \cdot)$ is a ring, which is called the ***polynomial ring over*** R.

We determine the indeterminate. The specification $x \notin R$ and the phrase "formal expression", which we used in the definition of polynomials, are mathematically correct, if "formal expression" means the following. With x and its powers only the operations specified in the definitions of $+$ and $\cdot$ are possible. This means in particular that two polynomials are equal if and only if their coefficients with the same index are the same. Then x is called an ***indeterminate*** over R, (or variable or also unknown—according to the word Unbekannte, often used in German). Somewhat more formally (see e.g. Körner [52]), an indeterminate over R is defined as an element of a *commutative overring of* R, which is *algebraically independent over* R, compare Sect. 9.5.

Basically, $R[x]$ is constructed in a similar way as $\mathbb{Q}(S)$ above. But now division is not possible. Therefore the result is only a ring and not a field, even if R is a field. This is symbolized by the fact that here square brackets are used for the adjoined element.

We denote the set of ***(formal) power series over*** R by

$$R[[x]] := \left\{ \sum_{i=0}^{\infty} a_i x^i \mid a_i \in R \right\}.$$

Obviously, we can consider polynomials as finite power series (that is there exists $n \in \mathbb{N}$ such that $a_i = 0$ for all $i > n$). If we extend the definitions of addition and multiplication in the obvious way to $R[[x]]$, then $R[[x]]$ is a ring. The quotient field of $K[[x]]$, K a field, i.e. the field of *Laurent series*, consists of (formal) power series with positive and negative exponents

$$K((x)) := \left\{ \sum_{i=-\infty}^{\infty} a_i x^i \mid a_i \in K \right\}.$$

Ordered Rings

After the definition of an ordered algebraic structure with one binary inner operation, which we have given in Chap. 7, we will define ordered rings and fields accordingly.

If $(R, \oplus, \odot)$ is a ring, then $(R, \oplus, \odot, \leq)$ is called an ***ordered ring***, if $(R, \oplus, 0, \leq)$ is an ordered group with neutral element 0 and for all $a, b, c \in R$ we have

$$0 \leq c \text{ and } a \leq b \Rightarrow a \odot c \leq b \odot c \text{ and } c \odot a \leq c \odot b.$$

Note that in this case $\leq$ is only "partially" compatible with $\odot$.

If $(K, \oplus, \odot)$ is a field, then $(K, \oplus, \odot, \leq)$ is called an ***ordered field***, if $(K, \oplus, \odot, \leq)$ is an ordered ring in which $\leq$ is a total order.

Examples 8.10 (Ordered Rings)

(1) $(\mathbb{Z}, +, \cdot, \leq)$ is an ordered ring.
(2) $(\mathbb{Q}, +, \cdot, \leq)$ and $(\mathbb{R}, +, \cdot, \leq)$ are ordered fields.
(3) $(\mathbb{C}, +, \cdot)$ with $\mathbb{C} = \mathbb{R}^2$, as defined in the Subsection "Complex Numbers" in Sect. 3.2, is a field, also compare Theorem 8.5. If we use the component-wise order $\leq_k$ on $\mathbb{C} \cong \mathbb{R}^2$, then $(\mathbb{C}, +, \leq_k)$ is an ordered group, but it is not linearly ordered. Thus, $(\mathbb{C}, +, \cdot, \leq_k)$ is not an ordered ring and therefore not an ordered field.

Ideal, Quotient Ring, Homomorphism Theorem

We now specify what structure-preserving means for mapping between rings. If $(R, +, \cdot)$ and $(L, +, \cdot)$ are two rings and $f : R \to L$ is a mapping, then f is called ***ring homomorphism***, if for all x, y in R one has

$$f(x + y) = f(x) + f(y) \quad \text{and} \quad f(xy) = f(x)f(y).$$

The ring homomorphism f is called ***unity-preserving***, if R and L are rings with unit elements e_R and e_L and if $f(e_R) = e_L$.

A subset $I \subseteq (R, \oplus, \odot)$ of a (semi)ring, such that $(I, \oplus, \odot)$ itself is a (semi)ring or field is called ***sub(semi)ring*** or ***subfield***. A sub(semi)ring $(I, \oplus, \odot)$ is called ***right ideal***, ***left ideal*** or ***ideal of the (semi)ring*** R, if $I \odot R \subseteq I$, $R \odot I \subseteq I$ or $R \odot I \odot R \subseteq I$. Intuitively, we say again, a left/right ideal has the ***swallowing property*** or ***absorption property*** from the left/right.

Note how this definition differs from that for algebras with only one binary operation in the subsection "Subalgebras", Sect. 7.2.

Obviously, R and $\{0\}$ are always ideals of the ring R. Any other ideal of the ring R we call a ***proper ideal***.

8.2 Rings and Fields

Theorem 8.11 (Ideals in Fields) *No field has proper ideals.*[5]

Proof Let K be a field and $I \neq \{0\}$ an ideal. For $x \in I$ there exists $x^{-1} \in K$ and consequently $1 = xx^{-1} \in I$. Thus for every $k \in K$, $k = 1k \in I$, so $K \subseteq I$ and therefore $I = K$. □

Examples 8.12 (Subrings and Ideals)

(1) $(\mathbb{N}, +, \cdot) \subseteq (\mathbb{Z}, +, \cdot) \subseteq (\mathbb{Q}, +, \cdot) \subseteq (\mathbb{R}, +, \cdot) \subseteq (\mathbb{C}, +, \cdot)$ is a chain of subrings, except for $(\mathbb{N}, +, \cdot)$, which is only a semiring. The last implication $(\mathbb{R}, +, \cdot) \subseteq (\mathbb{C}, +, \cdot)$ needs interpretation: it must mean precisely $(\mathbb{R} \times \{0\}, +, \cdot) \subseteq (\mathbb{R} \times \mathbb{R}, +, \cdot)$, where $+$ and $\cdot$ are addition and multiplication in $\mathbb{C}$. Note that the multiplication in $\mathbb{C} = \mathbb{R} \times \mathbb{R}$ restricted to $\mathbb{R} \times \{0\} \cong \mathbb{R}$ coincides with the multiplication in $\mathbb{R}$. The remaining structures except $(\mathbb{Z}, +, \cdot)$ are fields, but according to the previous theorem, none of the rings is an ideal in a larger one.
(2) The subring of even numbers $2\mathbb{Z} = \{2z \mid z \in \mathbb{Z}\}$ of the ring $(\mathbb{Z}, +, \cdot)$ is an ideal in $(\mathbb{Z}, +, \cdot)$.
(3) The subring $(\{0, 2\}, +, \cdot) \cong (\mathbb{Z}_2, +, \cdot)$ of $(\mathbb{Z}_4, +, \cdot)$ is an ideal in $(\mathbb{Z}_4, +, \cdot)$, see Example 7.6 (2).

Also for rings we get: every ring homomorphism f provides a congruence ϱ_f and conversely, every congruence ϱ on a ring provides a canonical surjection π_ϱ. Like groups, rings are algebraic structures in which congruences can be described by subobjects, namely the ideals, compare Theorem 7.36.

So again we define: For rings $(R, +, \cdot)$ and $(L, +, \cdot)$ and a ring homomorphism $f : R \to L$, we set $\ker f := \{r \in R \mid f(r) = 0\}$ and call it **kernel of** f. Compare Lemma 7.23 and the corresponding remarks before Theorem 7.36.

The following theorem again indicates the connection:

Theorem 8.13 (Kernels, Ideals, Congruences) *Let $(R, +, \cdot)$ and $(S, +, \cdot,)$ be rings, both of whose zeros being denoted by 0, $f : R \to S$ a ring homomorphism and ϱ a congruence on the ring $(R, +, \cdot)$. Then*

1. *the set $[0]_\varrho$ is an ideal of R and $\ker f = [0]_{\varrho_f}$;*
2. *if $I \subseteq R$ is an ideal of R, then for $r, s \in R$ by*

$$r\varrho_I s :\Leftrightarrow r - s \in I$$

*we get a congruence on R, the **congruence induced by** I, for which I is precisely the equivalence class of the neutral element $0 \in R$. The **canonical surjection***

---

[5] In German this theorem reads as "Kein Körper hat echte Ideale". This might be considered as a slightly trivialized "wisdom of life".

induced by I, given by $\pi_I := \pi_{\varrho_I} : R \to R/_{\varrho_I}$ *is a ring homomorphism with* $\ker \pi_I = I$;

3. *the assignment* $I \mapsto \varrho_I$ *between the set* $\mathcal{I}(R)$ *of all ideals of* R *and the set* $\mathcal{C}(R)$ *of all congruences on* R, *and also* $\varrho \mapsto [0]_\varrho$ *in the converse direction, define order-preserving bijections with respect to the order* $\subseteq$.

If I is an ideal of R, we write for the factor structure $R/_{\varrho_I}$ simply $R/_I$ and speak of the ***factor ring***, the name ***quotient ring (by*** I**)** is also common. This name can lead to misunderstandings, because it could be confused with the "ring of quotients", which is also called ***field of quotients***. The field of quotients of $\mathbb{Z}$ is $\mathbb{Q}$, factor rings are for example $\mathbb{Z}_n$.

When we have found all ideals of a ring, we know all its congruences and thus also (up to isomorphism) all its homomorphic images. Hence, we get the next homomorphism theorem. It follows from Theorem 7.27.

Theorem 8.14 (Homomorphism Theorem for Rings) *Let R and S be rings, $f : R \to S$ a homomorphism and I an ideal in R with $I \subseteq \ker f$. Then the homomorphism*

$$f' : \begin{cases} R/_I \to S \\ [a]_I \mapsto f(a) \end{cases}$$

is unique such that the diagram

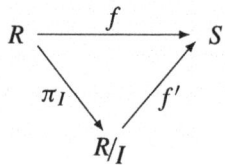

*is commutative.
Furthermore*

(a) f *is surjective* $\Rightarrow$ f' *is surjective;*

(b) $\quad I = \ker f \Rightarrow f'$ *is injective.*

In particular, $R/_{\ker f} \cong f[R]$ *as rings.*

Finite Fields and Prime Fields

We now specify the statements about residue fields from the beginning of this chapter.

Theorem 8.15 (Finite Fields) *For every natural number $n \geq 2$ one has that $\mathbb{Z}_n$ is a ring. Moreover, $\mathbb{Z}_p$ is a field if and only if p is a prime number.*

Proof By Lemma 7.2 we know that the operations on $\mathbb{Z}_n$ are well-defined. There we just showed the independence of representatives for the definition of addition. So take now $a, x, b, y \in \mathbb{Z}$ with $[a]_n = [x]_n$ and $[b]_n = [y]_n$, i.e., $n|x-a$ and $n|x-b$. Then $n|((x-a)b = xb - ab)$ and $n|(x(y-b) = xy - xb)$, so $n|((xy-xb)+(xb-ab) = xy-ab)$. Thus $[xy]_n = [ab]_n$. From the Homomorphism Theorem for rings, it follows that the ring properties of $(\mathbb{Z}, +, \cdot)$ are transferred to the set of residue classes.

We leave the proof of the statement about $\mathbb{Z}_p$ to the reader, with the following hints.

- If $x, y \neq 0$, but $xy = 0$, then x and y are called *zero divisors*. A field contains no zero divisors (because every $x \neq 0$ has an inverse element x^{-1} and thus from $xy = 0$ it follows that $y = x^{-1}xy = 0$).
- Every element $[a]_n \in \mathbb{Z}_n$ is a zero divisor in $\mathbb{Z}_n$ if and only if $\gcd(a, n) \neq 1$. If n is not a prime number, then $\mathbb{Z}_n$ contains zero divisors. In $\mathbb{Z}_6$ for example, $[2]_6 [3]_6 = [6]_6 = [0]_6$.
- If n is a prime number, then the n elements

$$[0]_n, [1]_n, [2]_n, \ldots, [n-1]_n$$

are pairwise different. Then these are all elements of $\mathbb{Z}_n$, and therefore every element $[a]_n \neq [0]_n$ in $(\mathbb{Z}_n, \cdot)$ is invertible, i.e. $(\mathbb{Z}_n, +, \cdot)$ is a field.

□

The fields $(\mathbb{Q}, +, \cdot)$ and $(\mathbb{Z}_p, +, \cdot)$ are called *prime fields*.
Without proof, we state

Theorem 8.16 (Prime Field) *Every field either contains $(\mathbb{Q}, +, \cdot)$ or $(\mathbb{Z}_p, +, \cdot)$ for exactly one prime number p as a subfield.*

The smallest finite field is $(\mathbb{Z}_2, +, \cdot)$ with the operation tables for addition and multiplication as follows:

+	0	1
0	0	1
1	1	0

·	0	1
0	0	0
1	0	1

The smallest finite ring with more than one element, which is not a field, is $(\mathbb{Z}_4, +, \cdot)$, cf. Example 7.6 (2) and Example 8.12 (3). We emphasize: $(\mathbb{Z}_1, +, \cdot) = (\{0\}, +, \cdot)$ is not a field, because $\mathbb{Z}_1 \setminus \{0\} = \{0\} \setminus \{0\} = \emptyset$ and therefore it cannot be a group.
In Theorem 9.16 we will describe all finite fields.

Products of Rings

In analogy with Sect. 7.5 we can now also form *products of rings*. We do not formulate the definition. Instead we consider one positive and one negative example:

- In $(\mathbb{R} \times \mathbb{R}, +, \cdot) = (\mathbb{R}, +, \cdot) \times (\mathbb{R}, +, \cdot)$, define both operations component-wise. Then this is the product of the ring $(\mathbb{R}, +, \cdot)$ with itself. It is a ring, but not a field.
- In $\mathbb{C} = \mathbb{R}^2$ multiplication is defined as $(x_1, y_1) \odot (x_2, y_2) = (x_1 x_2 - y_1 y_2, x_1 y_2 + x_2 y_1)$. That is not defined component-wise. Together with the component-wise addition, $\mathbb{C} = \mathbb{R}^2$ is a field. But it is thus not the product of the field $(\mathbb{R}, +, \cdot)$ with itself.

Exercise 8.17 (Equivalent Fractions) Consider the relation $\sim \subseteq (\mathbb{Z} \times (\mathbb{Z} \setminus \{0\}), +, \odot) \times (\mathbb{Z} \times (\mathbb{Z} \setminus \{0\}), +, \odot)$, defined by $(x, y) \sim (u, v) := xv = yu$. Show that this is a congruence on $(\mathbb{Z} \times \mathbb{Z} \setminus \{0\}, +, \odot)$. Give the following classes: $[(0, 1)]_\sim, [(1, 2)]_\sim, [(-2, 1)]_\sim, [(5, 9)]_\sim$. Explain what the title of the exercise has to do with it. Justify that the operations in $(\mathbb{Z} \times \mathbb{Z} \setminus \{0\}, +, \odot)$ define the usual addition and multiplication on $\mathbb{Q}$.

Exercise 8.18 (One-sided Ideals) Find in the ring of 2×2 matrices with entries from the field $(\mathbb{Z}_2, +, \cdot)$ right ideals that are not left ideals. Show that these right ideals do not induce congruences on this matrix ring.

Exercise 8.19 ($\mathbb{C}$ Lexicographically) Consider the questions discussed in Example 8.10 for $\mathbb{C}$ with the lexicographic order $\leq_l$. Is $(\mathbb{C}, +, \leq_l)$ an ordered group, is it linearly ordered? Is it an ordered ring or even an ordered field?

8.3 Universal Algebras

Universal (general) algebra, algebraic structure, signature, type, homomorphism, Boolean algebra, lattices, idempotent, variety of algebras

So far, we have studied unary and binary operations on sets and have encountered many algebraic structures in the examples. In this part, we will generalize the definitions and results introduced so far as follows.

8.3 Universal Algebras

A *universal algebra* (also: *general algebra*) is a set M with a set $\mathcal{V} = \{v_1, v_2, \ldots, v_n\}$ of (possibly partial) operations. We write this as a pair $(M, \mathcal{V})$ or as an $(n + 1)$-tuple $(M, v_1, v_2, \ldots, v_n)$. The adjective *universal* suggests a generalization of the algebras from Sect. 7.1. It serves to distinguish from the (classical) algebras in Sect. 9.1, but it is often omitted. The set $\mathcal{V}$ of operations is called *algebraic structure on M*. When $v_i \in \mathcal{V}$ is an s_i-ary operation, then the tuple $(s_i)_{i=1,\ldots,n}$ of the arities of the operations is called the *signature* or also *type* of the algebra.

Here too, we define for two universal algebras of the same signature $(A, \mathcal{V})$ and $(B, \mathcal{W})$ homomorphisms. A mapping $f : A \to B$ is called a *homomorphism* (*structure-preserving* or *compatible mapping*), if there is a bijective mapping $\varphi : \mathcal{V} \to \mathcal{W}$ such that

(a) for each s-ary operation $v \in \mathcal{V}$ the image $w := \varphi(v) \in \mathcal{W}$ is also s-ary,
(b) if v is defined on $A' \subseteq A$, then $\varphi(v)$ is also defined on $f(A')$, and
(c) for arbitrary elements $a_i \in A$, $1 \le i \le s$, s the arity of v, one has

$$f(v(a_1, a_2, \ldots, a_s)) = w(f(a_1), f(a_2), \ldots, f(a_s)).$$

Examples 8.20 (Number Systems as Universal Algebras)

(1) The algebra $(\mathbb{R}, +, \cdot, ^{-1}, 0, 1)$ of real numbers has the binary operations $+$ and $\cdot$, the partial unary operation $^{-1}$, as well as the distinguished elements 0 and 1, which are considered as 0-ary operations. So it is of type $(2, 2, 1, 0, 0)$.
(2) The algebra $(\mathbb{Z}, +)$ has the signature (2). But we can also consider the integers as an algebra $(\mathbb{Z}, +, -(\ ), 0)$ with signature $(2, 1, 0)$.

Lattices

We recall from Subsection "Semilattices and lattices" in Sect. 4.3: In every lattice $(V, \le)$, the mappings

$$\sup : \begin{cases} V \times V \to V \\ (x, y) \mapsto \sup\{x, y\} \end{cases} \quad \text{and} \quad \inf : \begin{cases} V \times V \to V \\ (x, y) \mapsto \inf\{x, y\} \end{cases}$$

are defined everywhere. We can therefore interpret them as binary operations on V.
We set

$$x \sqcup y := \sup\{x, y\} \quad \text{and} \quad x \sqcap y := \inf\{x, y\}$$

for the supremum and the infimum of two elements $x, y \in V$. These are associative over each other and commutative operations on V. They possess the following additional properties, also called *absorption laws*

$$(\forall x, y \in V): x \sqcup (x \sqcap y) = x = x \sqcap (x \sqcup y).$$

This implies in particular that both operations are *idempotent* (i.e. $(\forall x \in V): x \sqcup x = x$).

Therefore we can define a lattice as an algebraic structure $(V, \sqcap, \sqcup)$ with these properties. Similarly, we can describe each semilattice as a commutative semigroup, whose operation is idempotent. However, this loses the information whether it was an upper or a lower semilattice.

Conversely, we can show: If $(V, \sqcap, \sqcup)$ is an algebraic structure, whose operations $\sqcap$ and $\sqcup$ are associative, commutative and idempotent and have the mentioned additional property, then by $x \leq y :\Leftrightarrow x \sqcap y = x$ an order relation $\leq$ is defined on V, so that for any two elements $x, y \in V$ the supremum is equal to $x \sqcup y$ and the infimum is $x \sqcap y$.

We summarize.

Theorem 8.21 (Characterization of Lattices) *A lattice is a set with two commutative, associative operations that fulfill the two absorption laws and vice versa.*

Boolean algebras

The term Boolean algebra covers the commonalities between propositions and sets that we have observed in Chaps. 1 and 2.

A ***Boolean algebra*** is an algebraic structure $(B, +, \cdot, ', 0, 1)$, such that

1. $(B, +, 0)$ is a commutative monoid.
2. $(B, \cdot, 1)$ is a commutative monoid.
3. $\cdot$ is distributive over $+$ and vice versa.
4. $(\forall a \in B)(\exists a' \in B)\quad a + a' = 1$ and $a \cdot a' = 0$.

If $(B, +, \cdot, ', 0, 1)$ is a Boolean algebra, then $(B, +, \cdot, 1)$ and $(B, \cdot, +, 0)$ are commutative semirings with identity.

The power set lattice of any set M is a Boolean algebra $(\wp(M), \cup, \cap, \bar{\ }, \emptyset, M)$.

Varieties of Algebras

A *variety of algebras* is understood to be a class of algebraic structures of the same signature that satisfy given identities, i.e. equations. Equivalently: a class of algebraic structures of the same signature that is closed under the formation of homomorphic images, of subalgebras and of direct products.

The equivalence of these two definitions was proven by G. Birkhoff.[6]

Example 8.22 (Varieties) The class of all semigroups forms a class of algebras of signature (2). A defining equation is the associative law

$$x(yz) = (xy)z.$$

The class of all groups forms a class of algebras of signature (2,1,0). The three operations are multiplication, inversion, and identity. Defining equations are

$$x(yz) = (xy)z$$

$$1x = x1 = x$$

$$xx^{-1} = x^{-1}x = 1.$$

Exercise 8.23 (Lattice of Divisors) Draw the Hasse diagram of the lattice of divisors of 60. Observe that $x \mid y$ is an order relation on $\mathbb{N}$. Specify the algebraic structure defined by its operation tables and verify the two absorption laws.

Exercise 8.24 (Lattice of Subsets) Draw the Hasse diagram of the lattice of subsets of $\{x, y, z\}$. Specify the algebraic structure defined by its operation tables and verify the absorption laws. Verify that the corresponding lattice of the power set is a Boolean algebra.

---

[6] Garrett Birkhoff US-American mathematician, 1911–1996.

Act, Vector Space, Extension

9

In this chapter, we further explore the theory of algebraic structures by introducing outer (or external) operations. The sections on vector spaces, field extensions, and coding serve as applications of the theory, but are presented concisely, primarily using examples. These examples provide an initial overview of the topics and serve as a brief review for those already familiar with the content.

On the Literature On the subject of acts and semiautomata see also Chapter 1 of Kilp et al. [45]. If questions from computer science are in the focus of interest, Eilenberg [26] could be useful. For Linear Algebra and thus as an introduction to the theory of vector spaces Fischer [29] is suitable. For coding theory and for finite fields we found Ki Hang Kim and Fred W. Roush [46] as well as Barnett [6] useful. Also compare the references to Chap. 7.

9.1 Outer Operations

(Partial) outer (external) operation, operator domain, operand domain, R-act, groupoid act, R-set, R-operand, semi(group) automaton, module, vector space, algebra, max-plus algebra, evaluation mapping, left/right cosets modulo, input alphabet, state set, transfer function, multisorted systems (= heterogeneous algebras), semilinear

Let R and X be sets. A (partial) mapping

$$\bullet : \begin{cases} R \times X \to X \\ (r, x) \mapsto r \bullet x \end{cases}$$

is called *(partial) outer (or external) operation on X with operator domain R and operand domain X*.

As for inner operations, we also use terms like composition, everywhere defined and (not) closed.

From Automaton to Vector Space and R-Algebra

Similar to Chaps. 7 and 8, we proceed formally and combine different algebraic structures on operator domain and on operand domain. We present the most common structures with definition and name in Table 9.1.

For an outer operation $\bullet : R \times X \to X$ the equations given in Table 9.1 should be fulfilled for all $r, s \in R$ and $x, y \in X$. The word "additionally" refers to the validity of the equations already required for the weaker structures (those above and further to the left in the table). Empty fields mean that a separate name is not common.

Left R-acts are also called *R-sets* or *R-operands*.

The name *semi(group)automata* is also very common. In this case X is the set of states, changes of states are performed by the inputs, i.e. by the action of elements from R, without considering the outputs, see Example 9.1 (7).

Examples 9.1 (Acts)

(1) Every semigroup $(X, \oplus)$ is a left $(\mathbb{N} \setminus \{0\}, \cdot)$-act, if for $n \in \mathbb{N}$ and $x \in X$, we define:

$$n \bullet x := x \oplus \cdots \oplus x \ (n\text{-times}).$$

(2) Every commutative (= abelian) group $(X, +)$ is a $(\mathbb{Z}, +, \cdot)$-module, if for $z \in \mathbb{Z}$ and $x \in X$, we define:

Table 9.1 From left groupoid act to R-linear algebra

	X		
R	X set	$(X, +)$ commutative group	$(X, +, \cdot)$ ring
$(R, \cdot)$ groupoid	**Left groupoid act**		
$(R, \cdot)$ semigroup	**Left R-act**, if $(rs) \bullet x = r \bullet (s \bullet x)$		
$(R, +, \cdot)$ ring		*R-module*, if additionally $(r+s) \bullet x = r \bullet x + s \bullet x$ $r \bullet (x+y) = r \bullet x + r \bullet y$	*R-algebra*, if additionally $r \bullet (xy)$ $= (r \bullet x)y$ $= x(r \bullet y)$
$(R, +, \cdot)$ field		*R-vector space*, if additionally $1 \bullet x = x$	*R-linear algebra*

9.1 Outer Operations

$$zx := \begin{cases} x + \cdots + x \ (|z|\text{-times}), & \text{if } z > 0, \\ -x - \cdots - x \ (|z|\text{-times}), & \text{if } z < 0, \\ 0, & \text{if } z = 0. \end{cases}$$

(3) Every inner (internal) operation $\tau : M^2 \to M$ can be considered as an outer operation with operator domain M and operand domain M: $m \bullet n := \tau(m, n)$, e.g. every semigroup $(R, \cdot)$ is a left $(R, \cdot)$-act, if for $r \in R$ and $x \in R$, we define: $r \bullet x := rx$, and analogously every ring $(R, +, \cdot)$ is a left-$(R, +, \cdot)$ module.

(4) The *evaluation mapping*, also called *evaluation function*, of a mapping $f : M \to M$:

$$\text{eval} : \begin{cases} M^M \times M \to M \\ (f, x) \mapsto f \bullet x := f(x) \end{cases}$$

is an outer operation on M with operator domain M^M. In this way, we obtain the left $(M^M, \circ)$-act M.

(5) *Rotations* δ_φ of the plane $\mathbb{R}^2$ by an angle φ, where a given point, such as $(0, 0)$ remains fixed, can be linked with integers:

$$\bullet : \begin{cases} \mathbb{Z} \times D \to D \\ (z, \delta_\varphi) \mapsto z \bullet \delta_\varphi := \delta_{z\varphi}. \end{cases}$$

So we have a $(\mathbb{Z}, +)$-act.

(6) If G is a group and U is a subgroup of G, then

$$x \equiv_U y :\Leftrightarrow xU = yU$$

is a left congruence on G. The equivalence classes with respect to $\equiv_U$ are called *left cosets of G modulo U*. The set $_G U := \{ xU \mid x \in G \}$ of left cosets of G modulo U is a left G-act with respect to the outer operation

$$\bullet : \begin{cases} G \times {_G U} \to {_G U} \\ (r, xU) \mapsto r \bullet xU := (rx)U. \end{cases}$$

With analogous notation for right cosets modulo U in G, U_G becomes a right G-act. If $aU = Ua$ for every $a \in G$, i.e. if U is a normal subgroup of G (see the corresponding part in Sect. 7.4) then $_G U = U_G = G/U$.

(7) [From semiautomaton to Act] A triple $\mathcal{H} = (E, Z, \delta)$ with a set E, called *input alphabet*, a set Z, called *state set* and a mapping, called *"transfer function"*,

$$\delta : \begin{cases} E \times Z \to Z \\ (e, z) \mapsto \delta(e, z), \end{cases}$$

is called **semiautomaton**.

If E and Z are not too large, the *behavior* of the semiautomaton can be represented by a graph, whose vertex set is Z and whose edges are labeled with the elements from E, as the following sketch suggests:

$$z \xrightarrow{e} ez = \delta(e, z).$$

A semiautomaton is converted into a left E^*-act as follows:

On the set E^* of all words with finitely many letters from E together with the empty word ε (cf.Subsection "Formal languages" in Sect. 1.1) we define by

$$* : \begin{cases} E^* \times E^* \to E^* \\ ((a_1 \cdots a_m,), (b_1 \cdots b_n)) \mapsto (a_1 \cdots a_m) * (b_1 \cdots b_n) \\ \qquad := a_1 \cdots a_m b_1 \cdots b_n \end{cases}$$

and $\varepsilon a = a\varepsilon = a$ for all $a \in E^*$ a binary inner operation $*$, the concatenation of words *(concatenation)*, also section "Formal Languages" in Chap. 1.1. The transfer function δ is recursively extended to

$$\delta^* : \begin{cases} E^* \times Z \to Z \\ (\varepsilon, z) \mapsto \delta(\varepsilon, z) := z \\ (xa, z) \mapsto \delta^*(xa, z) := \delta(x, \delta^*(a, z)) \end{cases}$$

for $x \in E$, $z \in Z$ and $a \in E^*$.

Thus, δ^* is an outer operation on Z with operator domain $(E^*, *)$, i.e. for all $a \in E^*$ and all $z \in Z$ we have $a \bullet z := \delta^*(a, z)$.

Examples 9.2 (Operations on Vectors)

(1) The multiplication with scalars in $\mathbb{R}^3$ is an everywhere defined outer operation on $\mathbb{R}^3$ with operator domain $\mathbb{R}$:

$$\cdot : \begin{cases} \mathbb{R} \times \mathbb{R}^3 \to \mathbb{R}^3 \\ \left(\alpha, \begin{pmatrix} v_1 \\ v_2 \\ v_3 \end{pmatrix} \right) \mapsto \alpha \begin{pmatrix} v_1 \\ v_2 \\ v_3 \end{pmatrix} = \begin{pmatrix} \alpha v_1 \\ \alpha v_2 \\ \alpha v_3 \end{pmatrix}. \end{cases}$$

In Example 7.4 the component wise addition of elements from $\mathbb{R}^3$ is described. With both operations together, $\mathbb{R}^3$ becomes an $\mathbb{R}$-vector space. For every field

9.1 Outer Operations

(K, +, ·) these definitions can be generalized to K^n, also for $n > 3$. In the following Sect. 9.2 we go into more detail about vector spaces.

(2) **Max-plus algebra.** As in the preceding example (1), n-tuples of elements from the semifield ($\mathbb{R} \cup \{-\infty\}$, max, +), cf. Example 8.3 (4), scalar multiplication and addition, with the addition max and multiplication + from the semifield. In this way, one obtains a "max-plus algebra", also known as *tropical semiring*. This, for example, is useful for modeling timetables with transfer relations, cf. Braker [14] and Cunninghame-Green [17] and lies the foundation for "tropical geometry", see Mikhalkin [59].

(3) The scalar product in $\mathbb{R}^3$, which already served us as a non-example for inner operations (Example 7.4), is not an outer operation either.

Structure Preserving Mappings and Multisorted Systems

Groupoid acts over a fixed groupoid R are also related to each other by structure preserving mappings:

Let X and Y be two left groupoid acts for a groupoid R. A mapping $f : X \to Y$ is called *act homomorphism*, if for all $r \in R, x \in X$ one has $f(r \bullet x) = r \bullet f(x)$. If X and Y themselves have inner operations, it is additionally required that f is compatible with these.

In particular, for R-modules and R-algebras, as well as for K-vector spaces and K-linear algebras, homomorphisms must additionally be compatible with the addition on the abelian group $(X, +)$. They are then called *R-homomorphisms*.

Now again explicitly: For two vector spaces V, W over the field K, a mapping $f : V \longrightarrow W$ is called *linear mapping* or *vector space homomorphism* or also *K-homomorphism*, if for all $v, v' \in V, \lambda \in K$ one has

$$f(v) + f(v') = f(v + v') \text{ and } \lambda f(v) = f(\lambda v).$$

The first property is called *additivity*, the second *K-linearity*.

Of course, in each case there is a Homomorphism Theorem, which we will only formulate for R-modules in the next section.

A further generalization in the sense of universal algebra, as described at the end of Chap. 8, are so-called *multisorted systems (= heterogeneous algebras)*. Here, one considers Cartesian products of several sets with different algebraic structures as a new structure. The simplest example is the K-vector space V, written as $((K, +, \cdot), (V, +), \cdot)$. Addition and multiplication in K, addition in V and scalar multiplication are listed in this order. It is obvious that two such systems can be related to each other by corresponding tuples (here pairs) of mappings. It will be required that the mappings are compatible for all corresponding compositions. Compare with Birkhoff and Lipson [8] or Bloom and Wagner [10].

In the case of vector spaces, we speak of semilinear homomorphisms. They allow the transition from a K-vector space V to a K'-vector space V' while preserving structures: A pair of mappings (φ, f) is called a *semilinear homomorphism*, if

$\varphi : K \to K'$ is a field homomorphism, $f : V \to V'$ is a group homomorphism, and for all $k \in K$, $v \in V$ one has $f(kv) = \varphi(k)f(v)$.

Submodule, Factor Module, Homomorphism Theorem

In this subsection we reconsider Lemmas 7.23 and 7.26 now for R-modules.

Again, every homomorphism f provides a congruence ϱ_f, and every congruence ϱ provides a canonical surjection π_ϱ. And again, congruences can be described by submodules, compare Theorem 7.36.

We define: For an R-module V, a subset $U \subseteq V$ is called a **submodule**, if U itself is an R-module or more precisely, if the natural embedding $\iota : U \to V$ is an R-homomorphism. And we call $\ker f := f^{\leftarrow}(0)$ the **kernel of** f. Compare Lemma 7.23 and the corresponding remarks before Theorems 7.36 and 8.2.

Theorem 9.3 (Kernels, Submodules and Congruences) *Let $(V, +)$ and $(W, +)$ be R-modules, whose zeros are both denoted by 0.*

Let $f : V \to W$ be an R-homomorphism and ϱ a congruence on the R-module $(V, +)$, i.e. ϱ is relation compatible with $+$, so that additionally from $u \varrho v$ for $u, v \in V$ it follows that $ku \varrho kv$ for all $k \in R$.

1. *The set $[0]_\varrho$ is a submodule of V and $\ker f = [0]_{\varrho_f}$.*
2. *If $U \subseteq V$ is a submodule of V, then for $v, w \in V$ by $v \varrho_U w :\Leftrightarrow v - w \in U$ we get a congruence on V, the **congruence induced by** U. For this U is the equivalence class of the neutral element 0 of V. The **canonical surjection induced by** U, given by $\pi_U := \pi_{\varrho_U} : V \to V/_{\varrho_U}$ is an R-homomorphism with $\ker \pi_U = U$.*
3. *The assignments $U \mapsto \varrho_U$ between the set $\mathcal{U}(V)$ of all submodules of V and the set $\mathcal{C}(V)$ of all congruences on V or $\varrho \mapsto [0]_\varrho$ in the converse direction define order-preserving bijections with respect to $\subseteq$.*

If U is a submodule of the R-module V, we simply write $V/_{\varrho_U}$ as $V/_U$ for the factor structure and call $V/_U$ the *factor module* or *factor module of V after U*. The name *quotient module* is also common.

When we have found all submodules of a module, we know all its congruences and thus (up to isomorphism) all its homomorphic images. That is, the next theorem follows from the Homomorphism Theorem for groups (Theorem 7.38).

Theorem 9.4 (Homomorphism Theorem for R-Modules) *Consider R-modules V and W, an R-homomorphism $f : V \to W$ and a submodule U of V with $U \subseteq \ker f$. Then the R-homomorphism*

$$f' : \begin{cases} V/U \to W \\ [a]_U \mapsto f(a) \end{cases}$$

is unique such that the diagram

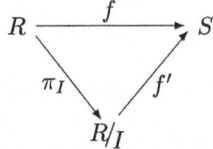

is commutative.
Furthermore,

(a) f is surjective $\Rightarrow f'$ is surjective;

(b) $U = ker f \Rightarrow f'$ is injective.

In particular, $V/_{ker f} \cong f[V]$ as R-modules.

If we replace the words submodule by subspace, factor module by factor space, R by K, and module by vector space, we get statements about vector spaces.

If we only consider semigroup actions, the Homomorphism Theorem is somewhat "weaker" because there $ker f$, the *kernel congruence*, cannot generally be described by a subact. See the footnote to Lemma 7.23.

9.2 Vector Spaces

K-vector space (= linear space), scalar, linear mapping (= vector space homomorphism), additive, linear, linear hull, polynomial, Laurent series, geometric interpretation, arrow, position vector, submodule, factor module, kernel, subspace, induced congruence/canonical surjection, factor space, Homomorphism Theorem

Vectors likely first appeared in physics as "directed quantities" without the consideration of the overarching structure of vector spaces. This structural perspective later proved invaluable for solving linear systems of equations and in analytic geometry, quickly spreading to numerous disciplines both within and outside mathematics.

Here, we provide some basic definitions and properties to give a first impression. We begin with the definition of a vector space, which can also be found in the table at the beginning of Sect. 9.1. We then define the associated structure preserving mappings, known as linear mappings.

Let $(K, +, \cdot)$ be a field. A commutative group $(V, +)$ is called a *K-vector space*, also: *vector space over K* or: *K-linear space*, if an outer operation $\bullet : K \times V \to V$ is defined, so that for all $r, s \in K$ and all $v, w \in V$ we get

$$(rs) \bullet v = r \bullet (s \bullet v)$$
$$(r+s) \bullet v = r \bullet v + s \bullet v$$
$$r \bullet (v+w) = r \bullet v + r \bullet w$$
$$1 \bullet v = v.$$

The field K is called the *scalar field*, its elements are called *scalars*. The elements of V are called *vectors* and the outer operation $\bullet$ is called *scalar multiplication*. The multiplication of a vector v with a scalar s assigns the *scalar multiple* $s \bullet v$ to the vector, which in turn is a vector.

Note We already know the term scalar product, see Example 9.2, but this has a completely different meaning than the scalar multiplication here.

What Is a Vector? – Geometric Interpretation We can now answer this obvious question as follows: A vector is an element of a vector space. However, this at first looks like a tautology.[1] And it does not correspond to our prior knowledge, which usually comes from physics and from the literal meaning.

Therefore, we now provide a geometric interpretation. We know that for $K = \mathbb{R}$ and $n = 2$, each vector $(x, y) \in \mathbb{R}^2$ corresponds exactly to a point in the two dimensional coordinate system. Vectors are encountered in physics as directed quantities, visually represented as *arrows*. – How can we bring together these two aspects? The arrow from the *origin* $(0, 0)$ of the coordinate system to the point (x, y) is called the *position vector* of (x, y) and serves as a representative for the set of all arrows of the same length and direction, not just those that start in $(0, 0)$. These sets form a partition on the set of all arrows in $\mathbb{R}^2$.

Each such class of arrows can be seen as an *action* on the set of all points $(a, b) \in \mathbb{R}^2$. The vector $(x, y) \in \mathbb{R}^2$ acts on $\mathbb{R}^2$ by moving the point (a, b) to the point $(a, b) + (x, y) = (a+x, b+y)$. This explains the name "vector". In other words: $(\mathbb{R}^2, (\mathbb{R}^2, +))$ is a group act with $+$ as the outer operation.

Examples 9.5 (Vector Spaces)

(1) The *standard vector space* K^n. For every field $(K, +, \cdot)$, K^n is a K-vector space in an obvious way, see Example 7.4.
   Linear Algebra shows that every finite dimensional K-vector space is *essentially* equivalent to such a vector space K^n, i.e., isomorphic to it. For this reason, we call K^n standard vector space. However, in applications such as coding, the different representations of isomorphic vector spaces are essential.

---

[1] This is not the case at all: In a K-vector space, an element, or vector, is only defined by satisfying the four equations noted above when combined with other elements of the vector space and scalars.

9.2 Vector Spaces

(2) The smallest non-trivial standard vector space is the $\mathbb{Z}_2$-vector space $(\mathbb{Z}_2)^2$. It is described by two operation tables, for addition (inner operation) and scalar multiplication (outer operation).

+	(0, 0)	(0, 1)	(1, 0)	(1, 1)
(0, 0)	(0, 0)	(0, 1)	(1, 0)	(1, 1)
(0, 1)	(0, 1)	(0, 0)	(1, 1)	(1, 0)
(1, 0)	(1, 0)	(1, 1)	(0, 0)	(0, 1)
(1, 1)	(1, 1)	(1, 0)	(0, 1)	(0, 0)

•	(0, 0)	(0, 1)	(1, 0)	(1, 1)
0	(0, 0)	(0, 0)	(0, 0)	(0, 0)
1	(0, 0)	(0, 1)	(1, 0)	(1, 1)

Note that the additive group of $((\mathbb{Z}_2)^2, +)$ is isomorphic to the Klein four group. Formally, the $\mathbb{Z}_2$-vector space $(\{0\}, +)$ with only one single vector is smallest, should be written as $(\mathbb{Z}_2)^1$, but is apparently not very interesting. The operation tables become somewhat more interesting for fields $\mathbb{Z}_p$ with larger p.

Further examples of such vector spaces over finite fields follow in Sect. 9.5 on extension fields.

(3) The **vector space of $m \times n$ matrices over a field** K. A rectangular scheme in which elements from K are arranged in m rows and n columns is called an $m \times n$-***matrix over*** K. For this construction, K does not even have to be a field, see Chap. 6. Such a matrix generally is presented in two ways like this:

$$A = (a_{ij})_{1 \leq i \leq m, 1 \leq j \leq n} = \begin{pmatrix} a_{11} & a_{12} & \cdots & a_{1n} \\ a_{21} & a_{22} & \cdots & a_{2n} \\ \vdots & \vdots & & \vdots \\ a_{m1} & a_{m2} & \cdots & a_{mn} \end{pmatrix}.$$

Examples for $m = 2$ and $n = 3$ as well as $m = 3$ and $n = 2$ can be found in Sect. 6.5. Now, if K is a field, we can add matrices over K elementwise, as also described there. And in analogue with K^n we define scalar multiplication by

$$s \bullet A = s \bullet (a_{ij})_{1 \leq i \leq m, 1 \leq j \leq n} := (sa_{ij})_{1 \leq i \leq m, 1 \leq j \leq n}.$$

The set of $m \times n$ matrices over K is a K-vector space. Attentive readers may have noticed that this is merely a special case of K^{mn}, where now a vector is written as an $m \times n$ matrix and not as an mn-tuple.

For $n = m$ we even get a non-commutative K-*linear algebra* with the usual matrix multiplication, compare also Sect. 6.5.

(4) The **vector space $K[x]$ of polynomials over a field** K. In the corresponding part of Sect. 8.2 we have represented a polynomial as a formal expression of the form

$$P := \sum_{i=0}^{n} a_i x^i := a_0 + a_1 x + a_2 x^2 + \cdots + a_n x^n$$

with $n \in \mathbb{N}$ and $a_0, \ldots, a_n \in K$, $a_n \neq 0$. In analogy to (1) we can also consider P as an infinite sequence $(a_0, a_1, a_2, \ldots)$ of elements from K, all of which are equal to zero from a certain index (in the above P from $n+1$) on. This has the advantage that we can adopt the definition of sum and scalar multiplication in K^n also from (1). Translated back into the notation of polynomials, we get as sum

$$\sum_{i=0}^{m} a_i x^i + \sum_{j=0}^{n} b_j x^j := \sum_{k=0}^{\max\{m,n\}} (a_k + b_k) x^k$$

for polynomials $P = \sum_{i=0}^{m} a_i x^i$ and $Q = \sum_{j=0}^{n} b_j x^j$ in $K[x]$. And for each $s \in K$ we get as scalar multiplication $\odot$:

$$s \odot \sum_{i=0}^{m} a_i x^i := \sum_{i=0}^{m} s a_i x^i.$$

With the usual polynomial multiplication, compare also Sect. 8.2, we even get a K-*linear algebra*.

(5) **Power series.** Again from Sect. 8.2 we know the ring $R[[x]]$ of formal power series over a ring R. If we extend the outer operation defined in the vector space $K[x]$ of polynomials over a field K in the obvious way to $R[[x]]$, then $R[[x]]$ is an R-*algebra* and, if R is a field, also a R-*linear algebra*.

9.3 Generating System, Basis, Linear (In)dependence

Standard vector space, standard basis (= canonical basis), generating system, linear combination, dimension, linearly (in)dependent

We now consider an arbitrary field K and standard vector spaces K^n for any $n \in \mathbb{N}$. We call the vectors

$$e_1 = (1, 0, \ldots, 0), e_2 = (0, 1, \ldots, 0), \ldots, e_n = (0, 0, \ldots, 1)$$

standard basis or **canonical basis of** K^n. The reason is:

1. Every vector $v = (x_1, x_2, \ldots, x_n) \in K^n$ can be written as a *linear combination* of the vectors $e_1, e_2, \ldots, e_n$, i.e.

$$v = (x_1, x_2, \ldots, x_n) = x_1 e_1 + x_2 e_2 + \cdots + x_n e_n.$$

In particular, the elements $e_1, e_2, \ldots, e_n$ form a generating system of V with respect to K.

9.3 Generating System, Basis, Linear (In)dependence

2. The set $\{e_1, e_2, \ldots, e_n\}$ is minimal as a *generating system* of V with respect to K. Because, if we were to omit one of these vectors, say e_n, then this omitted one could not be written as a linear combination of the remaining vectors. That is the remaining vectors would no longer be a generating system of K^n.

We now define in general:

A vector v in the K-vector space V is called a **linear combination** of the vectors $\{v_1, v_2, \ldots, v_k\} \subseteq V$, if there exist **linear factors** $\lambda_1, \lambda_2, \ldots, \lambda_k \in K$ such that $v = \lambda_1 v_1 + \lambda_2 v_2 + \cdots + \lambda_k v_k \in V$. The set of all linear combinations

$$LH_K(\{v_1, v_2, \ldots, v_k\}) := \{\lambda_1 v_1 + \lambda_2 v_2 + \cdots + \lambda_k v_k | \lambda_1, \lambda_2, \ldots, \lambda_k \in K\}$$

is called the **linear hull** of $\{v_1, v_2, \ldots, v_k\} \subseteq V$.

If $LH_K(v_1, v_2, \ldots, v_k) = V$, we call $\{v_1, v_2, \ldots, v_k\}$ a **generating system of** V. Indeed, every vector of V can be represented as a linear combination of the elements of the generating system.

A minimal generating system $v_1, v_2, \ldots, v_n$ of V is called a **basis** of V.

Now it is clear that $\{e_1, e_2, \ldots, e_n\}$ is a basis of K^n. It is called the standard basis, because the vectors look particularly simple.

Please note: **Only standard vector spaces can have standard bases!**

Of course, there are many other bases of K^n, e.g.

$$\{(1, 1, \ldots, 1), (0, 1, \ldots, 1), \ldots, (0, 0, \ldots, 1)\}.$$

Every vector space V has a generating system, namely V itself. If there is no finite generating system for a vector space, and V' is an infinite generating system, then this means: for each $v \in V$ there exists a finite subset $\{v_1, v_2, \ldots, v_k\} \subset V'$, such that $v \in LH_K(\{v_1, v_2, \ldots, v_k\})$. The vector space $\mathbb{R}[x]$ of polynomials in x for example has no finite generating system.

In Linear Algebra it is shown:

Theorem 9.6 (Existence of Bases) *Let V be a K-vector space, which has a finite generating system. Then V has a basis. All bases of V have the same cardinality. With respect to a given basis $\{v_1, v_2, \ldots, v_n\}$, every $v \in V$ is uniquely representable, i.e. $\lambda_1 v_1 + \lambda_2 v_2 + \cdots + \lambda_n v_n = \mu_1 v_1 + \mu_2 v_2 + \cdots + \mu_n v_n$ implies $\lambda_i = \mu_i \in K$ for all $i = 1, \ldots, n$.*

Remark Corresponding statements can be made about all vector spaces, including those that do not have a finite generating system. However, the proof requires *Zorn's Lemma*, which is equivalent to the axiom of choice. This means that, unlike the finite case, the proof does not provide a basis, so it is not constructive. It even turns out that the statement "Every vector space has a basis" is equivalent to the axiom of choice. The classic example of a vector space for which no basis can be specified is the $\mathbb{Q}$-vector space $\mathbb{R}$.

The number of elements in a basis of V is called the **dimension** of V over K. Then, for example, K^n has dimension n. We write $dim_K K^n = n$ and correspondingly in general: $dim_K V = n$. A family $\{v_1, v_2, \ldots, v_k\}$ of vectors from a K-vector space V is called **linearly dependent** over K if $\lambda_1, \lambda_2, \ldots, \lambda_k \in K$ exist such that $\lambda_1 v_1 + \lambda_2 v_2 + \cdots + \lambda_k v_k = 0 \in V$ and at least one of the λ's is not $0 \in K$. It is called **linearly independent** over K if only for $\lambda_1 = \lambda_2 = \cdots = \lambda_k = 0 \in K$ one has $\lambda_1 v_1 + \lambda_2 v_2 + \cdots + \lambda_k v_k = 0 \in V$.

Linear dependence of the set $\{v_1, v_2, \ldots, v_k\}$ occurs for example, if $v_k = v_{k-1}$. Then we can choose $\lambda_1 = \lambda_2 = \cdots = \lambda_{k-2} = 0 \in K$ and $\lambda_{k-1} = -\lambda_k = 1 \in K$. For example, the vectors of each basis of K^n, as well any of its s subsets is linearly independent.

The following lemma, usually called **Basis completion lemma**, indicates how to construct a basis from a linearly independent set of vectors. This works for every vector space, because the empty set is linearly independent(!).

Lemma 9.7 (Basis Completion Lemma) *Start with an independent set $\emptyset \subseteq U \subseteq V$ of the K-vector space V. If there is a linearly independent set $U' \subseteq V$ that contains more elements than U, then there is an element $u' \in U'$ such that $U \cup \{u'\}$ is linearly independent in V. If there is no such set, U is already a maximal linearly independent set. Thus it is a basis.*

In Linear Algebra, the following theorem is proven:

Theorem 9.8 (Basis) *For a family $B = \{v_1, v_2, \ldots, v_n\}$ of vectors from the K-vector space V, the following statements are equivalent.*

(i) B is a basis of V, i.e., a minimal generating system of V.
(ii) B is a linearly independent generating system of V.
(iii) B is maximally linearly independent in V.

Examples 9.9 (Dimensions) Consider again the vector spaces from Example 9.5:

(1) is n-dimensional,
(2) is 2-dimensional,
(3) is mn-dimensional,
(4) has countably infinite dimension,
(5) is a case where the existence of a basis can only be theoretically proven, as for the $\mathbb{Q}$-vector space $\mathbb{R}$.

The $\mathbb{R}$-vector space $\mathbb{C}$ is 2-dimensional with basis $\{1, i\}$. It is an $\mathbb{R}$-linear algebra with the known multiplication, cf. Theorem 8.5.

The $\mathbb{R}$-vector space $\mathbb{H}$ is 4-dimensional with basis $\{1, i, j, k\}$. It is an $\mathbb{R}$-linear algebra with the known multiplication, cf. Theorem 8.6.

The $\mathbb{R}$-vector space $\mathbb{O}$ is 8-dimensional with basis $\{1, i, j, k, l, m, n, o\}$. It is also an $\mathbb{R}$-linear algebra with a corresponding multiplication, cf. Theorem 8.7.

9.4 Linear Mappings and Matrices

Principle of linear extension, representing matrix, linear (in-)homogeneous equation system, column rank, extended matrix, elimination method, solution, uniquely/ universally solvable

Let $f : K^n \longrightarrow K^m$ be a linear mapping. Every vector $v \in K^n$ can be uniquely represented with respect to the standard basis $E = \{e_1, e_2, \ldots, e_n\}$ of K^n. So take $v = \lambda_1 e_1 + \lambda_2 e_2 + \cdots + \lambda_n e_n$. As f is a linear mapping we obtain the image vector $f(v) = \lambda_1 f(e_1) + \lambda_2 f(e_2) + \cdots + \lambda_n f(e_n)$. Thus, it suffices to know the image vectors $f(e_1), f(e_2), \ldots, f(e_n)$ is sufficient to specify $f(v)$ for every $v \in V$.

Conversely, by specifying vectors $w_1, w_2, \ldots, w_n \in W$ we can define a linear mapping $f : K^n \longrightarrow K^m$ by setting:

$$f(e_1) = w_1, f(e_2) = w_2, \ldots, f(e_n) = w_n.$$

Then for any $v \in V$ with $v = \lambda_1 e_1 + \lambda_2 e_2 + \cdots + \lambda_n e_n$ we get

$$f(v) = \lambda_1 w_1 + \lambda_2 w_2 + \cdots + \lambda_n w_n$$

according to the so-called ***principle of linear extension***.

Now we look at this interplay in more detail. Denote the standard basis of K^m by $E' = \{e'_1, e'_2, \ldots, e'_m\}$.

A linear mapping $f : K^n \longrightarrow K^m$ then provides the following scalars in K

$$a_{11}, a_{21}, \ldots, a_{m1}, a_{12}, a_{22}, \ldots, a_{m2}, \ldots, a_{1n}, a_{2n}, \ldots, a_{mn}.$$

They are obtained through

$$\begin{aligned} f(e_1) = w_1 &= a_{11} e'_1 + a_{21} e'_2 + \cdots + a_{m1} e'_m \\ f(e_2) = w_2 &= a_{12} e'_1 + a_{22} e'_2 + \cdots + a_{m2} e'_m \\ &\vdots \\ f(e_n) = w_n &= a_{1n} e'_1 + a_{2n} e'_2 + \cdots + a_{mn} e'_m. \end{aligned}$$

Here each $w_i, i \in \{1, 2, \ldots, m\}$, is written as a linear combination of the basis vectors from E'. We write these scalars in matrix form

$$A = (a_{ij})_{1\le i\le m,\, 1\le j\le n} = \begin{pmatrix} a_{11} & a_{12} & \cdots & a_{1n} \\ a_{21} & a_{22} & \cdots & a_{2n} \\ \vdots & \vdots & & \vdots \\ a_{m1} & a_{m2} & \cdots & a_{mn} \end{pmatrix}.$$

This matrix is called *representing matrix* of f with respect to the bases E and E' of K^n and K^m. We also use the following notation $M_E^{E'}(f) := A$.

Note *The first column of the representing matrix consists of the coefficients of the image of the first basis vector of K^n with respect to the chosen basis of K^m, and so on. Observe that the mapping f naturally has a different representing matrix with respect to other bases of K^n and K^m.*

Conversely, given a matrix

$$A = (a_{ij})_{1\le i\le m,\, 1\le j\le n} = \begin{pmatrix} a_{11} & a_{12} & \cdots & a_{1n} \\ a_{21} & a_{22} & \cdots & a_{2n} \\ \vdots & \vdots & & \vdots \\ a_{m1} & a_{m2} & \cdots & a_{mn} \end{pmatrix}$$

and bases of K^n and K^m, here the standard bases, we obtain a linear mapping $f : K^n \longrightarrow K^m$ by setting $f(e_i) = a_{1i}e'_1 + a_{2i}e'_2 + \cdots + a_{mi}e'_m$ for all $i \in \{1, 2, \ldots, n\}$. We write $L_E^{E'}(A) := f$.

We mention two properties that demonstrate the importance of the representing matrix.

Evaluation of $f(v)$ Let $f : K^n \longrightarrow K^m$ be a linear mapping and $v = (x_1, \ldots x_n) \in K^n$ a vector. Now we write v as a column, i.e., as a one column matrix with n rows. Then we get $f(v) = M_E^{E'}(f))v$, which is the product of the matrix $M_E^{E'}(f)$ with the one column matrix v (according to the usual matrix multiplication introduced in Sect. 6.5). As result we obtain a one column matrix with m rows. This we can therefore consider as a vector in K^m, which is the image of v under f.

As an example, take $K = \mathbb{R}$, $n = 2$, $m = 3$ and $M_E^{E'}(f) = \begin{pmatrix} 1 & 2 \\ 3 & 4 \\ 3 & 2 \end{pmatrix}$.

Then $f\left(\begin{pmatrix} x_1 \\ x_2 \end{pmatrix}\right) = \begin{pmatrix} x_1 + 2x_2 \\ 3x_1 + 4x_2 \\ 3x_1 + 2x_2 \end{pmatrix} = \begin{pmatrix} 1 & 2 \\ 3 & 4 \\ 3 & 2 \end{pmatrix} \begin{pmatrix} x_1 \\ x_2 \end{pmatrix}$.

9.4 Linear Mappings and Matrices

Matrix of fg Let $f : K^n \longrightarrow K^m$ and $g : K^m \longrightarrow K^r$ be linear mappings and A and B their respective representing matrices with respect to the standard bases in the three vector spaces. Then it follows that $gf : K^n \longrightarrow K^r$ with respect to the standard bases has the representing matrix BA. This was historically the reason why the matrix multiplication is defined as stated in Sect. 6.5. At that time, it was not yet known that this multiplication has many very useful interpretations for graphs.

As an example, choose $n = 2, m = 3, r = 2, K = \mathbb{R}$, take the above matrix as representing matrix A of f, and

$$B = \begin{pmatrix} 1 & 3 & 3 \\ 2 & 4 & 2 \end{pmatrix}$$

as representing matrix of g. Now one can easily compute $g(f(v)) = (gf)(v)$ as BAv, where we again set $v = \begin{pmatrix} x_1 \\ x_2 \end{pmatrix}$.

Systems of Linear Equations

Consider the following situation of matrices with entries from a field K:

$$\begin{pmatrix} a_1 & a_2 \\ a_3 & a_4 \\ a_5 & a_6 \end{pmatrix} \begin{pmatrix} x_1 \\ x_2 \end{pmatrix} = \begin{pmatrix} b_1 \\ b_2 \\ b_3 \end{pmatrix}.$$

We again interpret this as follows:

The vector $\begin{pmatrix} x_1 \\ x_2 \end{pmatrix}$ is mapped under $\begin{pmatrix} a_1 & a_2 \\ a_3 & a_4 \\ a_5 & a_6 \end{pmatrix}$ to $\begin{pmatrix} b_1 \\ b_2 \\ b_3 \end{pmatrix}$.

With the obvious abbreviations A, x, b we thus have $Ax = b$. If for given A and b such a vector x is called a **solution of the equational system** (A, b). And we call (A, b) a **system of linear[2] equations**. Here, A is called the **defining matrix**, b the **right hand side**. If $b = 0 \in K^3$, we speak of a **homogeneous** system, otherwise of an **inhomogeneous** system.

If we do not initially know such x_1, x_2, we use the same notation and call x_1, x_2 **unknowns**. In our specific case, we say (A, b) **consists of three equations with two unknowns**. Obviously, when writing out the matrix product using the unknowns

---

[2] The name is explained by the fact that x_1, x_2 do not also appear as squares or in higher powers.

x_1, x_2, we get three equations:

$$a_1x_1 + a_2x_2 = b_1$$
$$a_3x_1 + a_4x_2 = b_2$$
$$a_5x_1 + a_6x_2 = b_3$$

The above applies analogously for arbitrary matrices A with m rows and n columns, where then x is a column with n entries and b is a column with m entries. All of them come from an arbitrary field or from a ring or another suitable algebraic structure or are unknowns.

The following questions arise for a system of linear equations (A, b):

1. **Is there a solution?**
2. **How can we find such a solution?**
3. **Are there any other solutions?**
4. **How can we find all of them?**

We give answers without proofs:

 1. There is a solution exactly if the vector b can be written as a linear combination of the columns of the matrix A. The linear factors then represent, written as a column vector, a solution. One says: There is a solution for (A, b) exactly if the (column) rank of the matrix A is equal to the (column) rank of the so-called *extended matrix* (A, b), which is created by adjoining the column b to A. The **column rank** of a matrix is the maximum number of linearly independent columns of the matrix.

 2. There are different methods to solve the system (A, b). It can be shown that $Ax = b$ exactly if $BAx = Bb$ for every square matrix B with a suitable number of columns, all of whose columns are linearly independent. The *elimination method* according to Gauss allows the step by step construction of such a matrix B, so that BA has a triangular shape. A solution can then be read from the equation $BAx = Bb$. For details compare any book on Linear Algebra or also Schöning [66].

 4. If x and x' are solutions, then $A(x - x') = Ax - Ax' = b - b = 0$, i.e. $v := x - x'$ is a solution of the homogeneous system $(A, 0)$. Conversely, the same applies: if we know all solutions v of the homogeneous system $(A, 0)$ and at least one solution x of the inhomogeneous system (A, b), we obtain all solutions of (A, b) in the form $x + v$.

 3. One solution of the homogeneous system $(A, 0)$ is always the zero vector 0. If $(A, 0)$ has not more than one solution, then (A, b) has at most one solution. If the system (A, b) actually has exactly one solution, it is called **uniquely solvable**. If the matrix A is such that (A, b) has a solution for every b, which has as many components as the matrix A has rows, we say, the matrix A determines a **universally solvable** system.

Exercise 9.10 (Systems of Linear Equations) Write the linear equation systems defined by the following (A,b) in the usual form with unknowns. Indicate which of the equation systems are solvable, unsolvable, uniquely solvable, universally solvable. Find all solutions.

$A = \begin{pmatrix} 1 & 1 \\ 1 & 1 \end{pmatrix}$ or $\begin{pmatrix} 1 & 1 \\ 0 & 0 \end{pmatrix}$ or $\begin{pmatrix} 1 & 1 \\ 0 & 1 \end{pmatrix}$ and $b = \begin{pmatrix} 1 \\ 2 \end{pmatrix}$ or $\begin{pmatrix} 0 \\ 1 \end{pmatrix}$.

9.5 Field Extensions

Root, (extension) field, degree, irreducible, minimal polynomial, irreducibility criterion (by Eisenstein), algebraically (in)dependent

We return to the constructions of fields between $\mathbb{Q}$ and $\mathbb{R}$; see the relevant section in Sect. 8.2. The underlying question is how to solve polynomial[3] equations of the form $x^n + a_{n-1}x^{n-1} + \ldots + a_0 = 0$. This requires finding the **roots** of the polynomial $x^n + a_{n-1}x^{n-1} + \ldots + a_0$ with coefficients $a_0, \ldots, a_{n-1}$ from a field K in a (larger) field L.

We revisit $\mathbb{Q}(\sqrt{2})$ and $\mathbb{Q}(\sqrt[3]{5})$. We now interpret $\mathbb{Q}(\sqrt{2})$ as a 2-dimensional $\mathbb{Q}$-vector space with basis $\{1, \sqrt[2]{2}\}$. Similarly, the field $\mathbb{Q}(\sqrt[3]{5})$ forms a 3-dimensional $\mathbb{Q}$-vector space with basis $\{1, \sqrt[3]{5}, \sqrt[3]{25}\}$. We can multiply the vectors just like elements from $\mathbb{R}$, and we have obtained fields that are also $\mathbb{Q}$-algebras.

We recall that the inverses have the same form (please check!):
$(a + b\sqrt{2})^{-1} = -a/(2b^2 - a^2) + (b/(2b^2 - a^2))\sqrt{2} \in \mathbb{Q}(\sqrt{2})$.

We generalize this and define: A field L is called **extension** or **extension field** of a field K, written $L : K$, if $K \subseteq L$ is a subfield of L. Automatically, L is in particular a K-vector space. The **degree of the extension** $[L : K]$ is the dimension of L as a K-vector space. So here we have $[\mathbb{Q}(\sqrt{2}) : \mathbb{Q}] = 2$ and $[\mathbb{Q}(\sqrt[3]{5}) : \mathbb{Q}] = 3$. If $[L : K] = n$ is finite, then every element $q \in L$ must satisfy a polynomial equation of the form $x^n + a_{n-1}x^{n-1} + \ldots + a_0 = 0$ with coefficients in K. As elements of an n-dimensional K-vector space, the $n + 1$ elements $q^0, q^1, \ldots, q^n$ must be linearly dependent. In particular, every $q \in L$ is algebraic over K (see Chap. 3).

We call a polynomial $\neq 0$ of degree n in $K[x]$ **irreducible**, if it is not the product of two polynomials of lower degree. The normalized polynomial p of lowest degree that has $q \in L$ as a root is unique (if there were two, consider their difference). It cannot be a product of two polynomials of lower degree (otherwise one of them would have q as a root). Consequently it is irreducible. This polynomial is called **minimal polynomial of q over K**.

For example, $x^2 - 2$ is the minimal polynomial of $\sqrt{2}$ over $\mathbb{Q}$, the minimal polynomial of $\sqrt[3]{5}$ over $\mathbb{Q}$ is $x^3 - 5$.

---

[3] Which are evidently not linear.

Theorem 9.11 (Extension Field) *Let $g \in K[x]$ be an irreducible normalized polynomial of degree n over the field K. Then*

$$L := K[x]/\bigl(g(x)\bigr) := \{b_0 + b_1 x + \ldots + b_{n-1} x^{n-1} \mid b_0, b_1, \ldots, b_{n-1} \in K\}$$

is a field. In particular, L is an extension of K with basis $1, x, x^2, \ldots, x^{n-1}$ over K and g is the minimal polynomial of x over K. In L we calculate as in a polynomial ring, however modulo $g(x)$, i.e. if $g(x) = a_0 + a_1 x + \ldots + a_{n-1} x^{n-1} + x^n$, then $x^n = -a_0 - a_1 x - \cdots - a_{n-1} x^{n-1}$.

Proof (*sketch*) We first note that the above definition of $K[x]/\bigl(g(x)\bigr)$ actually hides the meaning of the notation $K[x]/\bigl(g(x)\bigr)$. It denotes the factorring of $K[x]$ by the (principal) ideal $g(x) K[x]$ generated by $g(x)$, compare Sect. 8.2. It turns out that this factorring consists exactly of the polynomials given in the above definition. Now, L is a ring that contains K, K is the subring of L consisting of the elements of L for which $b_1 = \ldots = b_{n-1} = 0$. We show: Every polynomial $0 \neq h \in K[x]/\bigl(g(x)\bigr)$ has a multiplicative inverse. We use the polynomial division with remainder (which we actually have not formally introduced). If the degree of h is greater than $n - 1$, then we have $h = gf + r$, where f and r are polynomials in $K[x]$, and the degree of r is less than n. Now we can use the Euclidean algorithm for polynomials, which gives us $s, q \in K[x]$ such that $qr + gs = 1$ (where it is to be used that g is irreducible (prime)). This then implies, that $qh = qgf + qr = qgf - gs + 1$. By substituting x we get $q(x) h(x) = 1$ and due to the commutativity of multiplication in the polynomial ring also $h(x) q(x) = 1$.

The remaining field properties follow from the fact that L is already a K-algebra. □

This theorem thus provides a method for constructing finite dimensional extensions of K which can also be applied multiple times. The only important thing is to find an irreducible polynomial over the respective K, i.e. in $K[x]$. The construction in Theorem 9.11 thus creates the roots of the respective irreducible polynomial in a field extension of K.

For the following example and the terms used there (irreducible, maximal ideal, principal ideal domain), also compare Chap. 10.

Example 9.12 (Complex Numbers) Suppose that the field K does not contain an element x with $x^2 = -1$. Then the polynomial $x^2 + 1$ is irreducible in $K[x]$, so the ideal generated by it is maximal in $K[x]$. (This implication holds in principal ideal domains.) Then $L = K[x]/(x^2 + 1)$ is a field extension of K that contains an element denoted in a usual way by i, whose square is -1, namely the residue class of x. In particular, $\mathbb{R}[x]/(x^2 + 1) \cong \mathbb{R}(i) \cong \mathbb{C}$. Similarly, $\mathbb{Z}[x]/(x^2 + 1) \cong \mathbb{Z}[i]$, where $\mathbb{Z}$ is not a field, but a principal ideal domain. The elements of $\mathbb{Z}[i]$ are called *Gaussian integers*. Obviously, $\mathbb{Z}[i] \subseteq \mathbb{R}(i)$.

9.5 Field Extensions

Table 9.2 Multiplication table of $\mathbb{Q}[x]/(x^3+2x^2+2x+2)$

	1	x	x^2
1	1	x	x^2
x	x	x^2	$-2x^2-2x-2$
x^2	x^2	$-2x^2-2x-2$	$2x^2+2x+4$

To test for irreducibility, we formulate without proof

Theorem 9.13 (Irreducibility Criteria)

(Irr1) If an $f \in \mathbb{Z}[x]$ is a product of two polynomials of lower degree in $\mathbb{Q}[x]$, then it is also a product of two polynomials of lower degree in $\mathbb{Z}[x]$. The first and the last coefficients of the factors divide the corresponding coefficients of f.

(Irr2) **(Eisenstein's Irreducibility Criterion**[4]**)** If $f(x) \in \mathbb{Q}[x]$ is normalized of degree n, and there is a prime number $p \in \mathbb{Z}$ that divides the coefficients of $x^{n-1}, x^{n-2}, \ldots, x^1$ and x^0, but p^2 does not divide the constant coefficient (i.e. the one of x^0), then f is irreducible.

(Irr3) A polynomial over K of degree 2 or 3 is irreducible if it has no root in K.

Example 9.14 (Field Extension $\mathbb{Q}[x]/(g(x))$) The polynomial $g(x) = x^3+2x^2+2x+2$ is irreducible in $\mathbb{Q}[x]$ according to (Irr2), since the prime number 2 divides the coefficients of x^2, x^1 and x^0, but 2^2 does not divide the constant coefficient 2. In $\mathbb{Q}[x]/(g(x))$ we then have the basis $(1, x, x^2)$. The products of these basis elements are given in Table 9.2.

Note that in $\mathbb{Q}[x]/(g(x))$ we have $x^3 = -2x^2 - 2x - 2$ and consequently $x^4 = -2x^3 - 2x^2 - 2x = -2(-2x^2 - 2x - 2) - 2x^2 - 2x = 2x^2 + 2x + 4$. With this, we can calculate the inverses. For example, $(x^2+x+1)(x+1) = g(x) - 1$, which in $\mathbb{Q}[x]/(g(x))$ is $= -1$, i.e. $(x^2+x+1)^{-1} = -(x+1)$ in $\mathbb{Q}[x]/(g(x))$.

If g is a reducible polynomial, i.e. if $g = fh$ and f and h have degrees smaller than that of g, the same construction works. Then $L = K[x]/(g(x))$ is not a field, but only a ring with zero divisors. Example 9.15 illustrates the situation.

From now on, $\mathbb{F}_p$ denotes a finite field with p elements.

Example 9.15 (Factor Ring, not a Field) In Table 9.3 we provide the multiplication table of the elements $\neq 0, 1$ of the factor ring $\mathbb{F}_2[x]/(x^3-1)$ of $\mathbb{F}_2[x]$, which is obviously not a field.

Theorem 9.16 (Finite Fields) *For every normalized irreducible polynomial $g(x) \in \mathbb{F}_p[x]$ of degree n, $\mathbb{F}_p[x]/(g(x))$ is an extension of degree n over $\mathbb{F}_p$, i.e., a field with p^n elements, if p is a prime number. All finite fields can be obtained this way. All finite fields with p^n elements are isomorphic, they are denote by $\mathbb{F}_{p^n}$.*

---

[4] Ferdinand Gotthold Max Eisenstein, German mathematician, 1823–1852.

Table 9.3 Multiplication table of $\mathbb{F}_2[x]/(x^3 - 1) \setminus \{0, 1\}$

	x	$x+1$	x^2	x^2+1	x^2+x	x^2+x+1
x	x^2	x^2+x	1	$x+1$	x^2+1	x^2+x+1
$x+1$	x^2+x	x^2+1	x^2+1	x^2+x	$x+1$	0
x^2	1	x^2+1	x	x^2+x	$x+1$	x^2+x+1
x^2+1	$x+1$	x^2+x	x^2+x	$x+1$	x^2+1	0
x^2+x	x^2+1	$x+1$	$x+1$	x^2+1	x^2+x	0
x^2+x+1	x^2+x+1	0	x^2+x+1	0	0	x^2+x+1

Table 9.4 Addition and multiplication in $\mathbb{F}_{2^2}$

+	0	1	x	$1+x$
0				
1		0	$1+x$	x
x		$1+x$	0	1
$1+x$		x	1	0

$\cdot$	0	1	x	$1+x$
0				
1				
x			$1+x$	1
$1+x$			1	x

It can be problematic to find an irreducible polynomial if only criterion (Irr3) from Theorem 9.13 can be applied. However, it is often possible to find irreducible polynomials over $\mathbb{F}_p$ by trial and error.

Example 9.17 ($\mathbb{F}_4$) In $\mathbb{F}_2$, the polynomial $x^2 + x + 1$ has no roots, so it is irreducible according to (Irr3). In Table 9.4, we calculate sums and products in $\mathbb{F}_{2^2} = \mathbb{F}_2[x]/(x^2+x+1)$ by reducing all occurring polynomials with $x^2+x+1 = 0$, i.e., using $x^2 = x + 1$.

The missing entries result from the role of 0 for $+$ and $\cdot$, and from the role of 1 for $\cdot$. We have found the smallest field that is not a prime field.

Due to the formal parallel with linear dependence, we introduce here the concept of *algebraic dependence*. Let $L : K$ be a field extension, where L is a commutative ring with 1. Let $v_1, \ldots, v_n$ be elements of L. If there exists a polynomial $f \neq 0$ in n variables with coefficients in K, i.e. $f \in K[x_1, \ldots, x_n] \setminus \{0\}$, such that

$$f(v_1, \ldots, v_n) = 0,$$

then $v_1, \ldots, v_n$ are called **algebraically dependent**. From this, the definition of **algebraically independent** follows. We will use the term later for the definition of algebraic matroids, see Theorem 11.12.

Example 9.18 (Algebraically Independent) From the definition it immediately follows that the pairwise different variables $x_1, \ldots, x_n$ are algebraically independent over any field.

A transcendental number over $\mathbb{R}$, such as π or e, is obviously algebraically independent over $\mathbb{R}$. So π, e are also algebraically independent. But the numbers π

and π^2 are algebraically dependent over $\mathbb{R}$: they "satisfy" the polynomial $x^2 - y \in \mathbb{R}[x, y]$.

Exercise 9.19 (Factorization of $\mathbb{Q}[x]$) Determine the multiplication table for the elements $1, x, x^2, x^3, x^2 - 2$ in $\mathbb{Q}[x]/((x^2 - 2)^2) = \mathbb{Q}[x]/(x^4 - 4x^2 + 4)$.

Exercise 9.20 (Groups in $\mathbb{F}_4$) The additive group of $\mathbb{F}_{2^2}$ is isomorphic to Klein's four group $V_4 \cong (\mathbb{Z}_2, +) \times (\mathbb{Z}_2, +)$, the multiplicative group $\mathbb{F}_{2^2}^* := \mathbb{F}_{2^2} \setminus \{0\}$ is isomorphic to $(\mathbb{Z}_3, +)$.

Exercise 9.21 (Irreducible Polynomials) Over $\mathbb{F}_{2^2}$ find an irreducible polynomial $g(x)$ of degree 2 (with proof) and calculate addition and multiplication tables for $\mathbb{F}_{2^2}[x]/(g(x))$. Similarly for an irreducible polynomial $g(x)$ of degree 3 over $\mathbb{F}_2$. Investigate analogously $\mathbb{F}_3(\sqrt{2})$.

9.6 Coding

> Message word, coding/decoding function, code(word), Hamming distance, linear/cyclic code, generated by, next neighbor decoding principle, BCH code, Reed Solomon code, Fire code, Burst error

Coding theory explores methods for representing and transmitting data such that errors occurring during representation or transmission can be detected and possibly corrected. "Correcting" means that the recipient can deduce the original data. Consider the following scenario: if all data consists only of combinations of 0 and 1, it is clear that any different character indicates an error. However, we cannot determine whether the correct character was 0 or 1.

Now, if we know that the allowed characters (codewords) are only 000 and 111, we can interpret a received word 001 as 000 and 011 as 111. This correction is possible if we are certain that at most one error occurs. If we know that at most two errors can occur, these errors are still detectable, but we can only recognize that "something is not correct" without being able to correct the errors.

Messages

The set of *characters* S with $|S| = q$ contains the elements that are used for transmission.

The *message* (information) then consists of a set of *message words*, which are m-tuples of elements from S. These form a set M, the so-called blocks.

An *encoding* consists in replacing the m-tuples (message words) with n-tuples (then called codewords) of elements from S with $n > m$.

The additional characters are used to increase the "reliability", they would be superfluous if we could be sure that no errors occur.

Message Words as Vectors

Now we use the tools of vector spaces over finite fields.

Set $S := \mathbb{F}_q$, q a prime power, i.e. the characters are the elements of a field.

Choose $M \subseteq V(n,q) := (\mathbb{F}_q)^n$, i.e. the set M of message words is considered as a subset of $V(n,q)$.

An *encoding function* is an injective mapping $cod : M \to V(n,q)$.

The set $C := cod(M) \subseteq V(n,q)$ is called *code*, an element of C is called *codeword*.

A *decoding function* for cod is a left inverse mapping $dec : V(n,q) \to M$, such that $dec(cod(v)) = v$ for all $v \in M$.

An (n,m)-*code* over $\mathbb{F}_q$ is a subset $cod(V(m,q)) \subseteq V(n,q)$, where $V(m,q)$ is a vector space over $\mathbb{F}_q$ and cod is a coding function.

Hamming Distance

For two codewords $u = (u_1, \ldots, u_n)$ and $v = (v_1, \ldots, v_n)$, i.e. vectors from $(\mathbb{F}_q)^n$, the *Hamming*[5] *distance* is defined as

$$d_H(u,v) := |\{i \in \{1, \ldots n\} \mid u_i \neq v_i\}|.$$

Example 9.22 (Hamming Distance) We take the field $\mathbb{F}_2$. For $m = 1, n = 3$ and $q = 2$ we get $V(1,2) = \{0,1\}$. And $cod(V(1,2)) = \{(0,0,0), (1,1,1)\} \subseteq V(3,2)$ is a $(3,1)$-code with Hamming distance 3 between each pair of codewords (here there are only two codewords).

As we have already noted, $n - m$ positions are not necessary for the message, but only serve to increase security of the transmission. The larger the difference $n - m$ is, the more errors can be detected or corrected. On the other hand, the transmission takes longer (and is therefore more expensive).

Theorem 9.23 (Detecting and correcting errors) *A code can detect up to k errors, exactly if the Hamming distance d between any two codewords is at least $k + 1$. A code can correct k errors, exactly if the Hamming distance between any two codewords is at least $2k + 1$.*

Proof See, for example, Theorem 6.4.1 in Kim and Roush [46]. □

From this it also follows that it is advantageous if all codewords pairwise have the same Hamming distance d.

---

[5] Richard Wesley Hamming, US-American mathematician, 1915–1998.

Linear and Cyclic Codes

A code $C = cod(M)$ is called **linear**, if C is a subspace of $(\mathbb{F}_q)^n$.

A code is called **cyclic**, if every cyclic permutation of the letters of a codeword $u \in C$ again yields a codeword.

Theorem 9.24 (Cyclic Codes and Ideals) *Cyclic codes are exactly the ideals I in the ring $R := \mathbb{F}_q[x]/((x^n - 1))$. It can be considered as a subspace of $(\mathbb{F}_q)^n$ consisting of polynomials of degree $\leq n - 1$.*

Each such ideal I in R is generated by a uniquely determined normalized polynomial of lowest degree $g(x) \in I$, i.e. $I = g(x)R$ with $g(x) \mid (x^n - 1)$.

*If $k = n - m$ is the degree of $g(x)$, we get the (n, m)-**code generated by** $g(x)$.*

Proof Compare, for example, Propositions 6.5.1 and 6.5.2 in Kim and Roush [46].
□

Example 9.25 (1-Error Correcting) For $q = 2, n = 3$ consider the ideal generated by $g(x) = x^2 + x + 1$ in $\mathbb{F}_2[x]/(x^3 + 1)$, cf. Example 9.15. We again obtain the $(3, 1)$-code consisting of the codewords 0 and $x^2 + x + 1$. The coefficients of these two polynomials, interpreted as 3-dimensional vectors, thus again yield the code 000 and 111, which is 1-error correcting.

Coding/Decoding

The question remains how to find possible, simple, beautiful, suitable coding and decoding functions. A simple coding function transforms the message word $f(x) \in M$ into the codeword (polynomial) $f(x)g(x)$, (i.e. $cod(f(x)) = f(x)g(x) \in C = cod(M)$). This can be decoded by dividing by the polynomial $g(x)$, but only if no errors occur.

It is clear that for the above $(3, 1)$-code an error correcting decoding is very simple by the so-called *next neighbor decoding principle*. For this the receiver has to have a complete list of the codewords. For larger codes, however, this method is less suitable.

Known codes with "good" decoding algorithms are the **BCH Codes** by Bose and Chaudhuri,[6] and independently by Hocquenghem,[7] developed in 1959.

---

[6] Raj Chandra Bose, Indian-American mathematician 1901–1987, Dwijendra Kumar Ray-Chaudhuri *1933, Bangladeshi-American mathematician.

[7] Alexis Hocquenghem, French mathematician 1908–1990.

Special BCH Codes are the **Reed Solomon Codes**.[8] These codes are used in CD players. There, $q = 2^8 = 256$ is used, and two Reed Solomon codes with $n = 28, m = 24, d = 5$ and with $n = 32, m = 28, d = 5$ are cleverly linked. Here n denotes the length of the codewords, m the length of the message words and d the Hamming distance between any two codewords.

E. Berlekamp[9] developed a fast, *error correcting algorithm* for BCH codes in 1968. For special cases, there are special decoding algorithms, for example an algorithm for 2-error correcting BCH codes in $\mathbb{F}_{16}$ (cf. for example Barnett [6]).

Finally, the **Fire** Codes[10] should be mentioned, which are particularly suitable for correcting errors such that several consecutive symbols are not correct. These are errors, that do not occur randomly, but are system related (so-called burst errors) cf. for example, Kim and Roush [46].

---

[8] Irving Stoy Reed, US-American mathematician, 1923–2012, Gustave Solomon, US-American mathematician, 1930–1996, another reading assumes one person: Reed Solomon.

[9] Elwyn Ralph Berlekamp, US-American mathematician, 1940–2019.

[10] Developed by Philip Fire in 1959.

Rings and Modules 10

In this chapter, we will explore some special aspects of the theory of rings and modules, mostly without proofs. These concepts were introduced in Chaps. 7 and 8. Rings and modules are significant in the fields of Commutative Algebra and Algebraic Geometry. We will provide some insight into these areas, though a full exploration is beyond the scope of this book.

Our focus will primarily be on commutative rings with unity and without zero divisors, although there are many other types of rings, as we will briefly discuss. Recall that if $x, y \neq 0$ but $xy = 0$, then x and y are called **zero divisors**. Each element $[a]_n \in \mathbb{Z}_n$ is a zero divisor in $\mathbb{Z}_n$ if and only if $\gcd(a, n) \neq 1$. Thus, $\mathbb{Z}_n$ contains zero divisors if n is not a prime number. For example, in $\mathbb{Z}_6$, $2 \cdot 3 = 6 = 0$. The ring $(2\mathbb{Z}, +, \cdot)$ has no identity element, and a non-trivial matrix ring is not commutative.

Literature Aluffi [4] provides a comprehensive and detailed introduction, especially in Chapter V. Hutchins [39] offers many unusual examples of various rings and modules.

10.1 Rings

Unit, associated, indecomposable, (left/right) principal ideal, prime, maximal, reducible, irreducible, Gaussian Integers, Noetherian, factorial, ZPR-ring, principal ideal domain, chain condition, Hilbert's Basis Theorem, Algebraic Geometry, algebraic curve, cissoid, Cartesian leaf, Euclidean domain

To start, we will analyze some of the seemingly self-evident properties of the "primal ring" $(\mathbb{Z}, +, \cdot)$ in more detail. These properties do not follow directly from the abstract definition of a ring, which leads to the creation of numerous new definitions. Here, we will present only the ones we consider to be the most important.

Special Elements and Ideals

We first describe a method for generating ideals in a ring R. The *left/right principal ideal* generated by an element $s \in R$ is Rs/sR, and the *(two-sided) principal ideal* generated by s is $(s) := RsR$. For a subset $A \subseteq R$ and left/right/2-sided ideals $J \subseteq R$ we call

$$(A) := \bigcap_{A \subseteq J} J$$

the *left/right/2-sided ideal generated by* A. It is the smallest left, right, 2-sided ideal in R that contains A.

If R has a unit element 1, then in the 2-sided case

$$(A) = \{r_1 a_1 s_1 + \cdots + r_n a_n s_n \mid r_i, s_i \in R, a_i \in A, n \geq 1\}.$$

Otherwise there may exist $a \in A$ which cannot be generated this way. Analogously for left/right ideals. If R is also commutative, this reduces to

$$(A) = \{r_1 a_1 + \cdots + r_n a_n \mid r_i \in R, a_i \in A, n \geq 1\}.$$

Perhaps the most important property of $(\mathbb{Z}, +, \cdot)$ is the unique decomposability into irreducible numbers or prime numbers (both terms coincide here). This decomposition is associated with the Euclidean algorithm, i.e., the possibility of division with remainder, compare Theorem 3.8.

So, in general, we investigate prime and irreducible elements and ideals separately.

Let R be a ring with 1. An element $e \in R$ is called a ***unit***, if e is invertible, i.e., there exists $e' \in R$ with $ee' = 1$. Two elements $a, b \in R$ are called ***associated***, if they differ only by a unit e, i.e., if $a = be$. Obviously, two elements a, b are associated, if and only if $a|b$ and $b|a$, i.e., a divides b and vice versa. For the definition of $a|b$ compare the definition of $a|b$ in $\mathbb{Z}$ (Subsection "Integers" in Sect. 3.2).

We see: in $\mathbb{Z}$ only 1 and -1 are units, and exactly a and $-a$ are associated for any $a \in \mathbb{Z}$.

Now we look at the prime factor decomposition, or in other words, the decomposition into prime elements. In analogy to the situation in $(\mathbb{Z}, +, \cdot)$: an element $p \in R$ is called ***prime element***, if p is neither 0 nor a unit, and $p|ab$ implies that $p|a$ or $p|b$ for $a, b \in R$.

If $a|p$ for a prime element p, then $a \in R$ is a unit or associated to p.

10.1 Rings

A two-sided ideal $\{0\} \neq I \subsetneq R$ of a ring R is called *prime ideal* or *prime*, if for all ideals $A, B \subseteq R$ we get that $AB \subseteq I$ implies $A \subseteq I$ or $B \subseteq I$.

An element $p \in R \backslash \{0\}$ is a prime element exactly if the principal ideal $(p) := RpR$ is a prime ideal.

We get the transition from elements to ideals, when we replace elements from R by principal ideals of R, "is divided by" by $\subseteq$ and $\cdot$ by $\cap$.

In analogy to the situation in $(\mathbb{Z}, +, \cdot)$ and as already done for polynomial rings, we now define in general: An element $r \in R$ is called *reducible* if r is a unit or $r = ab$, where $a, b \in R$ are not units. Correspondingly, $r \in R$ is called *irreducible* if r is not a unit and from $r = ab$ it follows that $r | a$ or $r | b$.

And parallel for ideals: An ideal I is called *irreducible*, if it is not the intersection of two ideals, which properly contain I.

Prime elements are irreducible if the ring has no zero divisors, the converse is not true in general.

Example 10.1 (Irreducible $\neq$ Prime) In the ring $\mathbb{Z}_6$, which has zero divisors, 2 is prime but reducible, as $2 \cdot 4 = 2$.

In the ring $\mathbb{Z}[\sqrt{-5}]$ the elements $2, 3, 1 + \sqrt{-5}, 1 - \sqrt{-5}$ are irreducible, but not prime. We have $2 \cdot 3 = (1 + \sqrt{-5})(1 - \sqrt{-5}) = 6$. This proves both properties. For the construction $\mathbb{Z}[\sqrt{-5}]$, compare the construction of $\mathbb{Q}[\sqrt{2}]$ in Sect. 8.2.

If every ideal in a ring is principal, this cannot happen.

An ideal $I \subseteq R$ in the ring R is called a *maximal ideal* if for every ideal $I \subsetneq J \subseteq R$, we get $J = R$. So maximal means maximal with respect to inclusion.

Example 10.2 (Prime = Max in $\mathbb{Z}$) Prime ideals $0 \neq P \subseteq \mathbb{Z}$ are also maximal in the ring $\mathbb{Z}$. Conversely, here maximal ideals are also prime.

We see: $2\mathbb{Z} \subseteq \mathbb{Z}$ is a prime and a maximal ideal in $\mathbb{Z}$, $4\mathbb{Z} \subseteq \mathbb{Z}$ is neither a prime ideal nor a maximal ideal in $\mathbb{Z}$.

Example 10.3 (Prime $\neq$ Max in General) The maximal ideal $4\mathbb{Z}$ in the ring $2\mathbb{Z}$ is not prime, since obviously $2\mathbb{Z}2\mathbb{Z} \subseteq 4\mathbb{Z}$, but $2\mathbb{Z} \not\subseteq 4\mathbb{Z}$. This is because $2\mathbb{Z}$ does not have an identity. Conversely: Prime ideals are generally not maximal. In the ring of polynomials $\mathbb{Z}[x, y]$ in two indeterminates over $\mathbb{Z}$ the ideal (x) is a prime ideal, which is not maximal since $(x) \subseteq (\{x, y\}) \subsetneq \mathbb{Z}[x, y]$.

Here we again glance on the ring of Gaussian integers, compare also Example 9.12. It has many peculiarities. So certain prime numbers from $\mathbb{Z}$ are no longer prime in $\mathbb{Z}[i]$. And even more impressing, Fermat's[1] Theorem can be proved in this context.

Lemma 10.4 (Prime in $\mathbb{Z}[i]$) *The element $q = a + bi \in \mathbb{Z}[i]$, with $a, b \neq 0$, is prime if and only if there exists a prime $p \in \mathbb{Z}$ with $a^2 + b^2 = p$ or $a^2 + b^2 = p^2$.*

---
[1] Pierre de Fermat, French mathematician and lawyer, 1607–1665.

For a proof see Aluffi [4], Section Gaussian Integers, Lemma 6.7 in Chapter V.

So, for example, $1+3i$ is not prime, since $1^2+3^2 = 10$. Indeed, $(1+i)(2+i) = 1+3i$. And, $2+3i$ is prime, since $2^2+3^2 = 13$. Also, $1+5i$ is not prime, since $(1+i)(3+2i) = 1+5i$. According to Lemma 6.9 in Chapter V of Aluffi [4], the following elements are no longer prim in $\mathbb{Z}[i]$. Indeed
$2 = 1^2 + 1^2 = (1+i)(1-i)$,
$5 = 1^2 + 2^2 = (1+2i)(1-2i)$,
$13 = 2^2 + 3^2 = (2+3i)(2-3i)$,
$17 = 1^2 + 4^2 = (1+4i)(1-4i)$.

Along the line of Gaussian integers it is possible to prove

Theorem 10.5 (Fermat's Little Theorem) *A positive odd prime $p \in \mathbb{Z}$ is a sum of two squares if and only if $p = 1$ mod 4.*

(Theorem 6.11 in Chapter V of Aluffi [4]).

Special Rings

We start with Table 10.1, which provides an overview of some of the diverse concepts, their definitions, and their "inheritances" on polynomial- and factor rings. Here ID stands for *integral domain*, ZDfree for free of zero divisors and PID for *principal ideal domain*. Basically, we assume commutative rings with 1, although

Table 10.1 Ring classes

$(R, +, \cdot)$	$R[x]$	$R[x, y]$	R/I I prime	R/I I maximal
Commutative with 1	Commutative with 1	Comm. with 1	ID	Field
ID (integral domain) =ZDfree (zero divisor free)	ID	ID	ID	field
Noetherian every ideal finitely generated	Noetherian Hilbert's Basis Theorem	Noetherian	ID	Field
Factorial elements finitely decomposable into irreducible factors	Factorial	Factorial	ID	Field
PID(principal ideal domain) every ideal is principal	Factorial, Noetherian	Factorial, Noetherian	Field	Field
Euclidean	Factorial, Noetherian	Factorial, Noetherian	Field	Field
Field	Euclidean $g =$ degree	Factorial, Noetherian	Inapplicable	Inapplicable

10.1 Rings

most of the definitions given in the table can also be transferred to non-commutative rings and to rings without 1.

In addition to that, there is also the concept of the ***principal ideal ring***. This can have zero divisors. Examples are the rings $\mathbb{Z}_n$, where n is not a prime number.

Similar tables you find in Chaps. 8 and 9, Tables 8.1, and 9.1.

A ***Euclidean domain***, also called ***Euclidean ring***, is an integral domain R with a ***weight function*** - also called valuation function or degree function, $g : R\setminus\{0\} \to \mathbb{N}$, which fulfills the "division with remainder" property: For $a, b \in R, a \neq 0$ and $g(a) \leq g(b)$ there exist $z, r \in R$ with $b = az + r$, where $r = 0$ or $g(r) < g(a)$.

The weight function usually used for $\mathbb{Z}$ is the absolute value $||: \mathbb{Z} \to \mathbb{N}$, compare Example 7.20.

The following implications show the relations of the concepts among each other, as they are also represented in the table. However, the "and" in the theorem is not visible there.

Theorem 10.7 *Field $\implies$ Euclidean $\implies$ PID $\implies$ factorial and Noetherian*

Beyond the statements in the table we get

Theorem 10.8 *If $R[x]$ is a principal ideal domain, then R is a field.*

That the polynomial ring $R[x]$ of a factorial ring R is also factorial goes back to a theorem by Gauss.

We note that for factorial rings the decomposition into irreducible factors is unique only up to order and units. In addition to factorial rings there are also ***unique factorization domains (UFD)***. There every element has a unique decomposition into prime elements. In zero-divisor-free rings, UFD = factorial.

A Noetherian[2] ring was originally characterized by the *"ascending chain condition (ACC) for ideals"*, i.e., every ascending chain of ideals becomes stationary, so it is finite.

The inheritance of this condition on the polynomial ring $R[x]$ is proven with Hilbert's[3] Basis Theorem, compare for example Kunz [53], Chapter I, Theorem 2.3. By induction, the result can also be extended to polynomial rings $R[x_1, \ldots, x_n]$.

This has an important application (in *Algebraic Geometry*): If a subset of a vector space K^n or even the (free) $\mathbb{Z}$-module $\mathbb{Z}^n$ is described by infinitely many polynomial equations, then a finite number of polynomial equations are already sufficient.

Example 10.9 (Algebraic Curves) The solutions of polynomial equations are algebraic curves. The *cardioid* is described by the equation $(x^2 + y^2)^2 - 2ax(x^2 + y^2) - a^2y^2 = 0$. The *Cartesian leaf*, is described by the equation $x^3 + y^3 - 3axy = 0$.

---

[2] Emmy Noether, German mathematician, 1882–1935.

[3] David Hilbert, German mathematician, 1862–1943.

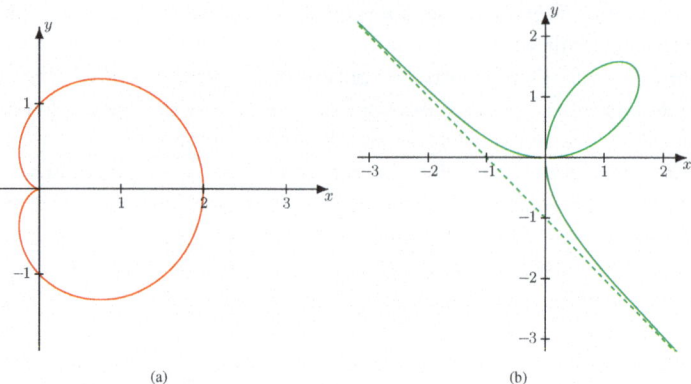

Fig. 10.1 The cardioid (**a**) and the Cartesian leaf (**b**)

Both in $\mathbb{R}[x, y]$. The solutions are subsets of $\mathbb{R}^2$. See Fig. 10.1. If we consider such equations over $\mathbb{Z}$, we get the integer solutions as subsets of $\mathbb{Z}^2$.

We continue with the investigation of the concepts from Table 10.1. The following examples illustrate these concepts and clarify invalid implications.

Example 10.11 The polynomial ring $\mathbb{R}[x_1, \ldots]$ in (countably) infinitely many indeterminates is obviously not Noetherian, but factorial.

The polynomial ring $\mathbb{Z}[x]$ is Noetherian, factorial, but not a principal ideal domain, also compare Example 10.3.

The ring $\mathbb{Z}[\sqrt{-5}]$ is Noetherian, but not factorial (cf. Example 10.1) and thus also not a principal ideal domain. This ring belongs to the not yet mentioned class of rings with not necessarily unique decomposition into irreducible factors.

There are also rings in which a decomposition into irreducible factors does not always exist, but is unique if it exists. An example is the ring of holomorphic functions on a domain in the complex plane $\mathbb{C}$.

Example 10.12 The 5 values $d = -1, -2$ or $-3, -7, -11$ are the only ones for which $\mathbb{Z}[\sqrt{d}]$ or $\mathbb{Z}[(1+\sqrt{d})/2]$ are Euclidean. Since these rings are all contained in $\mathbb{C} = \{a+bi \mid a, b \in \mathbb{R}\}$, the norm a^2+b^2 of the complex number $a+bi$ is a suitable weight function. For the values $d = -19, -43, -67, -163$ is $\mathbb{Z}[(1+\sqrt{d})/2]$ still a principal ideal domain. Compare for example Aluffi [4], p. 302.

The ring $\mathbb{Z}[(1+\sqrt{-19})/2]$ is therefore a non-Euclidean principal ideal domain. The same applies to $\mathbb{Q}[\sqrt{-19}]$.

We summarize the examples and relationships in Fig. 10.2. Note that the examples given there are always inserted in the "smallest" possible class. For example, $\mathbb{Z}[x]$ is not a principal ideal domain, or $K[x_1, x_2, \ldots]$ is not Noetherian.

10.2 Modules Over Rings

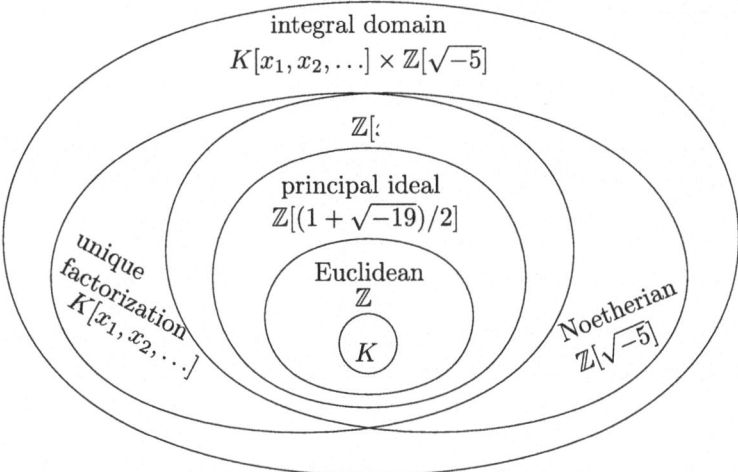

Fig. 10.2 Venn diagram of the considered ring classes

Exercise 10.14 (Proofs) Prove (some of) the statements in Examples 10.1, 10.3, 10.11 and 10.12. This will also prove the statements contained in Fig. 10.2.

10.2 Modules Over Rings

R-module, free module, basis, generating system, linearly independent, direct sum, torsion free, torsion module, homological algebra

Modules over rings can be viewed as analogous to vector spaces over fields, with the distinction that the underlying structure is a ring, not necessarily a field. The definition of modules is provided in Table 9.1, where the condition $1 \bullet x = x$ is typically required for modules over rings with a multiplicative identity 1.

Similar to how $(\mathbb{Z}, +, \cdot)$ is considered the "primal ring," abelian groups serve as the "primal modules," specifically as $\mathbb{Z}$-modules. It is straightforward to verify that all axioms and properties required for modules, as defined in Table 9.1, also apply to abelian groups considered as modules over $\mathbb{Z}$.

Every K-vector space is naturally a K-module for a field K, providing us with numerous examples of modules. Additionally, every ring R is inherently a left R-module and a right R-module. Similarly, every right or left ideal I of a ring R is a module over R. However, these examples do not require new terminology or theory.

The distinction between a module and a vector space lies in the nature of the ring versus the field. Specifically, it hinges on the absence of multiplicative inverses in rings compared to fields, where every nonzero element has a multiplicative inverse.

Bases of Modules

Every $\mathbb{R}$-vector space is also a $\mathbb{Z}$-module, as $(\mathbb{Z}, +, \cdot) \subseteq (\mathbb{R}, +, \cdot)$. Recall concepts and definitions of Sect. 9.3.

So consider the $\mathbb{R}$-vector space $\mathbb{R}^2$ as a $\mathbb{Z}$-module. Then it is obvious that $(\frac{1}{2}, \frac{1}{3}) \notin LH_{\mathbb{Z}}\{(1, 0), (0, 1)\}$, but $(\frac{1}{2}, \frac{1}{3}) \in LH_{\mathbb{R}}\{(1, 0), (0, 1)\}$. This means that $\{(1, 0), (0, 1)\}$ is not a generating system of the $\mathbb{Z}$-module $\mathbb{R}^2$, although it is even a basis (the standard basis) of the $\mathbb{R}$-vector space $\mathbb{R}^2$ as we know.

We now work our way towards a concept of basis for modules that includes that for vector spaces, but also shows the differences.

Let R be a ring with 1. A finite (non-empty) subset $\{m_1, \ldots, m_t\}$ of an R-module M is called *linearly dependent* if there exist elements $r_i \in R$ that are not all 0, such that $\sum_{i=1}^{t} r_i m_i = 0$. Otherwise, the subset is called *linearly independent*. In this case, $\sum_{i=1}^{t} r_i m_i = 0$ implies that $r_1 = \cdots = r_t = 0$.

An infinite set $X \subseteq M$ is called linearly independent if every finite subset of X is linearly independent. It is therefore convenient to define the empty set as linearly independent. In particular, every module then has a linearly independent subset, a fact we already used in vector spaces.

A subset $M' \subseteq M$ is called a *generating system* of M if every $m \in M$ can be written as a linear combination of finitely many elements $m_1, \ldots, m_t \in M'$.

We now call a linearly independent generating system M' of M a *basis* of M.

So far one cannot see a difference to vector spaces. The definition of a free module insinuates that there may exist modules which don't have a basis. This then is one of the differences to vector spaces.

A module M is called *free*, if it has a basis.

One can show that the free R-modules are of the form $\coprod_{i \in I} R$, also written as $\oplus_{i \in I} R$ or $R^{|I|}$. These are the $|I|$-fold direct sums of the R-module R with itself, compare Examples 12.12 (4) and 12.30 for vector spaces. As with infinite direct sums of vector spaces, the elements here can also be represented as columns of length $|I|$ with entries from R, of which only finitely many are $\neq 0$.

Example 10.15 The polynomial ring $R[x]$ is a free R-module. A basis is $\{x^0, x^1, x^2, \ldots\}$.

The product of free modules is generally not free. In $\mathbb{Z}^{\mathbb{N}}$, for example, $(1, 1, \ldots)$ cannot be represented as a linear combination of finitely many basis elements.

Another difference even between free modules and vector spaces arises in the next theorem. There free modules over non-commutative rings, in particular skew fields, can have bases of different lengths.

Theorem 10.16 *If R is a commutative ring, then $R^n \cong R^m$ implies that $n = m$, otherwise not.*

10.2 Modules Over Rings

Proof See Jantzen and Schwermer [42], page 165. There you will also find a counterexample for a non-commutative ring. □

Example 10.17 (Modules Are Different) In vector spaces, every generating system contains a basis. This is not generally the case for free modules: The free $\mathbb{Z}$-module $_\mathbb{Z}\mathbb{Z}$ is generated by $\{2, 3\}$, since the submodule generated by $\{2, 3\}$ contains $3 - 2 = 1$. But since $3 \cdot 2 - 2 \cdot 3 = 0$, the set $\{2, 3\}$ is linearly dependent over $\mathbb{Z}$. So it is not a basis and of course does not contain a basis, since neither ∅ nor any of the two elements of $\{2, 3\}$ generate $_\mathbb{Z}\mathbb{Z}$.

The same example also shows another peculiarity of free modules. If the vectors $v_1, \ldots, v_t$ are linearly dependent, not all 0, then each of the non-zero vectors can be represented as a linear combination of the others. This is of course due to the existence of multiplicative inverses in the field. For the set $\{2, 3\}$, which is linearly dependent over $\mathbb{Z}$, this is obviously not the case.

Finally, for principal ideal domains, an analogue to the Basis completion lemma (cf. Lemma 9.7) can be proven. It is shown that certain bases of a submodule $N \subseteq F$ can be derived from bases of the free module F. The proof basically takes up the entire Chapter 7 in Hartley and Hawkes [33].

Theorem 10.18 (Basis Completion Lemma, Analogue) *Let R be a principal ideal domain, F a free R-module with a finite basis of s elements, and let N be a submodule of F. Then there exists a basis $\{b_1, \ldots, b_s\}$ of F and elements $d_1, \ldots, d_s \in R$, such that the non-zero elements of $\{d_1 b_1, \ldots, d_s b_s\}$ form a basis of N and additionally: $d_1 | d_2 | \ldots | d_s$.*

Non-free Modules

From the discussion above, it is evident that very few modules have a basis and are therefore considered free. We now turn our attention to torsion modules, which are fundamentally different from free modules.

In an R-module M, an element $m \in M$ is called a **torsion element** if there exists a nonzero element $r \in R$ such that $rm = 0$. A module consisting entirely of torsion elements is called a **torsion module**.

Conversely, a module is labeled **torsion free** if it contains no nonzero torsion elements. The following theorem is Theorem 6.5 in Hartley and Hawkes [33], there with proof.

Theorem 10.19 (Torsion) *Let M be a module over an integral domain R. Denote the set of all torsion elements of M by T. Then T is a submodule of M and the factor module M/T is torsion free.*

Examples 10.20 (Torsion)

(a) The $\mathbb{Z}$-module $(\mathbb{Z}_4, +)$ is a torsion module. All elements $z \in \mathbb{Z}_4$ fulfill $4z = 0$.
(b) For every module, the submodule of its torsion elements is a torsion module.
(c) The $\mathbb{Z}$-module $\mathbb{Q}$ is torsion free, but not free.

Remark 10.21 (Homological Algebra) *A typical theorem of so-called Homological Algebra is the following:*
  All R-modules are free, if and only if R is a field.
Exercise 10.23 is an example. Another question of this type is:
  Over which rings are all torsion free modules free?
There are further homological specifications of module classes with the indicated implications. For example

$$\text{free} \implies \text{projective} \implies \text{flat} \implies \text{torsionfree}.$$

And over principal ideal rings

$$\text{free} \iff \text{projective} \text{ and } \text{flat} \iff \text{torsionfree}.$$

Definitions and corresponding homological classifications can be found, for example, in Anderson and Fuller [5] or in Faith [28].

The same concepts can also be formulated for acts over monoids. There, too, the corresponding questions of homological classification arise, compare for example the book "Monoids, Acts and Categories" by Kilp, Knauer, Mikhalev, [45].

Exercise 10.22 (Example 10.20) Prove the statements in Example 10.20.

Exercise 10.23 (Homological Classification) Show: All R-modules are free, if and only if R is a field.

Exercise 10.24 (Integral Domain Is Required) Form T and M/T for the $\mathbb{Z}_{12}$-module $(\mathbb{Z}_{12}, +)$ and show that M/T is not torsion-free.

Exercise 10.25 (Modules) Show that $(\mathbb{Z}_4, +)$ is not a $\mathbb{Z}_2$-module.

Matroids 11

Matroids are relatively recent objects in combinatorics, abstracting classes of structures that initially appear quite disparate. They provide a unifying framework that connects graphs, sets of vectors, and concepts of linear and algebraic independence.

Due to the variety of examples, matroids can be defined in multiple equivalent ways, each emphasizing different aspects. We explore characterizations such as independence systems, bases, rank functions, and circuits, which initially appear distinct but are interconnected.

Beyond their role as a generalization of fundamental structures in discrete mathematics, matroids have significant importance in computer science. They are characterized as set systems where greedy algorithms yield optimal solutions. Efficient algorithms can also find optimal solutions for intersections of two matroids, which is crucial in applications such as matching problems in bipartite graphs.

This chapter provides a glimpse into this rich theory. Therefore, proofs, which can be found in existing literature, are generally omitted.

Literature A seminal work on matroid theory is Oxley's book [61], which serves as a primary reference for notation and insights in this chapter. Other recommended texts include those by Aigner [2] and Welsh [73], which delve deeper into the subject from slightly earlier perspectives.

11.1 From Vectors, Graphs, and Fields to Matroids

Independent sets, vector matroids, bases, uniform matroids, partition matroids, transversal matroids, graphic matroids, algebraic matroids

In the mid-1930s, Whitney[1] and independently Nakasawa[2] introduced matroids. They can be described as a family of subsets of a finite ground set. A **matroid** $M = (E, \mathcal{U})$ is a pair consisting of a finite set E, called **ground set**, and a set $\mathcal{U} \subseteq \wp(E)$ of subsets, called **independent sets**, that satisfies the following axioms:

(I0) $\emptyset \in \mathcal{U}$ (not to be confused with $\emptyset \subseteq \mathcal{U}$, which would be trivial),
(I1) for all $X, Y \subseteq E$ with $Y \subseteq X$: $X \in \mathcal{U}$ implies $Y \in \mathcal{U}$ (that is, subsets of independent sets are independent),
(I2) for all $X, Y \in \mathcal{U}$ with $|X| < |Y|$ there exists $e \in Y \setminus X$, such that $X \cup \{e\} \in \mathcal{U}$ (in other words: Each independent set can be extended with elements from larger independent sets such that it remains independent).

Even if this definition seems to fall from the sky at first, it makes sense upon closer inspection, namely when replacing "independent" with "linearly independent" (see Sect. 9.3). More precisely: If E is a set of vectors of a vector space, then the empty set is linearly independent, (one can say anything about the empty set, just not that it contains an element). Every subset of a linearly independent set is linearly independent. Furthermore, with the Basis Extension Lemma (Lemma 9.7), every linearly independent set in E can be extended with elements from larger linearly independent sets to a larger linearly independent set.

We obtain the following class of matroids:

Theorem 11.1 (Vector Matroid) *Let K be a field, V a K-vector space, and $E \subseteq V$ a finite subset. Define $\mathcal{U}$ as the set of subsets of E that are linearly independent over K in V. The pair $M = (E, \mathcal{U})$ is a matroid, called a **vector matroid** or K-**vector matroid**.*

As an example, we take the $\mathbb{R}$-vector space $V = \mathbb{R}^5$, as the ground set the columns $E = \{e_1, \ldots e_{10}\}$ of the following matrix, and as independent sets the subsets of E that correspond to linearly independent column sets.

$$\begin{pmatrix} 1 & 0 & 0 & 0 & 0 & -1 & 1 & 0 & 0 & 1 \\ 0 & 1 & 0 & 0 & 0 & 1 & -1 & 1 & 0 & 0 \\ 0 & 0 & 1 & 0 & 0 & 0 & 1 & -1 & 1 & 0 \\ 0 & 0 & 0 & 1 & 0 & 0 & 0 & 1 & -1 & 1 \\ 0 & 0 & 0 & 0 & 1 & 1 & 0 & 0 & 1 & -1 \end{pmatrix}$$

The resulting matroid is called R_{10}. By definition, it is an $\mathbb{R}$-vector matroid.

Since finite subsets of a vector space can always be viewed as matrices, the name *matroid* is explained here as a generalization of *matrix*. However, different matrices

---

[1] Hassler Whitney, US-American mathematician, 1907–1989.
[2] Takeo Nakasawa, Japanese mathematician, 1913–1946.

11.1 From Vectors, Graphs, and Fields to Matroids

can represent the same matroid. For example, if we multiply any set of columns and rows in the above matrix by different scalars from $\mathbb{R} \setminus \{0\}$, the sets of independent columns remain the same. Even if we permute columns or rows, the independent sets essentially remain the same. To better formalize the latter, it is now time to define isomorphisms for matroids.

Let $M_1 = (E_1, \mathcal{U}_1)$ and $M_2 = (E_2, \mathcal{U}_2)$ be matroids. We say that M_1 and M_2 are *isomorphic*, written $M_1 \cong M_2$, if there is a bijective mapping $\phi : E_1 \to E_2$ such that $X \in \mathcal{U}_1 \Leftrightarrow \phi(X) \in \mathcal{U}_2$. Such a mapping ϕ is called a **matroid isomorphism**. It is clear that matroid isomorphisms arise from swapping and scaling rows and columns. We will find below another matroid isomorphism.

An $\mathbb{R}$-vector matroid is called *real* and an $\mathbb{F}_2$-vector matroid is called **binary** If a matroid is a K-vector matroid for every field K, it is called **regular**. Surprisingly, the following theorem can be proven:

Theorem 11.2 (Bland and Las Vergnas 1978 [9]) *A matroid is regular if and only if it is binary and real.*

It can be shown that R_{10} is binary, since we can represent it by the following matrix over $\mathbb{F}_2$. Hence R_{10} is also regular by Theorem 11.2.

$$\begin{pmatrix} 1 & 0 & 0 & 0 & 0 & 1 & 1 & 0 & 0 & 1 \\ 0 & 1 & 0 & 0 & 0 & 1 & 1 & 1 & 0 & 0 \\ 0 & 0 & 1 & 0 & 0 & 0 & 1 & 1 & 1 & 0 \\ 0 & 0 & 0 & 1 & 0 & 0 & 0 & 1 & 1 & 1 \\ 0 & 0 & 0 & 0 & 1 & 1 & 0 & 0 & 1 & 1 \end{pmatrix}$$

If we call the columns of this matrix $E = \{e_1, \ldots, e_{10}\}$, then the subsets of E that correspond to independent column sets coincide with those of the previous matrix. This means that both matrices represent isomorphic matroids.

We have therefore found a vector matroid isomorphism that arises not only from swapping and scaling rows and columns.

We now turn to another way of representing matroids.

We consider Fig. 11.1. The **Fano**[3] **matroid** depicted in Fig. 11.1a is to be understood as follows. Its ground set consists of the 7 vertices of the diagram, and the independent subsets are exactly the sets with at most 3 elements that lie neither on one of the straight lines nor on the circuit.

The **non-Pappus**[4] **matroid** depicted in Fig. 11.1b has 9 vertices as its ground set. Its independent sets are all subsets with at most 3 elements that do not form any of the depicted lines.

---

[3] Gino Fano, Italian mathematician, 1871–1952.
[4] Pappus of Alexandria, Greek mathematician, forth century AD.

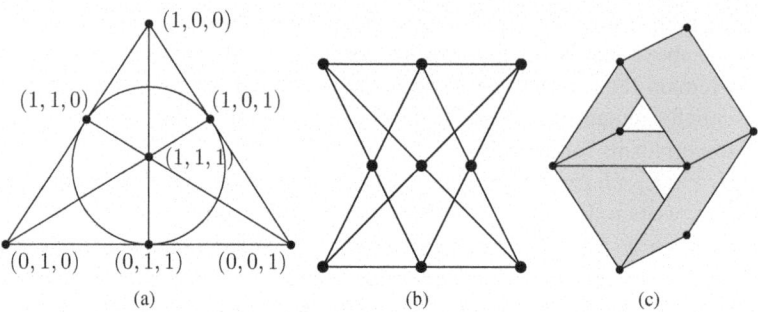

Fig. 11.1 (a) The Fano matroid. The vectors correspond to a representation as a binary matroid. (b) The non-Pappus matroid. (c) The Vámos matroid

Figure 11.1c represents the **Vámos[5] matroid**, whose ground set consists of the 8 vertices of the depicted cube, while the independent sets are exactly the subsets with at most 4 elements except the 5 gray sets.

Another example of such a diagram can be found in Fig. 11.5 at the end of this chapter. For a systematic introduction to the representation of matroids using this type of diagrams, we refer to Oxley's book [61].

The three examples in Fig. 11.1 are of different significance. In particular, it can be shown that the Fano matroid is binary (see Exercise 11.15), but not real (see [61] Proposition 6.4.8). Theorem 11.2 thus implies that the Fano matroid is not regular. We will discuss the non-Pappus and the Vámos matroid later.

After discussing vector matroids, the following question arises: If one can generalize linear independence so easily, what about bases? No problem! Given a matroid $M = (E, \mathcal{U})$, we define its **bases** $\mathcal{B}$ as the inclusion maximal elements of $\mathcal{U}$, that is, we choose exactly the independent elements that are not contained in any other independent set. We know that in the case of vector matroids this is correct, because the bases of a vector space are exactly the inclusion maximal linearly independent sets, see Theorem 9.8. More generally, one can now say when a set of sets is the set of bases of a matroid:

Theorem 11.4 (Bases) *A set $\mathcal{B} \subseteq \wp(E)$ is the set of bases of a matroid $M = (E, \mathcal{U})$ if and only if $\mathcal{B}$ satisfies the following axioms:*

(B0) $\mathcal{B} \neq \emptyset$,
(B1) *for $A, B \in \mathcal{B}$ and each $e \in A \setminus B$ there is an $e' \in B \setminus A$ such that $(A \setminus \{e\}) \cup \{e'\}$ is a basis.* *(basis exchange)*

We have just described how to obtain the bases from the independent sets. Of course, this operation can also be reversed. If the bases $\mathcal{B}$ of a matroid are given, we define

---
[5] Peter Vámos, English-Hungarian mathematician, 1940–2020.

11.1 From Vectors, Graphs, and Fields to Matroids

the independent sets $\mathcal{U}$ as the set of all subsets of bases from $\mathcal{B}$. In other words, there is a bijection between matroids $M = (E, \mathcal{U})$ and pairs $(E, \mathcal{B})$, consisting of sets of bases $\mathcal{B}$ on a ground set E. This leads to the fact that matroids can also be defined as pairs $(E, \mathcal{B})$ with $\mathcal{B} \subseteq \wp(E)$ that satisfy (B0), (B1).

Such an equivalence of different axiomatizations is called a *cryptomorphism*—a term first used by Birkhoff. This is to say that two axiomatizations are "isomorphic" in a non-obvious or even cryptic way. A formal definition follows in the next section, where we will also discuss more examples of cryptomorphisms.

Now we present some more classes of matroids.

Definition 11.5 (Uniform Matroids) Given a set E with $|E| = n$ and $r \in \mathbb{N}$ with $r \leq n$, set $\mathcal{U} := \{X \in E \mid |X| \leq r\}$. The pair $U_{r,n} := (E, \mathcal{U})$ is a matroid. It is called the **uniform matroid** of rank r on n elements.

The bases of $U_{r,n}$ are thus exactly given by $\mathcal{B} := \{X \in E \mid |X| = r\}$. Even though uniform matroids may seem very simple, they are an important element of the theory. They help to determine whether a given matroid is an K-vector matroid. It can be shown that $U_{2,n}$ is an K-vector matroid if and only if $|K| \geq n - 1$, see for example Oxley [61], Proposition 6.5.2. A special case of this statement is shown in Exercise 11.15, namely that even the small $U_{2,4}$ is not binary. On the other hand, every uniform matroid is a $\mathbb{R}$-vector matroid.

The next class is even a bit more general (cf. Fig. 11.4) and is based on partitions of a finite set. See also the corresponding part in Sect. 2.2.

Definition 11.6 (Partition Matroids) Take a finite set E with a partition $\mathcal{P} = \{A_i \subseteq E \mid i \in I\}$ and natural numbers $r_i \leq |A_i|$ for $i \in I$. Define $\mathcal{U} := \{X \in E \mid |X \cap A_i| \leq r_i\}$. The pair $M_\mathcal{P} := (E, \mathcal{U})$ is a matroid, which is called the **partition matroid** of $\mathcal{P}$.

A set $X \subseteq E$ is independent in $M_\mathcal{P}$ if and only if X has at most r_i elements in each partition class A_i. Of course, the uniform matroid $U_{r,n} =: M_P$ is a partition matroid, where $\mathcal{P} = \{A_1\}$ with $A_1 = E$ and $r_1 = r$.

Let us generalize further. If $r_i = 1$ for all i in a partition matroid, then the bases correspond exactly to the representative systems of the corresponding equivalence relation and the independent sets to their subsets. But if we now assume to have only a compatibility relation, we get a cover of the ground set instead of a partition, see Exercise 4.25. A partial representative system of a set of not necessarily disjoint sets is called a transversal. We even define transversals for families of sets, i.e., a set can occur several times. Let E be a finite set and $\mathcal{A} = (A_i \mid i \in I)$ a family of neither necessarily disjoint nor necessarily pairwise different subsets of E. A subset $T \subseteq E$ is called a **transversal** of $\mathcal{A}$ if there is an injective mapping $\phi : T \to I$ such that for all $e \in T$ one has that $e \in A_{\phi(e)}$.

Example 11.7 (Transversal) Take $E = \{b_1, \ldots, b_5\}$ and

$\mathcal{A} = (\{b_1, b_2\}, \{b_2, b_3\}, \{b_2, b_3\}, \{b_4, b_5\})$. Then $\{b_1, b_2, b_3, b_4\}$ and $\{b_1, b_2, b_3, b_5\}$ and all subsets of these sets are exactly the transversals of $\mathcal{A}$.

We denote the set of all transversals of a family $\mathcal{A}$ of sets by $\mathcal{T}_\mathcal{A}$. The following can be shown just using the definitions:

Theorem 11.8 (Transversal Matroids) *Let E be a finite set and $\mathcal{A} = (A_i \mid i \in I)$ a family of subsets of E. The set $\mathcal{T}_\mathcal{A}$ is the set of independent sets of a matroid with ground set E, called **transversal matroid**.*

Note that this justifies the name matroid in Definitions 11.5 and 11.6. In particular, the uniform matroid $U_{r,n}$ is a transversal matroid if the family $\mathcal{A}$ consists of r copies of the ground set E.

A useful way to imagine transversal matroids is by means of a bipartite graph G with vertex set $E \cup \mathcal{A}$, which is also the bipartition. An edge $\{e, A_i\}$ exists exactly if $e \in A_i$. The transversals are now the $X \subseteq E$, which are endpoints of a matching. In this way, every bipartite graph with a distinguished bipartition set provides a transversal matroid. In Fig. 11.2 there is such a representation of the transversals from Example 11.7.

The set of transversals is a completely different structure than the set of matchings in G. The latter generally does not form a matroid, see Exercise 11.27 and Fig. 11.2.

Matroids can also be defined based on the forests of a graph (see Sect. 6.4). The following is not hard if you are used to manipulate graphs:

Theorem 11.9 (Graphic Matroids) *Let $G = (V, E)$ be an undirected graph with vertex set V and edge set E—possibly with multiple edges, loops, and not necessarily connected. The edge sets of the forests of G are the independent sets of a matroid with ground set E. This is denoted by M_G and called **graphic matroid***

An example of a graphic matroid can be found in Fig. 11.2. The graph depicted there is bipartite and thus leads to two transversal and one graphic matroid. More precisely, we depict a bipartite graph $G = (A \cup B, E)$ with $A = \{a_1, \ldots, a_4\}$, $B = \{b_1, \ldots, b_5\}$ and $E = \{1, \ldots, 8\}$. Further, $\mathcal{T}_A$ is the basis of the transversal matroid

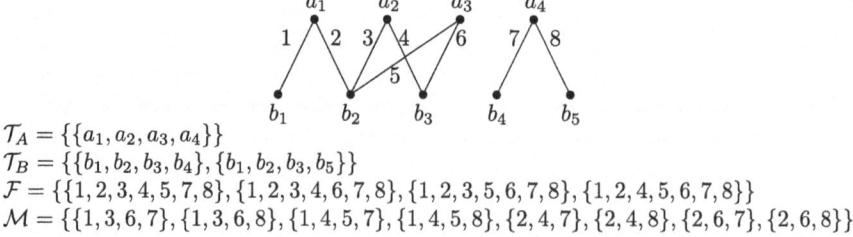

$\mathcal{T}_A = \{\{a_1, a_2, a_3, a_4\}\}$
$\mathcal{T}_B = \{\{b_1, b_2, b_3, b_4\}, \{b_1, b_2, b_3, b_5\}\}$
$\mathcal{F} = \{\{1, 2, 3, 4, 5, 7, 8\}, \{1, 2, 3, 4, 6, 7, 8\}, \{1, 2, 3, 5, 6, 7, 8\}, \{1, 2, 4, 5, 6, 7, 8\}\}$
$\mathcal{M} = \{\{1, 3, 6, 7\}, \{1, 3, 6, 8\}, \{1, 4, 5, 7\}, \{1, 4, 5, 8\}, \{2, 4, 7\}, \{2, 4, 8\}, \{2, 6, 7\}, \{2, 6, 8\}\}$

Fig. 11.2 A transversal matroid

11.1 From Vectors, Graphs, and Fields to Matroids

with distinguished bipartition set A and $\mathcal{T}_B$ are the bases of the transversal matroid with distinguished bipartition set B. It is the matroid from Example 11.7. The set $\mathcal{F}$ contains the bases of the graphic matroid M_G and $\mathcal{M}$ the maximal matchings in G.

Similar to the case of vector matroids, non-isomorphic graphs can represent isomorphic graphic matroids. For example, let T and T' be trees with n nodes. Then $M_T \cong M_{T'} \cong U_{n-1,n-1}$, since in both trees all sets are independent. See also Exercise 11.16. Even though, as we will see later, graphic matroids are very specific in the class of all matroids, all matroids with at most 3 elements are graphic. We have selected a representing graph for each of these matroids in Fig. 11.3. Note that there are of course more than 15 non-isomorphic graphs with at most 3 edges.

It can be shown that graphic matroids are binary and real by considering the linear independencies of the columns of the incidence matrix of G, see Chap. 6, compare Theorem 6.15. From Theorem 11.2 it follows that graphic matroids are regular. On the other hand, it can be shown that R_{10} is not graphic.

To introduce the last and most general example class, we return to algebra and consider algebraic independence, cf. Sect. 9.5.

Theorem 11.12 (Algebraic Matroids) *Let K' be an extension field of the field K and E a finite subset of K'. Let $\mathcal{U}$ be the family of subsets of E that are algebraically independent over K. Then $(E, \mathcal{U})$ is a matroid, which is called **algebraic**.*

It can be shown that all previous classes in this section are algebraic matroids.

There are matroids that are not vector matroids, but are algebraic. An example of this is the non-Pappus matroid in Fig. 11.1b. We cannot provide the proof for this here, but see Oxley [61], Proposition 6.1.11.

Since algebraic independence is a very general concept, one might think that perhaps all matroids are algebraic. However, this is not the case. The smallest example of a non-algebraic matroid is the Vámos matroid from Fig. 11.1c. This proof would go beyond the scope here, compare the article by Ingleton and Main [40].

See Fig. 11.4 for an overview of some matroid classes.

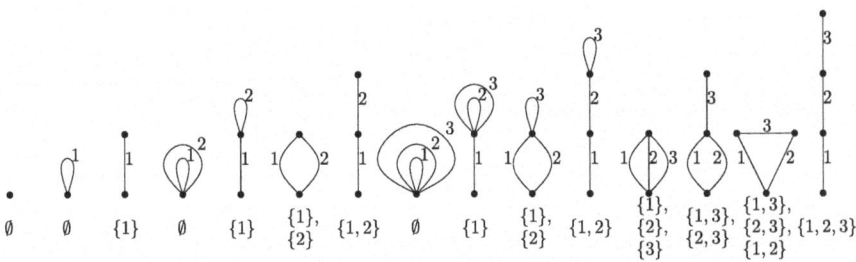

Fig. 11.3 Graphs representing all matroids with at most 3 elements. Below each graph is the list of bases of the corresponding matroid

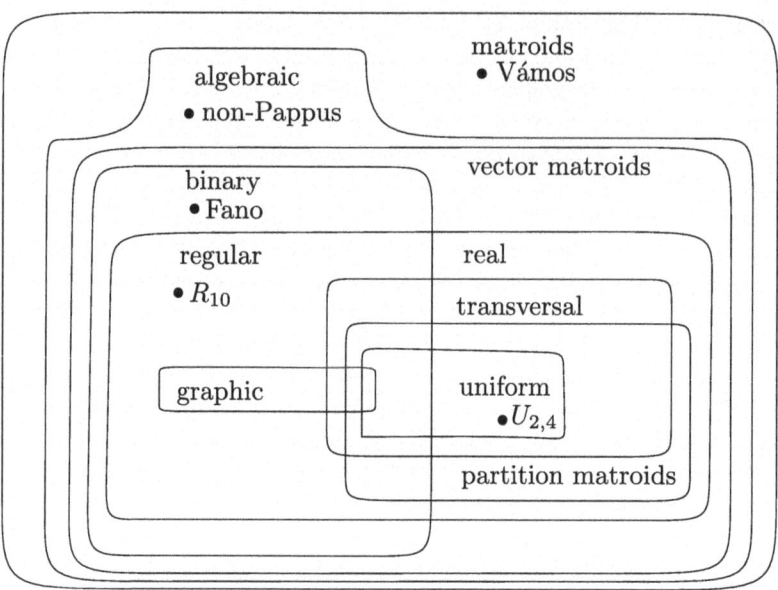

Fig. 11.4 Venn diagram of matroid classes

We have formally defined matroids, considered examples and example classes of matroids. However, none of these classes is large enough to exhaust the class of all matroids. The question remains:

What does it mean to be a matroid?

See also Oxley's survey article *What is a Matroid?* [60]. In the following section, we will discuss further ways to describe matroids in order to get closer to an answer to this question.

Exercise 11.14 (Bases)

(a) Show that for bases B_1, B_2 of a matroid we get
    $B_1 \subseteq B_2 \implies B_1 = B_2$ (Hint: Use (B1).)
(b) Show that all bases of a matroid have the same cardinality. (Hint: Use (U2).)

Exercise 11.15 (Binary Matroids)

(a) Show that the Fano matroid is represented by the vectors given in Fig. 11.1a over $\mathbb{F}_2$. Therefore it is binary.
(b) Show that there is no matrix with 4 columns and entries from $\mathbb{F}_2$ whose linearly independent column sets are exactly the at most 2-element column sets. This means that $U_{2,4}$ is not binary.

Exercise 11.16 (Graphical Matroids) Show that the graphical uniform matroids are exactly $U_{0,n}$, $U_{1,n}$, $U_{n-1,n}$ and $U_{n,n}$. Describe the set of non-isomorphic graphs that represent the respective uniform matroid.

11.2 Further Axiomatizations

Cryptomorphism, circuits, rank function, Greedy algorithms

In the previous section, we have already seen two different ways to axiomatize matroids: independent sets and bases. However, we have contented ourselves there with saying that both systems describe the same thing in a certain sense. But how can we formalize that two sets of axioms, which make statements about different objects, essentially describe the same class? Let $(P1), \ldots, (Pk)$ and $(Q1), \ldots, (Q\ell)$ be two systems of axioms that refer to set systems $\mathcal{P}$ and $\mathcal{Q}$ on a ground set E.[6] With $\mathbf{P}(E)$ and $\mathbf{Q}(E)$ we denote the set of set systems with ground set E, which fulfill the axioms $(P1), \ldots, (Pk)$ and $(Q1), \ldots, (Q\ell)$, respectively. The axioms $(P1), \ldots, (Pk)$ and $(Q1), \ldots, (Q\ell)$ are **cryptomorphic**, if there are mappings $\alpha : \mathbf{P}(E) \to \mathbf{Q}(E)$ and $\beta : \mathbf{Q}(E) \to \mathbf{P}(E)$ such that $\alpha \circ \beta$ and $\beta \circ \alpha$ are the identity. In this case, α and β are called **cryptomorphisms**. As a first example, we return to independent sets and bases. Let $\mathbf{U}(E)$ and $\mathbf{B}(E)$ be the sets of set systems with ground set E that satisfy (U0), (U1), (U2) and (B0), (B1) respectively. The cryptomorphism α from $\mathbf{U}(E)$ to $\mathbf{B}(E)$ maps each set system $\mathcal{U} \in \mathbf{U}(E)$ to the set of its maximal subsets. The cryptomorphism β from $\mathbf{B}(E)$ to $\mathbf{U}(E)$ maps each set system $\mathcal{B} \in \mathbf{B}(E)$ to the set of all subsets of elements in $\mathcal{B}$.

In this section, we will describe three axiom systems that are cryptomorphic to (U0), (U1), (U2): circuits, rank functions, and systems on which greedy algorithms are optimal.

Circuits

A *circuit* $M = (E, \mathcal{U})$ is an inclusion minimal **dependent** subset $C \subseteq E$. That is, C is not independent, but every proper subset of C is independent. The set $\mathcal{C}$ of circuits of $U_{r,n}$, for example, are exactly the $(r+1)$-element sets. The word circuit comes from the fact that the circuits of a graphic matroid M_G are exactly the sets of edges of the cycles, of G, see Sect. 6.4.

Theorem 11.17 (Circuits) *A set $\mathcal{C} \subseteq \wp(E)$ is the set of circuits of a matroid M with ground set E if and only if $\mathcal{C}$ satisfies the following axioms:*

---

[6] Instead of set systems, one can also consider functions from $\wp(E)$ to $\wp(E)$ or $\mathbb{N}$. For simplicity, however, we will restrict ourselves to set systems.

(C0) $\emptyset \neq \mathcal{C}$,
(C1) *for* $C_1, C_2 \in \mathcal{C}$, $C_1 \subseteq C_2 \implies C_1 = C_2$,
(C2) *for all* $C_1, C_2 \in \mathcal{C}$ *with* $C_1 \neq C_2$ *and* $e \in C_1 \cap C_2$ *there is* $C_3 \in \mathcal{C}$ *such that* $C_3 \subseteq C_1 \cup C_2 \setminus \{e\}$.

The cryptomorphisms between independent sets and circuits can be described as follows: A set $\mathcal{U}$ of independent sets is mapped to the minimal subsets of $\wp(E) \setminus \mathcal{U}$. A set $\mathcal{C}$ of circuits is mapped to the set of all proper subsets of elements of $\mathcal{C}$.

Rank Function

For a matroid $M = (E, \mathcal{U})$, we define its **rank function** rg : $\wp(E) \to \{0, \ldots, |E|\} \subseteq \mathbb{N}$ as $\mathrm{rg}(X) := \max\{|Y| \mid Y \subseteq X \text{ and } Y \in \mathcal{U}\}$. The **rank** of M is the number $\mathrm{rg}(E)$.

The Vámos matroid, for example, has rank 4. As another example, one can convince oneself that the uniform matroid of rank r on n elements $U_{r,n}$ indeed has rank r. In a vector matroid M with respect to a set E of vectors of a vector space, the rank of a subset $X \subseteq E$ can be defined as $\mathrm{rg}(X) := \dim(\mathrm{span}(X))$.

The rank of a graphic matroid is the number of edges of a spanning forest in a graph representing it. It can be shown that this is exactly the number of vertices minus the number of connected components of the graph. The rank function is not just a function to measure properties of a matroid. The class of matroids can also be characterized as the class of their rank functions. This means that each matroid is mapped to a rank function as described above, and thus the class of matroids is mapped to a class of rank functions. This mapping is a cryptomorphism and we can specify an axiomatization of its image:

Theorem 11.18 (Rank Function) *A function* rg : $\wp(E) \to \{0, \ldots, |E|\}$ *is the rank function of a matroid M with ground set E if and only if* rg *satisfies the following axioms:*

(R1) *For all* $X \subseteq E$, $\mathrm{rg}(X) \leq |X|$,
(R2) *if* $X \subseteq Y \subseteq E$, *then* $\mathrm{rg}(X) \leq \mathrm{rg}(Y)$,
(R3) *for all* $X, Y \subseteq E$, $\mathrm{rg}(X \cup Y) + \mathrm{rg}(X \cap Y) \leq \mathrm{rg}(X) + \mathrm{rg}(Y)$.

The cryptomorphism from (R1), (R2), (R3) to (I0), (I1), (I2) is as follows. Let rg be a function that meets the above axioms. The independent sets of the associated matroid are given by $\mathcal{U}_{\mathrm{rg}} := \{U \subseteq E \mid \mathrm{rg}(U) = |U|\}$.

Greedy Algorithms

An algorithmic characterization of matroids develops in the context of *combinatorial optimization*.

11.2 Further Axiomatizations

Let $E = \{e_1, \ldots, e_n\}$ be a finite ground set and $g : E \to \mathbb{R}_{\geq 0}$ a mapping, called *weight function*. We define the weight for subsets $X \subseteq E$ as $g(X) := \sum_{e \in X} g(e)$. If additionally $\mathcal{U}' \subseteq \wp(E)$ is given, then the triple $(E, \mathcal{U}', g)$ is called a ***combinatorial optimization problem***. A *solution* of $(E, \mathcal{U}', g)$ is an element $X_{\text{opt}} \in \mathcal{U}'$ with maximum weight.

The following algorithm describes a strategy to search for X_{opt}, and is called ***Greedy Algorithm***:

- Set $X := \emptyset$.
- Order E, so that $g(e_1) \geq g(e_2) \geq \ldots \geq g(e_n)$.
- Repeat for i from 1 to n:
  If $X \cup \{e_i\} \in \mathcal{U}'$, then set $X := X \cup e_i$.
- Set $X_{\text{opt}} := X$.

Note that the above description does not specify a unique procedure when there are elements of the same weight. Furthermore, it should be noted that it is not specified how the algorithm checks whether $X \cup \{e_i\} \in \mathcal{U}'$. Greedy algorithms are very simple and also fast, as you only have to look at each element once after ordering E.

An algorithm is called ***efficient*** or ***polynomial*** if the number of steps of the algorithm can be bounded by a polynomial in the input size $|E| = n$.

If we can check the statement $X \cup \{e_i\} \in \mathcal{U}'$ in one step and sort the set E in at most $n \log(n)$ steps, see for example Gersting [31], a greedy algorithm is efficient. Therefore, it is desirable to know whether a greedy algorithm actually delivers a good result. The following theorem states that this is the case for set systems that satisfy (U0) and (U1), if and only if they are a matroids:

Theorem 11.19 (Greedy Algorithms) *Let E be a finite set and $\mathcal{U}' \subseteq \wp(E)$ a set of subsets satisfying* (U0) *and* (U1). *The pair $(E, \mathcal{U}')$ is a matroid if and only if every greedy algorithm for every weight function $g : E \to \mathbb{R}_{\geq 0}$ finds a solution to the combinatorial optimization problem $(E, \mathcal{U}', g)$.*

An example of the usefulness of the above theorem is Kruskal's[7] algorithm, compare Example 6.5 (8). Let $G = (V, E)$ be a graph and $g : E \to \mathbb{R}_{\geq 0}$ a weight function on the edges. Kruskal's algorithm calculates a minimum weight spanning forest and is precisely a greedy algorithm on the graphic matroid M_G. More precisely, Kruskal's algorithm first sorts the edges of G in ascending order by weight and then adds the next heaviest edge to the already partial solution, provided this does not create a cycle.

---

[7] Joseph Kruskal, US-American mathematician, computer scientist, and psychometrician, 1928–2010.

Since the description of a matroid using the greedy algorithm is based on the same set system as the axiomatization by independent sets, the associated cryptomorphism is simply the identity mapping.

11.3 Operations on Matroids

Duality, union, direct sum, intersection, matchings, Hamilton cycles

We will now see how to construct new structures from one or more matroids. We will consider unary and binary operations and investigate when these in turn form matroids. For the intersection and the product, this will not be the case, but interesting algorithmic statements can still be made about intersections of two matroids.

Theorem 11.20 (Duality) *Given a matroid $M = (E, \mathcal{U})$ with bases $\mathcal{B}$. Set $\overline{\mathcal{B}} := \{X \subseteq E \mid E \setminus X \in \mathcal{B}\}$. The set $\overline{\mathcal{B}}$ are the bases of a matroid $M^* = (E, \overline{\mathcal{U}})$, called the **dual matroid** to M. Furthermore, $M = (M^*)^*$.*

It can easily be seen that $U_{r,n}^* = U_{n-r,n}$. That is, the class of uniform matroids is closed under duality. The same applies to the class of partition matroids (see Exercise 11.26), the class of K-vector matroids and thus also for regular matroids. Transversal and graphic matroids are not closed under duality. Whether the dual matroid of every algebraic matroid is algebraic is an open problem.

In the following, we describe binary operations on matroids. We start with the union. The proof of the following theorem requires a bit of work because it is not so obvious that $(U2)$ holds.

Theorem 11.21 (Union) *Given are matroids $M_1 = (E_1, \mathcal{U}_1)$, $M_2 = (E_2, \mathcal{U}_2)$. The pair $M_1 \cup M_2 := (E_1 \cup E_2, \mathcal{U}_{1,2})$ with $\mathcal{U}_{1,2} := \{X_1 \cup X_2 \mid X_1 \in \mathcal{U}_1, X_2 \in \mathcal{U}_2\}$ is a matroid, called the **union** of M_1 and M_2.*

From this it naturally follows that unions of finitely many matroids are also matroids. If matroids $M_1 = (E_1, \mathcal{U}_1)$, $M_2 = (E_2, \mathcal{U}_2)$ have disjoint ground sets, the union of M_1 and M_2 is called their **direct sum** and is written as $M_1 \oplus M_2$.

The direct sum is a very "tame" operation, in the sense that, apart from uniform matroids, all example classes from the first section are closed under the direct sum. Uniform matroids seem to be even tamer than the direct sum. For this, see Exercise 11.26.

Things get more difficult when considering the intersection of matroids. The definition is still as natural as in the preceding constructions: Let $M_1 = (E_1, \mathcal{U}_1), \ldots, M_k = (E_k, \mathcal{U}_k)$ be matroids. The **intersection** of $M_1, \ldots, M_k$, written as $M_1 \cap \ldots \cap M_k$, is defined as the pair

$$(E_1 \cap \ldots \cap E_k, \{X_1 \cap \ldots \cap X_k \mid X_1 \in \mathcal{U}_1, \ldots, X_k \in \mathcal{U}_k\}).$$

11.3 Operations on Matroids

Note that $\{X_1 \cap \ldots \cap X_k \mid X_1 \in \mathcal{U}_1, \ldots, X_k \in \mathcal{U}_k\} = \mathcal{U}_1 \cap \ldots \cap \mathcal{U}_k$, while the analogous statement for the union is not true. Many graph theoretic objects can be represented as the intersection of two matroids. An interesting example are matchings in bipartite graphs (compare the corresponding parts in Sect. 6.3):

Theorem 11.22 (Matchings) *Let $G = (V, E)$ be a bipartite graph. The set $\mathcal{M} \subseteq E$ of matchings of G is the set of independent sets of the intersection of two matroids.*

Proof We construct two partition matroids $M_1 = (E, \mathcal{U}_1)$, $M_2 = (E, \mathcal{U}_2)$ with ground set E as follows, compare Definition 11.6. For $v \in V$ take the set $A_v := \{e \in E \mid e \text{ incident to } v\}$. Now let $V_1 \cup V_2 = V$ be the two parts of the bipartition of V. For $i \in \{1, 2\}$ we define the partition matroids M_i by the partitions $P_i := \{A_v \subseteq E \mid v \in V_i\}$ with $r_v = 1$ for all $v \in V_i$. It is easy to see that $\mathcal{M} = \mathcal{M}_1 \cap \mathcal{M}_2$. □

Matchings of bipartite graphs are also an example of an intersection of two matroids, which is generally not a matroid, see Exercise 11.27.

Similar to the Greedy algorithm, one might ask whether in the intersection $M_1 \cap \ldots \cap M_k$ of several matroids, one can efficiently find a subset from $\mathcal{U}_1 \cap \ldots \cap \mathcal{U}_k$ of maximum weight. For this, a weight function g from $E_1 \cap \ldots \cap E_k$ to $\mathbb{R}_{\geq 0}$ is assumed. So, one wants to solve a combinatorial optimization problem $(E_1 \cap \ldots \cap E_k, \mathcal{U}_1 \cap \ldots \cap \mathcal{U}_k, g)$. A very important theorem in this context goes back to Edmonds:[8]

Theorem 11.23 (Edmonds 1970) *Given are matroids $M_1 = (E_1, \mathcal{U}_1), M_2 = (E_2, \mathcal{U}_2)$ and a weight function $g : E_1 \cap E_2 \to \mathbb{R}_{\geq 0}$. Then there is an efficient algorithm that finds a solution to the combinatorial optimization problem $(E_1 \cap E_2, \mathcal{U}_1 \cap \mathcal{U}_2, g)$.*

It should be noted that the efficient algorithm in the above theorem is not a greedy algorithm, because then, according to Theorem 11.19, the intersection of two matroids would be a matroid, which is not the case, see Exercise 11.27. In terms of the matchings of a bipartite graph, the above theorem, together with Theorem 11.22, now provides an efficient algorithm for finding a matching of maximum weight in a bipartite graph. Unfortunately, the above theorem cannot be generalized to intersections of more than two matroids. For this, we consider Hamilton cycles in graphs. See also the corresponding part in Sect. 6.7.

Theorem 11.24 (Hamilton Cycles) *Finding a Hamilton cycle in a graph with n vertices is equivalent to finding an independent set with n elements in the intersection of three matroids.*

Proof Let G be a graph with n vertices. Replace each edge with two oppositely directed edges. Now replace any vertex v with v_1 and v_2 and let all outgoing edges

---

[8] Jack Edmonds, Canadian mathematician and computer scientist, *1934.

from v start at v_1 and the incoming edges end at v_2. We denote the constructed directed graph as G'. Let $N^+(u)$ be the set of outgoing edges of the vertex u and $N^-(u)$ the set of incoming edges of the vertex u. Define three matroids:

- M_1 is the graphic matroid based on the undirected graph $\underline{G'}$, which we obtain by forgetting the orientation from G',
- M_2 is the partition matroid on the ground set of directed edges of G' partitioned by the sets $N^+(u)$ for all vertices u and $r_u = 1$ for all u,
- M_3 is the partition matroid on the ground set of directed edges of G' partitioned by the sets $N^-(u)$ for all vertices u and $r_u = 1$ for all u.

It is now easy to convince oneself that G contains a Hamilton cycle if and only if $M_1 \cap M_2 \cap M_3$ contains an independent set of size n. □

The above theorem shows: Finding an independent set of large weight in the intersection of three matroids is at least as hard as checking whether a graph contains a Hamilton cycle. It is assumed that the latter problem is not efficiently solvable, as it is NP-complete, and the same applies to the combinatorial optimization problem $(K_1 \cap K_2 \cap K_3, \mathcal{U}_1 \cap \mathcal{U}_2 \cap \mathcal{U}_3, g)$. In this sense, one cannot expect that Edmonds' theorem can be generalized to the intersection of more than two matroids.

We conclude with the (Cartesian) product under which matroids are not generally closed.

Example 11.25 (Products) Given are matroids $M_1 = (K_1, \mathcal{U}_1)$, $M_2 = (K_2, \mathcal{U}_2)$. The pair $M_1 \times M_2 := (K_1 \times K_2, \mathcal{U}_1 \times \mathcal{U}_2)$ is generally not a matroid. If you take, for example, $U_{2,4} \times U_{2,4}$, then $\{(1, 1), (1, 2), (2, 1), (2, 2)\} \in \mathcal{U}_1 \times \mathcal{U}_2$. But this is not the case for the subset $\{(1, 1), (1, 2), (2, 1)\}$.

Exercise 11.26 (Partition Matroids)

(a) Show that partition matroids are closed under duality.
(b) Show that partition matroids are exactly the direct sums of uniform matroids.
(c) Show: Partition matroids, which are direct sums of uniform matroids of rank 1, are transversal.

Exercise 11.27 (Matchings) Show that the matchings of a bipartite graph are generally not the independent sets of a matroid. (Hint: Use Exercise 11.14.)

11.4 Structure-Preserving Mappings

Topological space, continuous, open, weak mapping, independent mapping

Before we delve into structure preserving mappings of matroids, we turn to topological spaces. Their mappings may serve as models here. Very similar to a

11.4 Structure-Preserving Mappings

matroid, a ***topological space*** is a pair $(E, \mathcal{O})$. It consists of a ground set E and a set of subsets $\mathcal{O}$, called ***open sets***, which fulfill a series of axioms:

(O0) $\emptyset, E \in \mathcal{O}$ (the empty set and the entire set are open),
(O1) for $X, Y \in \mathcal{O}$ one has $X \cap Y \in \mathcal{O}$ (the finite intersection of open sets is open),
(O2) for $\mathcal{O}' \subseteq \mathcal{O}$ one has $\bigcup_{X \in \mathcal{O}'} X \in \mathcal{O}$ (the arbitrary union of open sets is open).

A crucial difference to matroids is that in topological spaces, the finiteness of the ground set E is not fundamentally assumed. Apart from that, the above definition of topological spaces has quite similar flavor to the definition of a matroid as system of independent sets. Just like in the case of matroids, the definition of a topological space is motivated by a large set of natural example classes. The easiest one is probably the following:

Theorem 11.28 (Open Intervals) *Let $E = \mathbb{R}$ and $\mathcal{O}$ be the set of open intervals and their arbitrary unions in $\mathbb{R}$. Then $(\mathbb{R}, \mathcal{O})$ is a topological space.*

We use the analogy of matroids and topological spaces to introduce structure preserving mappings of matroids here. Typically, the structure preserving mappings of topological spaces are the continuous mappings. Let $(E, \mathcal{O})$ and $(E', \mathcal{O}')$ be topological spaces. Then a mapping $\phi : E \to E'$ is called ***continuous*** if $X \in \mathcal{O}'$ implies $\phi^{-1}(X) \in \mathcal{O}$ for all $X \in \mathcal{O}'$. That is, preimages of open sets are open.
Analogously, weak mappings of matroids are defined. Let $(E, \mathcal{U})$ and $(E', \mathcal{U}')$ be matroids. A mapping $\phi : E \to E'$ is called ***weak*** if for all $X \in \mathcal{U}'$ one gets $\phi^{-1}(X) \in \mathcal{U}$. That is, preimages of independent sets are independent. One can convince oneself of the following statement:

Theorem 11.29 (Isomorphism) *Let $(E, \mathcal{U})$ and $(E', \mathcal{U}')$ be matroids. A bijective mapping $\phi : E \to E'$ is a matroid isomorphism (see the definition in Sect. 11.1) if and only if both ϕ and ϕ^{-1} are weak.*

Another way to define structure preserving mappings for topological spaces are open mappings. A mapping ϕ from a topological space $(E, \mathcal{O})$ to a topological space $(E', \mathcal{O}')$ is called ***open*** if for all $X \in \mathcal{O}$ one gets $\phi(X) \in \mathcal{O}'$. In other words: images of open sets are open.
Analogously, we define another type of structure preserving mappings between matroids. Let $(E, \mathcal{U})$ and $(E', \mathcal{U}')$ be matroids. We call a mapping $\phi : E \to E'$ ***independent*** if for all $X \in \mathcal{U}$ one gets $\phi(X) \in \mathcal{U}'$. Independent mappings are, as far as we know, defined here for the first time. An investigation of the properties of these mappings could be an interesting direction of research.

In matroid literature, there are two more definitions of structure preserving mappings: ***strong mappings*** and ***co-maps***, see the chapters by Joseph Kung in the book by Neil White [75]. We will not go into detail here.

Fig. 11.5 The non-Fano matroid

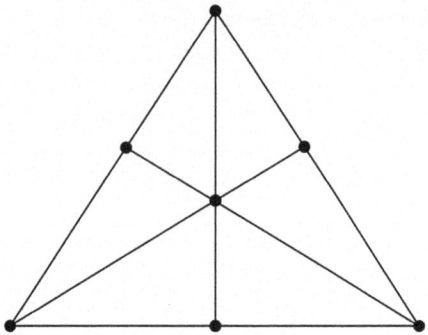

Exercise 11.31 (Structure Preserving Mappings) Find mappings ϕ, ϕ' from the non-Fano matroid (Fig. 11.5) to the Fano matroid (Fig. 11.1a), such that ϕ is weak, but not independent and ϕ' is independent, but not weak.

Categories   12

In the first chapters of this book we set up axioms and calculation rules. We used them for elements of sets, mostly numbers, with concretely given operations. In Chaps. 7–10 we have formulated these axioms and calculation rules for elements of general algebraic structures. We applied them to elements of semigroups, rings and fields or acts, modules and vector spaces. For graphs or ordered sets the situation is different. These sets are not structured by operations of the elements, but by more general relations, namely edges or order relations. Nevertheless, results such as the Homomorphism Theorem also apply to graphs or even ordered sets.

In this chapter, we venture into higher levels of abstraction. Until now, our focus has been on sets with elements whose structure is defined by relations among these elements. Categories consist of objects and morphisms—an abstraction of the elements of sets and mappings. Objects, in particular may be sets, but don't even have to be sets anymore, i.e. don't have to have elements. They then will be structured by appropriate homomorphisms (hereafter called morphisms). Consequently if the objects don't have elements, morphisms cannot be mappings. The morphisms between the objects can be represented as arrows between the objects. The examples in this chapter illustrate the diverse interpretations these arrows can take.

In Computer Science, categories play a crucial role in defining abstract data types and graph substitution systems.

On the Literature The importance of Category Theory at least for theoretical Computer Science is presented in Ehrig [25]. We recommend the books by Herrlich and Strecker [35] or Adamek et al. [1]. Also very popular among computer scientists is the book by MacLane [58]. For an introduction, Blyth [11] or Preuß [63] are also suitable.

Parts of this chapter can also be found in Kilp et al. [45].

12.1 Basic Concepts

Category, object, morphism, source, target, composition (defining properties), associative law, identical morphism, construct (= concrete category), underlying set/mapping, small/large category, dual category, functor

A *category* $\mathcal{K}$ has the following components:

(Ob) A class $\text{Ob}\,\mathcal{K}$, the $\mathcal{K}$-*objects*.
If A is a $\mathcal{K}$-object, we write $A \in \text{Ob}\,\mathcal{K}$ or simply $A \in \mathcal{K}$.

(Mor) A set $\mathcal{K}(A, B)$ for each pair (A, B) of $\mathcal{K}$-objects, such that

$$\mathcal{K}(A, B) \cap \mathcal{K}(C, D) = \emptyset$$

for all $A, B, C, D \in \mathcal{K}$ with $(A, B) \neq (C, D)$.
The elements of $\mathcal{K}(A, B)$ are called $\mathcal{K}$- *morphisms from* A *to* B. For $f \in \mathcal{K}(A, B)$, we call A the *source* or *domain* and B the *target* or *codomain* of f. We write $f : A \longrightarrow B$.

(Comp) A *composition rule* for morphisms as follows:
for any three objects $A, B, C \in \mathcal{K}$, there is given a mapping

$$\circ : \begin{cases} \mathcal{K}(A, B) \times \mathcal{K}(B, C) \to \mathcal{K}(A, C) \\ (f, g) \mapsto g \circ f, \end{cases}$$

such that

  (ass) the *associative law* $h \circ (g \circ f) = (h \circ g) \circ f$ holds, whenever all involved compositions are defined, and
  (id) an *identical morphism* $id_B \in \mathcal{K}(B, B)$ exists for each object $B \in \mathcal{K}$. It behaves like a neutral element with respect to the composition of morphisms, i.e., for all $f \in \mathcal{K}(A, B)$, $g \in \mathcal{K}(B, C)$

$$g \circ id_B = g \quad \text{and} \quad id_B \circ f = f.$$

Explanations

(Ob): Here we explicitly allow the possibility that the totality of the considered objects no longer forms a set, but a real class. This applies e.g. for all graphs, all semigroups, all $\mathbb{R}$-vector spaces. In particular, the objects themselves do not have to be sets.

(Mor): This ensures that two morphisms are considered different if they have different sources or different targets. We also have ensured this in the definition of mappings in Sect. 5.1.

12.1 Basic Concepts

(Comp): The partial composition of $\mathcal{K}$-morphisms, which fulfills (ass) and (id), is also called the *defining property for $\mathcal{K}$-morphisms*. This way the set $\mathcal{K}(A, B)$ may be called a *set of morphisms*. Morphisms may not be mappings, even if the objects are sets.

Hence we make the following definitions:
A category $\mathcal{K}$ is called a *construct* or *concrete category* or *set concrete*, when its objects are (structured) sets and its morphisms are (structure compatible) mappings. These sets are then called *underlying sets*, the mappings *underlying mappings*.

A category $\mathcal{K}$ is called *small*, if $\text{Ob}\,\mathcal{K}$ is a set, otherwise it is called *large*.

Theorem 12.1 (Dual Category) *If $\mathcal{K}$ is a category, then $\mathcal{K}^{op}$ is also a category, where*

$$\text{Ob}\,\mathcal{K}^{op} := \text{Ob}\,\mathcal{K},$$
$$\mathcal{K}^{op}(A, B) := \mathcal{K}(B, A) \text{ and}$$
$$g \bullet f := f \circ g \quad \text{for } f \in \mathcal{K}^{op}(A, B) = \mathcal{K}(B, A)$$
$$\text{and } g \in \mathcal{K}^{op}(B, C) = \mathcal{K}(C, B).$$

The category $\mathcal{K}^{op}$ is called the *category dual to* $\mathcal{K}$ and is obtained from $\mathcal{K}$ by "reversing all arrows".

Example 12.2 (Large, Not Concrete) The category **Rel** has all sets as objects, and **Rel**$(A, B) := \wp(A \times B)$ is the set of binary relations between the sets A, $B \in$ **Rel**. So these are not mappings. The composition in **Rel** is the composition of relations. This category is **large and not concrete**.

Examples 12.3 (Large, Concrete) The structures that serve as a pattern for the concept of a category are the **large constructs,** some of them are listed below. Table 12.1 contains in the first column the name of the category, in the second the objects and in the third the morphisms; the composition is always the composition of the underlying mappings.

Examples 12.4 (Miscellaneous)

- The set **Gra**$_4$ of all graphs with 4 vertices with the graph morphisms is a **small concrete category**.
- The following four categories are **small and not set-concrete**.
  - For a monoid $(M, \cdot, 1)$ set $\text{Ob}\,\mathcal{M} := \{1\}$ and $\mathcal{M}(1, 1) := M$, i.e. the category $\mathcal{M}$ has exactly one object and each monoid element is a morphism. The composition in $\mathcal{M}$ is the multiplication in M.
  - Objects of the category $\mathbb{Z}$-**Mat** are all natural numbers $m, n > 0$, morphisms from m to n are all $m \times n$ matrices over $\mathbb{Z}$, composition of morphisms is matrix multiplication.

Table 12.1 Categories

Set	Sets	Mappings
Sgr	Semigroups	Semigroup homomorphisms
Mon	Monoids	Monoid homomorphisms
Grp	Groups	Group homomorphisms
Ab	Abelian groups	Group homomorphisms
Rng	Rings	Ring homomorphisms
Rng$_1$	Rings with One	Ring homomorphisms preserving one
Field	Fields	Field homomorphisms
S-Act	Left S-Acts, $S \in$ **Sgr**	Act homomorphisms
Act-S	Right S-Acts, $S \in$ **Sgr**	Act homomorphisms
R-Mod	Left R-Modules, $R \in$ **Rng**	Module homomorphisms
Mod-R	Right R-Modules, $R \in$ **Rng**	Module homomorphisms
K-Vec	K-Vector spaces, $K \in$ **Field**	Linear mappings
Top	Topological spaces	Continuous mappings
Top^o	Topological spaces	Open mappings
Ord	Ordered sets	Isotone mappings
Gra	Graphs	Graph morphisms
SGra	Graphs	Strong graph morphisms
EGra	Graphs	Graph endomorphisms
WMad	Matroids	Weak mappings
IMad	Matroids	Independent mappings

- For a set X take Ob$\mathcal{P} := \mathcal{P}(X)$ and $\mathcal{P}(A, B) := \{(A, B)\}$, if $A \subseteq B$, otherwise $\emptyset$. The composition of morphisms is $(A, B) \circ (B, C) := (A, C)$.
- For each ordered set $(P, \leq)$ the objects of the category **P** are the elements of the set P and the morphisms are the pairs (x, y) with $x \leq y$. The previous example is the special case with $(P, \leq) = (\mathcal{P}(X), \subseteq)$.
- Ordered sets with antitone mappings $(x \leq y \Rightarrow f(x) \geq f(y))$ and the composition of mappings do **not form a category**, (since the composition of two antitone mappings is antitone).

Remark Just as mappings belong to sets and morphisms belong to objects, functors belong to categories. Functors establish relationships between objects and sets of morphisms across different categories. For a detailed understanding of functors, we refer to the aforementioned books on category theory. A mere definition might not suffice here, and space constraints prevent us from delving deeper into this topic.

12.2 Special Objects and Morphisms

Epi/monomorphism, (co)retract(ion), section, isomorphism, sub/factor object, (right/left) zero morphism, initial/terminal/null object

12.2 Special Objects and Morphisms

At this point, we generalize the terms surjective (to epi) and injective (to mono) in such a way that they make sense in every category, i.e., even if the considered morphisms are not mappings. Compare Sect. 5.3. It also turns out, that in every category, the definition of objects and morphisms already determines how subobjects and factor objects must be defined.

Let $\mathcal{K}$ be a category. A morphism $f \in \mathcal{K}(A, B)$ is called

- *epimorphism*, if it is right cancellable, i.e., if for all morphisms g and h

$$g \circ f = h \circ f \Rightarrow g = h.$$

- *monomorphism*, if it is left cancellable, i.e., if for all morphisms g and h

$$f \circ g = f \circ h \Rightarrow g = h.$$

- *retraction*, if it is right invertible, i.e., if there is a morphism $f' \in \mathcal{K}(B, A)$ such that $f \circ f' = id_B$.
- *section* (also: *coretraction*), if it is left invertible, i.e., if there is a morphism $f' \in \mathcal{K}(B, A)$ such that $f' \circ f = id_A$.
- *isomorphism*, if it is invertible, i.e., if there is a morphism $f^{-1} \in \mathcal{K}(B, A)$ such that $f^{-1} \circ f = id_A$ and $f \circ f^{-1} = id_B$.

Let A be a $\mathcal{K}$-object. A $\mathcal{K}$-object U is called

- *factor object of* A, if $\mathcal{K}(A, U)$ contains an epimorphism;
- *subobject of* A, if $\mathcal{K}(U, A)$ contains a monomorphism;
- *retract of* A, if $\mathcal{K}(A, U)$ contains a retraction (which is equivalent by definition to $\mathcal{K}(U, A)$ containing a section.)

The following theorems provide an overview of important properties of morphisms. Proofs can be found, for example, in Herrlich and Strecker [35].

Theorem 12.6 (Morphisms)

(a) For the category **Set** *we get*

$$retraction = surjection = epimorphism$$
$$section\ \ \ = injection\ \ = monomorphism.$$

(b) In every construct, the we have following implications for every mapping

$$retraction \Rightarrow surjection \Rightarrow epimorphism$$
$$section\ \ \ \Rightarrow injection\ \ \Rightarrow monomorphism.$$

(c) In every category, we have the following implications for every morphism

$$retraction \Rightarrow epimorphism$$
$$section \Rightarrow monomorphism.$$

The situation is even a bit more complicated. In the concrete category **Mon**, the embedding of $(\mathbb{N}, +)$ into $(\mathbb{Z}, +)$ is a non-surjective epimorphism, which is also a monomorphism, but not an isomorphism. This shows that the second implications in Theorem 12.6 are not reversible. And it shows that even for algebras, epimorphism and monomorphism do not imply isomorphism. Similar examples exist in **Rng** and in **Sgr**, see also Herrlich and Strecker [35] or Kilp et al. [45].
For **Set**, we have also already proven the following two theorems in Chap. 4.

Theorem 12.7 (Isomorphisms) *For every morphism f, the following statements are equivalent:*

(i) f is an isomorphism.
(ii) f is a monomorphism and a retraction.
(iii) f is an epimorphism and a section.

The statement of Theorem 12.7 takes into account the fact that the characterization of isomorphisms from Theorem 7.19 as bijective homomorphisms is not true in every category. In Example 6.7, we saw a bijective graph morphism that is not an isomorphism, because its inverse mapping does not preserve edges.

Theorem 12.8 (Compositions) *In every category, the following implications apply to all composable morphisms f and g:*

f, g epimorphisms	$\Rightarrow f \circ g$ is an epimorphism
	$\Rightarrow f$ is an epimorphism;
f, g monomorphisms	$\Rightarrow f \circ g$ is a monomorphism
	$\Rightarrow g$ is a monomorphism;
$f \circ g$ is a retraction	$\Rightarrow f$ is a retraction;
$f \circ g$ is a section	$\Rightarrow g$ is a section;
$f \circ g$ is an isomorphism	$\Rightarrow f$ is a retraction and g is a section.

Mnemonics The terms in Table 12.2 correspond to each other, column wise by dualizing and row wise (1st to 4th column) from right to left by specializing, i.e., a retraction is a special epimorphism. The bracketed part of the second column only makes sense in concrete categories, i.e., if the morphisms are actually mappings. The

12.3 Products and Coproducts

Table 12.2 Mnemonics

Retraction	(Surjective =) right invertible	Right cancellable	Epi	First factor
Section	(Injective =) left invertible	Left cancellable	Mono	Second factor

last column describes the heredity of the respective property from the composition of morphisms to the corresponding factor.[1]

To conclude this section, we provide two triples of definitions that have significance in applications. For all $A, B, C, D \in Ob(\mathcal{K}), g, h \in \mathcal{K}(C, A)$ or $g, h \in \mathcal{K}(B, D)$, a morphism $f : A \to B$ in a category $\mathcal{K}$ is called

- *left zero morphism*, if $f \circ g = f \circ h$;
- *right zero morphism*, if $g \circ f = h \circ f$;
- *zero morphism*, if f is both a left and right zero morphism.

An object $A \in \mathcal{K}$ is called

- *initial object*, if $|\mathcal{K}(A, D)| = 1$ for all $D \in Ob(\mathcal{K})$;
- *terminal object*, if $|\mathcal{K}(C, A)| = 1$ for all $C \in Ob(\mathcal{K})$;
- *zero object*, if it is both initial and terminal.

Exercise 12.10 (Initial—Terminal)

(a) In **Set**, ∅ is the initial object and any single element set is the terminal object. In **EGra**, the graph with only one vertex and no edges is the terminal object, initial objects do not exist.
(b) If $A \in Ob(\mathcal{K})$ is an initial object, then every morphism $f \in \mathcal{K}(A, D)$ is a right zero morphism for all $D \in Ob(\mathcal{K})$.
(c) What can we say about terminal objects in analogy to (b)?

12.3 Products and Coproducts

Product, projection, factors of the product, coproduct, injection, summands of the coproduct, duality of product and coproduct, cross product, box cross product, direct sum/product

---

[1] Unfortunately the key words epi and mono do not fit into the scheme. In the original table in German the first line is dominated by "e" (in rechts und erster), the second line by "n" (in links und hinterer). Then epi and mono also fit in.

Let $(A_i)_{i \in I}$ be a family of objects in a category $\mathcal{K}$, $P \in \mathcal{K}$ an object and let $\pi_i \in \mathcal{K}(P, A_i), i \in I$, be morphisms.

The pair $(P, (\pi_i)_{i \in I})$ is called **categorical product of** $(A_i)_{i \in I}$ **in** $\mathcal{K}$, if it is the solution of the following *universal problem* (compare Theorem 7.41):

For every $\mathcal{K}$-object Q with morphisms $q_i \in \mathcal{K}(Q, A_i), i \in I$, there exists exactly one morphism $q \in \mathcal{K}(Q, P)$, such that

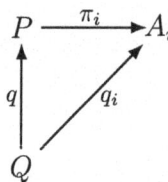

is commutative for every $i \in I$. Note that this definition is not constructive. In a given category one has to test, if a suitable pair exists. In Theorem 7.41, the above diagram is given for two algebras A_1 and A_2. The category of sets with mappings can also serve as example.

Without any specification, the **product** of $(A_i)_{i \in I}$ is usually denoted by $\prod_{i \in I} A_i$, the morphisms $\pi_i, i \in I$, are called **projections**, and the objects $A_i, i \in I$, are called **factors of the product**.

Next, we give the dual definition.

Let $(A_i)_{i \in I}$ be a family of objects in a category $\mathcal{K}$. The pair $((\iota_i)_{i \in I}, C)$, where $C \in \mathcal{K}$ is an object and $\iota_i \in \mathcal{K}(A_i, C), i \in I$, are morphisms in the category $\mathcal{K}$, is called **categorical coproduct of** $(A_i)_{i \in I}$ **in** $\mathcal{K}$, if it is the solution of the following universal problem:

For each $\mathcal{K}$ object D with morphisms $d_i \in \mathcal{K}(A_i, D), i \in I$, there exists exactly one morphism $d \in \mathcal{K}(C, D)$, so that

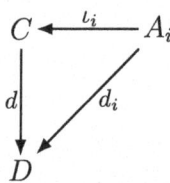

is commutative for each $i \in I$.

Again, this is not constructive.

Nevertheless, the **coproduct** of $(A_i)_{i \in I}$ is usually denoted by $\coprod_{i \in I} A_i$, the morphisms $\iota_i, i \in I$, are called **injections**, the objects $A_i, i \in I$, **summands of the coproduct**.

In the diagrams for product and coproduct, it can be seen that dualizing is accomplished by reversing all arrows. This also results in monomorphism and epimorphism being swapped.

The following statement follows directly from the definition of the dual category.

12.3 Products and Coproducts

Theorem 12.11 (Duality of Product and Coproduct) *The product $\prod A_i$ of objects A_i in a category $\mathcal{K}$ is equal to the coproduct of the same objects in the dual category $\mathcal{K}^{op}$ and vice versa.*

Examples 12.12 (Products and Co.) Not every category has products or coproducts, see, for example, Herrlich/Strecker [35]. However, they do exist in the common categories. And if we have a construct, the situation for the underlying sets and mappings is often the same as in **Set** with the canonical projections and injections.

Differences arise for the coproduct of vector spaces (4) and in the category **P** (5).

Note that categories with the same objects but different morphisms can also have different products and coproducts.

(1) In the category **Set** of sets with mappings, the categorical product is the Cartesian product with the canonical projections, and the categorical coproduct is the disjoint union with the canonical injections, see Examples 5.3 (7) and (8).
(2) In the category **Ord** of ordered sets with the isotone mappings, the categorical product is the Cartesian product with component wise order (Example 4.33), and the categorical coproduct is the disjoint union of the sets, where comparabilities in the summands are preserved and any two elements from different summands are incomparable.
(3) In the category **Gra** of simple, undirected graphs with the graph morphisms, the categorical product of two graphs $G_i = (V_i, E_i)$, $i = 1, 2$, is the graph
$G_1 \times G_2 = (V_1 \times V_2, E)$ with
$E = \{ \{(x_1, x_2), (y_1, y_2)\} \mid \{x_i, y_i\} \in E_i, i = 1, 2 \}$.
This is the so-called *cross product* (sometimes also called *Cartesian product*) of G_1 and G_2.
We introduced the box product of graphs in Exercise 6.41. It can be shown that this also has a categorical interpretation, namely as "tensor product" in **Gra** and in **EGra**. Definitions and details can be found in Knauer/Knauer, Algebraic Graph Theory, [50]. Note that the box product is sometimes also called Cartesian product.
The coproduct in **Gra** is the disjoint union of the graphs with their edges, without edges between the summands of the coproduct.
In the category **EGra**, the (categorical) product of two graphs $G_i = (V_i, E_i)$, $i = 1, 2$, is the so-called *boxcross product* $G_1 \boxtimes G_2 := (V_1 \times V_2, E^{\boxtimes})$ with

$$E^{\boxtimes} = \{ \{(x_1, x_2), (y_1, y_2)\} \mid \{x_i, y_i\} \in E_i, i = 1, 2 \text{ or}$$
$$x_1 = y_1, \{x_2, y_2\} \in E_2 \text{ or } \{x_1, y_1\} \in E_1, x_2 = y_2 \}.$$

In Fig. 12.1 on the left is the cross product of the paths P_1 and P_2 of lengths 1 and 2. The graph depicted on the right is the boxcross product of these graphs. Product and coproduct of any family of graphs are defined analogously.
Also for directed graphs, multigraphs and hypergraphs product and coproduct can be given. These of course also depend on the chosen morphisms.

Fig. 12.1 Cross and boxcross product

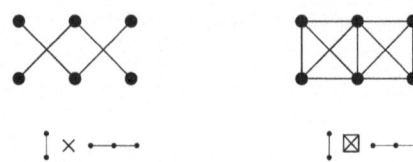

(4) In the category K-**Vec** of vector spaces over a field K with K-linear mappings, for every family $(V_i)_{i \in I}$ of K-vector spaces: The categorical product of $(V_i)_{i \in I}$ is equal to the ***direct product***, which is the Cartesian product of the sets $V_i, i \in I$, with addition and component wise outer multiplication:

$$(x_i)_{i \in I} + (y_i)_{i \in I} = (x_i + y_i)_{i \in I} \text{ and}$$

$$r(x_i)_{i \in I} = (rx_i)_{i \in I} \quad \text{for } x_i, y_i \in V_i, r \in K, i \in I.$$

The categorical coproduct of $(V_i)_{i \in I}$ is equal to the ***direct sum*** (usually denoted with $\oplus V_i$), which is the subspace of $\prod_{i \in I} V_i$, consisting of those elements $(x_i)_{i \in I}$ for which $x_i \neq 0$ for at most finitely many $i \in I$. The j-th injection is then

$$\iota_j : \begin{cases} V_j \to \coprod V_i, \\ x_j \mapsto (v_i)_{i \in I} \end{cases} : \begin{cases} x_j, \text{ if } i = j, \\ 0 \quad \text{otherwise}. \end{cases}$$

Since the division in the field is not used in this process, the categorical products and categorical coproducts are the same for R-modules, where R is any ring, and thus in particular for abelian groups.

Attention: the coproduct of vector spaces (or R-modules, or abelian groups) is therefore **not** the disjoint union of the summands, although these categories are set concrete. – It is, as one says, a *non-set concrete coproduct*.

(5) If **P** is the category whose objects are the elements of an ordered set $(P, \leq)$ and whose morphisms are the pairs (x, y) with $x \leq y$, then the categorical product of a family $(x_i)_{i \in I}$ is just their greatest lower bound $\inf\{x_i \mid i \in I\}$ and the categorical coproduct is their smallest upper bound $\sup\{x_i \mid i \in I\}$, if they exist. This example provides categories in which products and/or coproducts do not exist for every family of objects. – From $(P, =)$ we get a category **P**, in which no products or coproducts exist at all. The latter also applies to the category **SGra**, the graphs with strong graph morphisms.

The following two examples show what nice blossoms one can grow with categories. We use the terminology of the definitions of the coproduct and the product, as given at the beginning of this chapter.

Example 12.14 (Coproduct Is Initial) We take two objects A_1, A_2 in a category **C** and construct the new category $\mathbf{C}^{(A_1, A_2)}$. Its objects are triples (d_1, d_2, D), where

12.4 Free

d_1 and d_2 are morphisms in **C** that both end in D and start in A_1 and A_2 respectively. For two such triples (d_1, d_2, D) and (d_1', d_2', D'), a morphism in this category is a morphism f from the category **C**, such that $f d_1' = d_1$ and analogously with index 2. The coproduct (ι_1, ι_2, C) with $C = A_1 \coprod A_2$ is the *initial object* in this new category, because there exists exactly one d with $d\iota_1 = d_1$ and $d\iota_2 = d_2$.

And categorically dual:

Example 12.15 (Product Is Terminal) We take again two objects A_1, A_2 in **C** and and construct the new category $\mathbf{C}_{(A_1,A_2)}$, whose objects are triples (Q, q_1, q_2), where q_1 and q_2 are morphisms in **C** that both start in Q and end in A_1 and A_2 respectively. For two such triples (Q, q_1, q_2) and (Q', q_1', q_2'), a morphism in this category is a morphism f in the category **C**, such that $q_1' f = q_1$ and analogously with index 2. The product (P, π_1, π_2) with $P = A_1 \prod A_2$ is the *terminal object* in the new category, because there exists exactly one q with $\pi_1 q = q_1$ and $\pi_2 q = q_2$.

Theorem 12.16 (Product and Coproduct Are Unique) *If they exist, the product and coproduct are uniquely determined up to isomorphism.*

Proof This follows because there is at most one initial and one terminal object. □

Exercise 12.17 (Direct Sum of Vector Spaces) Prove that the direct sum of infinitely many $\mathbb{R}$-vector spaces fulfills the defining property of the coproduct in the category $\mathbb{R}$-**Vec**, but not that of the product. (Hint: You can take the 1-dimensional $\mathbb{R}$-vector space $\mathbb{R}$ itself infinitely many times.)

Exercise 12.18 (Products and Co.) Prove the, or some of the, statements in Example 12.12, in particular: in **SGra** there are neither categorical products nor categorical coproducts, except in trivial cases. This is despite the fact that cross products and unions can of course be formed.

Exercise 12.19 (Product in WMad) Is the product of matroids the categorical product in **WMad**?

Exercise 12.20 (Product and Co. in IMad) Investigate the product and coproduct in the category **IMad** of matroids with independent mappings.

12.4 Free

Free object, (free) generating system, uniqueness of free objects, sufficient free objects, basis, free (commutative) groupoid/monoid, free (commutative) semigroup/group, concatenation, word semigroup/monoid

Let $\mathcal{K}$ be a concrete category and take X in **Set**. A pair $(u, F(X))$, where $F(X) \in \mathcal{K}$ and $u : X \to F(X)$ is an injective mapping in **Set**, is called *X-free* (or *|X|-free* or *free*) *in* $\mathcal{K}$, if it is the solution of the following *universal problem*:

For each $A \in \mathcal{K}$ and each mapping $f : X \to A$ in **Set** there exists exactly one morphism $\overline{f} : F(X) \to A$ in $\mathcal{K}$, such that the following diagram is commutative

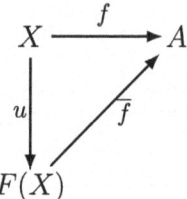

The analysis of the diagram shows that here the principle of linear extension from Linear Algebra (cf. Sect. 9.4) has been the godmother: Every mapping f, which is defined on the basis X of a K-vector space $F(X)$, can be uniquely extended to a linear mapping of the entire vector space. This means we know a linear mapping already when we only know how it maps the basis elements.

The following theorem corresponds in Linear Algebra to the statement that all K-vector spaces of the same dimension are isomorphic (cf. Theorem 9.6).

Theorem 12.21 (Uniqueness of Free Objects) *X-free objects are unique up to isomorphism, i.e., if (u_1, F_1) and (u_2, F_2) are both X-free in $\mathcal{K}$, then $F_1 \cong F_2$. We therefore speak of the X-free object.*

Proof The statement follows again, as already in Theorem 12.16, from the category constructed in the following example. The statement can also be proven directly. Substitutes (u_2, F_2) for (f, A) and (u_1, F_1) for $(u, F(X))$ and then (u_1, F_1) for (f, A) and (u_2, F_2) for $(u, F(X))$. Then one obtains $\overline{u_1} \circ \overline{u_2}$, which due to uniqueness must be id_{F_2}. Conversely, one obtains $\overline{u_2} \circ \overline{u_1} = id_{F_1}$. This shows that $\overline{u_2}$ and $\overline{u_1}$ are isomorphisms, thus $F_1 \cong F_2$. □

Similar to the Examples 12.14 and 12.15, the free object, in a specially constructed category, can also be described as an initial object.

Example 12.22 (Free are Initial) We start with a set $X \in$ **Set** and a concrete category $\mathcal{K}$. The objects of the new category have the form (f, A) with $A \in Ob\mathcal{K}$ and mappings $f : X \to A$ in **Set**. A morphism $g : (f, A) \to (f', A')$ in the new category is a morphism $g : A \to A'$ in $\mathcal{K}$ with the property that $gf' = f$. Then the pair $(u, F(X))$ is the initial object in the new category.

In the same way, the field of quotients of a ring without zero divisors can be described categorically as a solution of a universal problem.

12.4 Free

Example 12.23 (Fields of Quotients are Initial) We start with a ring without zero divisors R and the category **Field** of fields. The objects of the new category have the form (f, A), where A is a field and $f : R \to A$ is a ring homomorphism. A morphism $g : (f, A) \to (f', A')$ is a homomorphism $g : A \to A'$ in **Field** with the property that $gf' = f$. Then the pair $(u, Q(R))$ is the initial object in the new category, where $Q(R)$ is the field of quotients of R and $u : R \to Q(R)$ is the natural embedding. The standard example is $Q(\mathbb{Z}) = \mathbb{Q}$.

Let $A \in \mathcal{K}$, then $E \subseteq A$ is called a *generating system of* A, if there is an epimorphism $p : F(E) \to A$.

A minimal generating system B of $A \in \mathcal{K}$ is called a *free generating system* (in some categories, e.g. in K-**Vec**, as *basis*, cf. Sect. 9.3). If B is a free generating system of A, then every $a \in A$ is uniquely "representable" using the elements from B.

Theorem 12.24 (Existence of Generating Systems) *If in a concrete category for every set A an A-free object $(\iota, F(A))$ exists, then every object of this category has a generating system.*

Proof Every object generates itself, because $\bar{\iota} : F(A) \to F(A)$ is the extension of $\iota : A \to F(A), a \mapsto a$ to $F(A)$. □

In the following constructions and theorems, $u := \iota$ is always the canonical embedding of X into the object being constructed.

Let X be a set. A sequence of i elements, $i \in \mathbb{N} \setminus \{0\}$, with arbitrary (meaningful) bracketing is called a *groupoid word of length i over X*, written for example as $x_1(x_2 x_3) x_4 (x_5 ((\cdots) x_i))$.

With $(X)^i$ we denote the set of groupoid words of length i over X and set

$$(X)^+ := \bigcup_{i \in \mathbb{N} \setminus \{0\}} (X)^i.$$

We call $(X)^+$ with the concatenation of groupoid words the *free groupoid generated by X*.

Theorem 12.25 (Free Groupoid) *The groupoid $(X)^+$ with the canonical embedding*

$$u := \iota : X \to (X)^+, \ x \mapsto x$$

is X-free in the category of groupoids with groupoid homomorphisms.

Let X be a set and X^* the set of all words over X including the empty word ε (cf. Chap. 1). That is, $X^+ := X^* \setminus \{\varepsilon\}$ is the set of all non-empty words. The concatenation of words is an associative operation on X^* and on X^+. Therefore, $(X^+, \cdot)$ is a semigroup, called **word semigroup** and $(X^*, \cdot)$ is a monoid called **word monoid**.

Theorem 12.26 (Free Semigroups and Free Monoids) *The word monoid $(X^*, \cdot)$ is X-free in the category **Mon** of monoids and the word semigroup $(X^+, \cdot)$ is X-free in the category **Sgr** of semigroups each with their homomorphisms.*

Theorem 12.27 (Free Groupoid $\leadsto$ Semigroup) *Consider the equivalence relation α on the set $(X)^+$ of groupoid words, which identifies exactly those groupoid words that yield the same sequence of characters after dropping the brackets. Then α is a congruence on $((X)^+, \cdot)$. The factor groupoid of $(X)^+$ by α is a semigroup, written*

$$X^+ := (X)^+/_\alpha.$$

Theorem 12.28 (Free Commutative Semigroups and Monoids) *Consider the equivalence relation κ on X^+ (or on X^*), which identifies exactly those words that differ only by the order of their characters. Then κ is a congruence on X^+ (or on X^*) and the factor semigroup of X^+ (or of X^*) by κ is commutative. Now $X^+/_\kappa$ is X-free in the category of commutative semigroups and $X^*/_\kappa$ is X-free in the category of commutative monoids, each with the usual homomorphisms and with $u := \iota$.*

Theorem 12.29 (Free Groups) *Set $X^{-1} := \{x^{-1} \mid x \in X\}$. Consider the monoid $(X \cup X^{-1})^*$ with concatenation as operation. Let γ be the equivalence relation on $(X \cup X^{-1})^*$, which replaces any string of the form $x^{-1}x$ or xx^{-1}, $x \in X$, with the empty word ε. This is a congruence of monoids and the factor monoid of $(X \cup X^{-1})^*$ by γ is a group. Now $(X \cup X^{-1})^*/_\gamma, \cdot)$ is X-free in the category of groups with homomorphisms.*

Examples 12.30 (1-Free)

(1) $(\mathbb{N} \setminus \{0\}, +)$ is the 1-free semigroup, with $u : 1 \mapsto 1$,
(2) $(\mathbb{N}, +)$ is the 1-free monoid, with $u : 1 \mapsto 1$,
(3) $(\mathbb{Z}, +)$ is the 1-free group, with $u : 1 \mapsto 1$.
(4) Let K be a field and V an n-dimensional K-vector space with basis $X := \{x_1, x_2, \ldots, x_n\}$. With $u : \{1, 2, \ldots, n\} \to X$, $i \mapsto x_i$ one gets: V is n-free in K-**Mod** and in K-**Vec**. Compare also free R-modules in Sect. 10.2.
(5) The polynomial ring $(\mathbb{Z}[x], +, \cdot)$ is 1-free in the category of commutative rings with one and is generated by $\{x\}$, with $u : x \mapsto x$.
(6) The ring of polynomials with constant term equal to 0, i.e., $((\mathbb{Z}[x] \setminus \mathbb{Z}), +, \cdot)$, is 1-free in the category of all commutative rings. Here too, $\{x\}$ is a free generating system, and again $u : x \mapsto x$.
(7) In the category of fields, there is no 1-free object.

12.4 Free

Exercise 12.31 (Words) List all words with up to three letters over the alphabet $X = \{1, 2\}$ in the 2-free groupoid/semigroup/commutative monoid/commutative group.

Exercise 12.32 (Free Matroids) The uniform matroid $U_{n,n}$ is n-free in the category **WMad**. These matroids are also called *free matroids*. Compare also Aigner [2].

Bibliography

1. J. Adamek, H. Herrlich, G.E. Strecker, *Abstract and Concrete Categories* (John Wiley, New York, 1990)
2. M. Aigner, *Combinatorial Theory*. Classics in Mathematics (Springer, Berlin, 1987)
3. M. Aigner, *Discrete Mathematics* (American Mathematical Society, Providence, 2007). Translated from the German by David Kramer
4. P. Aluffi, *Algebra*, chap. 0 (American Mathematical Society, Providence, 2007)
5. F.W. Anderson, K.R. Fuller, *Rings and Categories of Modules* (Springer, New York, 1974)
6. S. Barnett, *Discrete Mathematics. Numbers and Beyond* (Addison Wesley Longman, Harlow, 1998)
7. A. Beutelspacher, *Das ist o.B.d.A. trivial!* (Vieweg, Braunschweig, 1991)
8. G. Birkhoff, J.D. Lipson, Heterogeneous algebras. J. Combin. Theory **8**, 115–133 (1970)
9. R.G. Bland, M. Las Vergnas, Orientability of matroids. J. Combinatorial Theory Ser. B. **24**, 94–123 (1978)
10. S. Bloom, E. Wagner, Many-sorted theories and their algebras with some applications to data types, in *Algebraic Methods in Semantics*, ed. by J.R.M. Nivat (Cambridge University Press, Cambridge, 1985), pp. 133–168
11. T.S. Blyth, *Categories* (Longman, London, 1986)
12. G. Böhme, *Fuzzy-Logik. Einführung in die algebraischen und logischen Grundlagen* (Springer, Berlin, 1993)
13. O. Borůvka, *Grundlagen der Gruppoid- und Gruppentheorie* (VEB Verlag der Wissenschaften, Berlin, 1960)
14. J.G. Braker, *Algorithms and Applications in Time Discrete Event Systems*. Thesis, Technische Universiteit Delft, 1993
15. R.H. Bruck, *A Survey of Binary Systems*. (Springer, Berlin, 1971)
16. G. Chartrand, L. Lesniak, *Graphs and Digraphs*, 4th edn. (Chapman and Hall, London, 2005)
17. R. Cunninghame-Green, *Minimax Algebra* (Springer, Berlin, 1979)
18. P. Damerow, R. Englund, H. Nissen, *Die ersten Zahlendarstellungen und die Entwicklung des Zahlbegriffs* (Spektrum der Wissenschaft, 1988), pp. 46–55
19. B.A. Davey, H.A. Priestley, *Introduction to Lattices and Order* (Cambridge University Press, Cambridge, 1990)
20. R. Dedekind, *Was sind und was sollen die Zahlen? und Stetigkeit und Irrationale Zahlen* (Vieweg, Braunschweig, 1969)
21. K. Denecke, *Algebra und Diskrete Mathematik für Informatiker* (Vieweg, Braunschweig, 2003)
22. G. Di Battista, P. Eades, R. Tamassia, I. Tollis, *Graph Drawing: Algorithms for the Visualization of Graphs* (Prentice Hall, Hoboken, 1999)
23. W. Dörfler, W. Peschek, *Einführung in die Mathematik für Informatiker* (Carl Hanser, München, 1988)

24. H. Ebbinghaus, H. Hermes, F. Hirzebruch, M. Koecher, K. Mainzer, A. Prestel, R. Remmert, *Numbers* (Springer, New York, 1991)
25. H. Ehrig, *Mathematisch-Strukturelle Grundlagen der Informatik* (Springer, Berlin, 1999)
26. S. Eilenberg, *Automata, Languages, and Machines, Vol. A and B.*, Academic Press, New York, 1976.
27. M. Erné, *Einführung in die Ordnungstheorie* (BI Wissenschaftsverlag, Mannheim, 1982)
28. C. Faith, *Algebra. Rings, Modules and Categories* (Springer, Berlin, 1987)
29. G. Fischer, *Lineare Algebra*, 11th edn. (Vieweg, Braunschweig, 1997)
30. B. Ganter, R. Wille, *Formal Concept Analysis. Mathematical Foundations* (Springer, Berlin, 1999)
31. J.L. Gersting, *Mathematical Structures for Computer Science* (W. H. Freeman and Company, New York, 1982) 1987(2).
32. F. Harary, *Graph Theory*. Addison-Wesley Series in Mathematics (Addison-Wesley Publishing Company, Reading, 1969)
33. B. Hartley, T.O. Hawkes, *Rings, Modules and Linear Algebra*, 1st edn. (Chapman and Hall, London, 1970/1994)
34. U. Hebisch, H.J. Weinert, *Semirings. Algebraic Theory and Applications in Computer Science*. Series in Algebra, vol. 5 (World Scientific Publishing, Singapore, 1998)
35. H. Herrlich, G.E. Strecker, *Category Theory* (Allyn and Bacon, Boston, 1973)
36. K. Houston, *How to Think Like a Mathematician. A Companion to Undergraduate Mathematics.* (Cambridge University Press, Cambridge, 2009)
37. J.M. Howie, *An Introduction to Semigroup Theory* (Academic Press, London, 1976)
38. K. Hrbacek, T. Jech, *Introduction to Set Theory* (Marcel Dekker, New York, 1978)
39. H. Hutchins, *Examples of Commutative Rings* (Polygonal Publishing House, Passai, 1981)
40. A.W. Ingleton, R.A. Main, Non-algebraic matroids exist. Bull. London Math. Soc. **7**, 144–146 (1975)
41. K. Jacobs, *Einführung in die Kombinatorik* (Walter de Gruyter, Berlin, 1983)
42. J.C. Jantzen, J. Schwermer, *Algebra* (Springer, Heidelberg, 2006)
43. D. Jungnickel, *Graphs, Networks and Algorithms*. Algorithms and Computation in Mathematics (Springer, Berlin, 2013)
44. M. Kaufmann, D. Wagner (ed.), *Drawing Graphs: Methods and Models* (Teubner, Stuttgart, 20000
45. M. Kilp, U. Knauer, A.V. Mikhalev, *Monoids, Acts and Categories* (Walter de Gruyter, Berlin, 2000)
46. H.K. Kim, F.W. Roush, *Applied Abstract Algebra* (Ellis Horwood Ltd., Chichester, 1983)
47. D. Klaua, *Kardinal- und Ordinalzahlen, Teil 1 und 2* (Vieweg, Braunschweig, 1974)
48. U. Knauer, *Diskrete Strukturen - kurz gefasst* (Spektrum, Berlin, 2001)
49. U. Knauer, K. Knauer, *Diskrete und algebraische Strukturen - kurz gefasst* (Springer Spektrum, Berlin, 2015)
50. U. Knauer, K. Knauer, *Algebraic Graph Theory: Morphisms, Monoids and Matrices* (Walter de Gruyter, Berlin, 2019)
51. B. Kolman, R.C. Busby, S.C. Ross, *Discrete Mathematical Structures* (Prentice Hall, Upper Saddle River, 2000)
52. O. Körner, *Algebra* (Akademische Verlags-Gesellschaft, Frankfurt a. M., 1974)
53. E. Kunz, *Introduction to Commutative Algebra and Algebraic Geometry* (Modern Birkhäuser Classics, New York, 2013). Translated from the German by Michael Ackerman, with a preface by David Mumford
54. G. Lallement, *Semigroups and Combinatorial Applications* (Wiley, New York, 1979)
55. F. Lorenz, *Algebra. Volume I: Fields and Galois Theory* (Springer, New York, 2006). Translated from the 1987 German edition by Silvio Levy. With the collaboration of the translator
56. H. Lugowski, *Grundzüge der universellen Algebra* (Teubner, Leipzig, 1982)
57. H. Lüneburg, *Gruppen, Ringe, Körper* (Oldenbourg, München, 1999)
58. S. MacLane, *Categories for the Working Mathematician* (Springer, New York, 1971)

59. G. Mikhalkin, Tropical geometry and its applications, in *Proceedings of the International Congress of Mathematicians (ICM), Madrid, Spain, August 22–30, 2006*. Invited Lectures, vol. II (European Mathematical Society (EMS), Zürich, 2006), pp. 827–852
60. J. Oxley, What is a matroid? Cubo Mat. Educ. **5**, 179–218 (2003)
61. J. Oxley, *Matroid Theory*. Oxford Graduate Texts in Mathematics, vol. 21, 2nd edn. (Oxford University Press, Oxford, 2011)
62. G. Pilz, *Near Rings – The Theory and Its Applications* (North-Holland, Amsterdam, 1983)
63. G. Preuß, *Grundbegriffe der Kategorientheorie* (BI Wissenschaftsverlag, Mannheim, 1975)
64. W.V.O. Quine, *Set Theory and Its Logic* (The Belknap Press of Harvard University Press, Cambridge, 1969)
65. E. Scheinerman, *Discrete Mathematics for Computer Scientists* (Brooks/Cole, Salt Lake City, 2000)
66. U. Schöning, *Algorithmen – kurz gefasst* (Spektrum, Heidelberg, 1997)
67. U. Schöning, *Logic for Computer Scientists*. Modern Birkhäuser Classics (Birkhäuser, Boston, 2008)
68. A. Steger, *Diskrete Strukturen*, vol. 1 (Springer, Berlin, 2001)
69. G. Stumme, R. Wille (eds.), *Begriffliche Wissenverarbeitung* (Springer, Berlin, 2000)
70. J. Truss, *Discrete Mathematics for Computer Scientists* (Addison Wesley, Harlow, 1999)
71. H. Wähling, *Theorie der Fastkörper* (Thales, Essen, 1987)
72. W. Wechler, *Universal Algebra for Computer Scientists* (Springer, Berlin, 1992)
73. D.J.A. Welsh, *Matroid Theory*. L. M. S. Monographs, vol. 8 (Academic Press/Harcourt Brace Jovanovich Publishers, London/New York, 1976)
74. H. Werner, *Einführung in die allgemeine Algebra* (BI Wissenschaftsverlag, Mannheim, 1978)
75. N. White (ed.), *Theory of Matroids*, Reprint of the 1986 hardback edn. (Cambridge University Press, Cambridge, 2008)
76. R. Wille, *Begriffliche Wissensverarbeitung: Theorie und Praxis* (Spektrum, Heidelberg, 2000)
77. K. Zhevlakov, A. Slinko, I. Schestakov, A. Schirschov, *Rings That Are Nearly Associative* (Academic Press, New York, 1982)

Symbols

{}, 30
{ }, 86
{ | }, 29
$^{-1}$, 141
$(\emptyset, \emptyset, B)$, 96
:⇔, 2
⊆, 50, 76
< , >, 144
>, 76
[ ], 86
⇔, 8
⇒, 2, 8
$\bigcap_{i=1}^{r}$, 34
$\bigcap_{i \in I}$, 34
$\bigcup_{i=1}^{r}$, 34
$\bigcup_{i \in I}$, 34
$\bigvee_x$, 18
$\bigwedge_x$, 18
⊠, 241
∩, 31
∘, 144
⊔, 242
∪, 31
$\dot{\bigcup} f_i$, 88
$\dot{\cup}$, 35
∅, 30
≥, 76
⌈ ⌉, 86
≤, 50, 76
⋖, 76
⌊ ⌋, 86
¬, 8
⊈, 30
⊕, 228
∥, 76
$\prod f_i$, 88
$\prod_{i=1}^{r}$, 34
$\prod_{i \in I}$, 34

\, 31
⊓, 179
⊔, 179
$\sqrt{-1}$, 55
□, 137
×, 31, 144
⊊, 30
∨, 8
∧, 8

0, 9
1, 9

A

A^*, 2
$\overline{A}$, 31
(A), 208
$A \subseteq B$, 2
Ab, 235
$(a, b) \in \varrho$, 64
$a \varrho b$, 64
Act−S, 235
$A(G)$, 118
A^I, 34
$(A_i)_{i \in I}$, 34
$a_{ij}^{(r)}$, 123
$(a_{ij}^{<r>})$, 125
$A^* \setminus L$, 4
A^{lb}, 79
$\aleph_0$, 102
aN, 162
A^r, 34, 123
$A^{<r>}$, 125
A^{ub}, 79
$A(x)$, 118

B
$B(G)$, 120
B^A, 96

C
$C(n, k)$, 42
$C^r(n, k)$, 43
$\mathbb{C}$, 55
$\mathcal{C}$, 31
$\bigsqcup_{i \in I} R$, 214
$\oplus_{i \in I} R$, 214
card(M), 36
C_n, 115
cos ⊲, 144
$\subset \cdot$, 78

D
$\mathcal{D}_A$, 14
$d(x, y)$, 112
Δ, 31
δ, 114
$\delta^{\leftarrow}$, 112
$\delta^{\rightarrow}$, 112
Δ, 70
Δ_A, 64
$D(G)$, 123
$d_H(u, v)$, 204
div, 141
D_n, 160
dom f, 85

E
$\in$, 2, 4, 28
$\notin$, 4, 28
ε, 2
e, 56
EGra, 235
E_ϱ, 130
$\exists$, 18
$\exists"$, 18
$\exists^1$, 18
eval, 185
exp, 141

F
$\mathbb{F}^*$, 203
$\forall$, 2, 18
$f(A)$, 85
$f[A]$, 85

(f, A, B), 84
$f^{\leftarrow}[B]$, 85
Field, 235
$\mathbb{F}_{p^n}$, 201
$(F(X)$, 243

G
$G/_\varrho$, 130
$\overline{G}$, 115
$[G]$, 135
$G(A)$, 119
gcd, 52
$G/_N$, 163
Gra, 235
$G/_{\varrho_N}$, 163
Grp, 235
${}_GU$, 185

I
i, 55
(IA), 23
id_B, 234
(IH), 23
Im f, 85
IMad, 235, 243
inf, 79
(IS), 23

K
$\mathcal{K}^{op}$, 235
$\mathcal{K}(A, B)$, 234
ker f, 98, 162, 175, 188
$K_{m,n}$, 115
K_n, 115
$\overline{K_n}$, 115
K–**Vec**, 235
$K((x))$, 171, 173
$K(x)$, 171
$K[[x]]$, 173
$K[x]$, 171, 191
$\mathbb{R}[[x]]$, 171

L
$\ell(P_n)$, 112
$L(G)$, 137
LH_K, 193
$L : K$, 199
$[L : K]$, 199

Symbols

M
$(M, \leq)$, 76
$\#M$, 36
$\mathcal{M}$, 235
$|M|$, 36
M^0, 140
M_a, 143
$\text{Map}(A'', B)$, 96
max, 78, 142
M_g, 143
min, 78
$M \subseteq N$, 30
$\equiv \text{mod}$, 72, 163
Mod$-R$, 235
Mon, 235
$(M, \mathcal{V})$, 179

N
$N(x)$, 114
$N^{\leftarrow}$, 112
$N^{\rightarrow}$, 112
$\mathbb{N}$, 48
$\mathbb{N}_{n_0}$, 22
n!, 41
∇, 64, 70
$\binom{n}{k}$, 43

O
Ob $\mathcal{K}$, 234
Ord, 235

P
P, 236
$P(n, k)$, 41
$P^r(n, k)$, 42
$\mathcal{P}$, 236
(p), 209
$\wp$, 32
π, 56
π_ϱ, 98
π_I, 176
π_N, 163
π_U, 188
P_n, 115
$\mathcal{PT}(A)$, 96

Q
$\mathbb{Q}$, 53
$\mathbb{Q}(\sqrt{2})$, 172
$\mathbb{Q}(\sqrt[3]{5})$, 172

Q_6, 156
Q_n, 127
$Q(R)$, 245
$\mathbb{Q}(S)$, 172

R
R_{10}, 218
$\mathbb{R}$, 54
$(R, \oplus, \odot, \leq)$, 174
ran f, 85
Rel, 235
$R(G)$, 124
$/\varrho$, 98
$\varrho(a)$, 64
$\varrho(x)$, 70
$\varrho[A]$, 64
$\varrho \cap \sigma$, 67
$\varrho \cup \sigma$, 67
$\varrho^{\leftarrow}(b)$, 64
$\varrho^{\leftarrow}[B]$, 64
$\varrho^{\leftarrow}$, 67
ϱ', 67
ϱ_f, 98
ϱ_I, 176
ϱ_N, 163
ϱ_U, 188
R/I, 176
$R-\textbf{Mod}$, 235
Rng, 235
Rng$_1$, 235
R/ϱ_I, 176
$R[[x]]$, 192

S
$S(n, k)$, 45
$\sum$, 23
$\mathcal{S}_n$, 96
Rs, 208
sR, 208
(s), 208
$S-\textbf{Act.}$, 235
Set, 235
Sgr, 235
$\sigma \circ \varrho$, 67
S_n, 155, 161
sup, 79
$\text{Sym}(A)$, 96

T
$\mathcal{T}(A)$, 96
$\tau(x_1, x_2, \cdots, x_n)$, 140

T^d, 10
θ, 96, 140

U
$(u, F(X))$, 243
U_G, 185

V
E^+, 124
$\| v \|$, 144
(V, E), 111
(V, E, p), 110
$V/_{\varrho_U}$, 188
$V/_U$, 188

W
WMad, 235

X
$(X)^+$, 245

X^*, 246
X^+, 246
$\downarrow x$, 78
$(X)^+/_\alpha$, 246
$(X)^i$, 245
$X^*/_\kappa$, 246
$X^+/_\kappa$, 246
$[x]_n$, 72
$(X \cup X^{-1})^*/_\gamma$, 246
xy, 111
(x, y), 111
$\{x, y\}$, 111
$x \equiv y(n)$, 72
$x \equiv y \bmod n$, 72

Z
$\mathbb{Z}$, 52
$\mathbb{Z}/n\mathbb{Z}$, 72
$\mathbb{Z}/_n$, 72
$\mathbb{Z}$-**Mat**, 235
$(\mathbb{Z}_4, +)$, 160
$\mathbb{Z}_n$, 72, 172

Index

A
Abelian group, 185
Absolute value, 211
Absolute value function, 86
Absorbing element, 146
Absorption laws, 180
Absorption property, 145, 174
Act, 184
Act homomorphism, 187
Addition of complex numbers, 55
Addition of natural numbers, 48
Addition of vectors, 190
Additivity, 187
Adjacency list, 118
Adjacency matrix, 118
Adjacent, 111
Adjunction, 172
Affine plane, 12
Aleph zero, 102
Algebra, 140
Algebra automorphism, 154
Algebra endomorphism, 154
Algebra homomorphism, 154, 179
Algebraically closed, 56
Algebraically dependent, 202
Algebraically independent, 173, 202
Algebraic curve, 211
Algebraic equation of degree, 56
Algebraic Geometry, 211
Algebraic matroid, 223
Algebraic number, 56
Algebraic structure, 140, 179
Algebra over R, 184
Algorithm, 59
Alphabet, 2
Angle between two vectors, 144
Antiparallel edge, 111
Antisymmetric relation, 67

Antitone mapping, 154
Arbitrary but fixed, 23
Argument, 89
Arithmetic mean, 143
Arity of a relation, 64
Arrow, 190
Ascending chain condition (ACC), 211
Assertion, 20
Assignment of variables, 8
Associated, 208
Associative law, 146
Associative law for mappings, 89
Associative law for morphisms, 234
Associative law for relations, 67
Associative law for sets, 33
Associative law in logic, 11
Associative law of addition, 49
Assumption, 20
Asymmetric relation, 67
Auxiliary theorem, 20
Axiom, 13
Axiomatic structure, 13
Axiom of choice, 27, 193
Axiom of separation, 37
Axiom system, 13

B
Bad students multiplication, 122
Basis, 193, 214, 220, 245
Basis completion lemma, 194
Basis completion lemma, analogue, 215
BCH code, 206
Bijection/bijective, 92
Binary composition, 139
Binary matroid, 219
Binary number, 58
Binary relation, 64

Binary/unary composition, 8
Binomial coefficient, 43
Binomial theorem, 43
Bipartite graph, 115
Blocks of a partition, 35
Boolean algebra, 180
Boolean function, 143
Boolean matrix, 125
Boolean operation, 125
Boolean power, 125
Boxcross product, 241
Box product, 137
Burst error, 206

C
Canonical basis, 192
Canonical extension, 90
Canonical projection, 98, 156
Canonical surjection, 98, 130, 156
Cantor's diagonal method first/second, 101
Cardinality, 36, 103
Cardinality of the power set, 36, 107
Cardinal number, 47, 103
Cardinal number/cardinality, 36
Cardioid, 211
Cartesian coordinate system, 31, 65, 89
Cartesian leaf, 211
Cartesian product, 31, 137
Cartesian product of graphs, 241
Categorical, 48
Categorical coproduct in K-**Vec**, 242
Categorical coproduct in **Gra**, 241
Categorical product in **EGra**, 241
Categorical product in **Gra**, 241
Categorical product in K-**Vec**, 242
Category, 234
Cauchy sequence, 54
Cayley octaves, 170
Ceiling function, 86
Chain, 76
Character, 203
Circle number π, 57
Circuit, 14, 112, 115
Circuit of a matroid, 225
Circuit with n edges, 115
Class, 38, 71, 104, 234
Clause, 14, 16
Closed, 140
Closure of a graph, 135
Code, 204
Code generated by, 205
Code, 1-error correcting, 205
Code word, 204

Codomain, 85
Coefficients of a polynomial, 172
Column rank, 198
Co-map, 231
Combinations with/without repetition, 42
Combinatorial optimization problem, 227
Commutative law for mappings, 67, 89
Commutative law for relations, 67
Commutative law for sets, 33
Commutative law in logic, 11
Compatibility set/relation, 73
Compatible mapping, 154, 179
Compatible relation, 153
Complement, 31
Complement graph, 115
Complement in logic, 11
Complement of a relation, 67
Complete axiom system, 13
Complete bipartite graph, 115
Complete graph, 115
Complete semilattice (lattice), 80
Complex number, 55
Complex product, 162
Componentwise order, 77
Composition, 139
Composition of cycles, 161
Composition of mappings, 88
Composition of morphisms, 234
Composition of partial mappings, 88
Composition of statements, 8
Composition on $\mathbb{Z}_n$, 142
Composition table, 150
Concatenation, 2, 186, 245
Conclusion, 8, 20
Concrete category, 235
Congruence and ideal, 176
Congruence and normal subgroup, 163
Congruence and submodule, 188
Congruence on an algebra, 153
Congruence on $\mathbb{Z}_n$, 164
Congruent modulo n, 72
Conjecture, 20
Conjunction, 8
Conjunction term, 16
Conjunctive normal form, 16
Connected, 115
Connected component, 113
Consistent axiom system, 13
Constant mapping, 86
Construct, 235
Continuous mapping, 231
Continuum hypothesis, 28, 103
Contradiction, 9, 15
Contraposition, 12

Convergent sequence, 19
Converse relation, 67
Coordinate system, 65
Coprime, 52
Coproduct in categories, 240
Coproduct, non-set concrete, 242
Coretract/ion, 237
Corollary, 20
Cosine, 144
Countability of algebraic numbers, 102
Countability of $\mathbb{N}$, $\mathbb{Z}$ and $\mathbb{Q}$, 100
Countable set, 100
Counterexample, 21
Covered by, 76
Cross product of graphs, 241
Cryptomorphism, 225, 226, 228
Cycle, 112, 115
Cycle notation, 161
Cycle of a permutation, 161
Cyclic code, 205
Cyclic group, 160
Cyclic group of order 4, 160

D
Decidable, 4
Decimal system, 58
Decision tree, 134
Decoding function, 204
Decoding principle, next neighbor, 205
Dedekind's cut, 54
Deductive reasoning, 21
Defining property of a set, 29
Defining matrix, 197
Defining property for morphisms, 235
Defining property of the imaginary unit, 55
Degree function, 211
Degree of a polynomial, 172
Degree of a vertex, 114
Degree of the extension, 199
de Morgans laws for sets, 33
de Morgan's rules in logic, 12
Derivation rule, 13
Diagonal, 64, 70
Difference set, 31
Digit, 58
Dihedral group, 160
Dimension, 194
Direct product of vector spaces, 242
Direct proof, 21
Direct sum of matroids, 228
Direct sum of vector spaces, 242
Directed graph, 110
Dirichlet's pigeonhole principle, 38

Disjoint, 32
Disjoint union, 35
Disjunction, 8
Disjunctive normal form, 14
Disjunctive term, 14
Distance, 112
Distance function, 87, 113
Distance matrix, 123
Distinguished element, 140
Distributive law, 167
Distributive law for sets, 33
Distributive law in logic, 11
Divide, 52
Division algebra, 169
Division with remainder, 52
Divisor relation, 76
Dodecahedron, 134
Domain, 85
Doubling the cube, 57
Downwards, 78
Dual category, 235
Duality of product and coproduct, 240
Dualizing of propositional formulas, 10
Dual matroid, 228
Dual number, 58
Dual propositional formula, 10
Dyadic number, 58

E
Edge graph, 137
Edge preserving mapping, 128
Edge sequence, 112, 115
Edge set, 110
Efficient algorithm, 227
Elliptic geometry, 13
Embedding, 87
Empty mapping, 96
Empty relation, 64
Empty set, 30
Empty word, 2
Encoding, 203
Encoding function, 204
Endomorphism graph, 128
Enumerable, 4
Epimorphism, 237
Equality of mappings, 84
Equality of relations, 64
Equality of sets, 29
Equation/equational form, 51
Equivalence class, 71
Equivalence relation, 69
Equivalence relation of a partition, 71
Equivalent propositional form, 15

Equivalent propositional formula, 9
Error correcting code, 204
Error detecting code, 204
Euclidean algorithm, 53
Euclidean domain, 210
Euclidean geometry, 12
Euclidean plane, 55
Euler graph/trail/cycle, 132
Eulerian number e, 57
Evaluation function, 185
Evaluation mapping, 185
Everywhere defined composition, 140
Exactly if, 8
Existential quantifier, 18
Exponentiation, 141
Extended matrix, 198
Extension field, 172, 199
Extension of partial mappings, 90
Extension of the range, 91
Extension of the source, 90
Extension of the target, 90

F

Factor algebra/structure, 156
Factor graph, 130
Factor group, 163
Factorial, 41, 210
Factor module, 188
Factor object in category, 237
Factor of the product, 240
Factor ring, 176
Factor set, 71
Family of sets, 34
Fano matroid, 220, 232
Fermat's theorem, 210
Field, 168
Field of fractions, 171
Field of quotients, 176, 245
Field of rational functions, 171
Finite field, 172
Finite ring, 172
Finite set, 36
Fire code, 206
First component, 31
Fixed point of a permutation, 161
Floor function, 86
Follows from, 8
For all, 18
Forest, 115, 123
Formal language, 2
Formal power series, 173
Free commutative monoid, 246
Free commutative semigroup, 246

Free generating system, 245
Free group, 246
Free groupoid, 245
Free matroid, 247
Free module, 211, 214
Free monoid, 246
Free object, 243
Free semigroup, 246
Free variable, 18
Free word monoid, 146
Full permutation group, 155
Full transformation monoid, 146
Function, 83
Function graph, 85
Functor, 236
Fundamental counting principle, 40
Fundamental Theorem of Algebra, 56
Fuzzy logic, 1, 7

G

Gauss bracket, 86
Gauss elimination method, 198
Gaussian integers, 200, 210
Gaussian number plane, 55
General algebra, 178
Generating element, 160
Generating system, 193, 214, 245
Geometric interpretation, 190
Geometric mean, 143
Goldbach conjecture, 21
Graph, 110
Graph automorphism, 128
Graph congruence, 129
Graph egamorphism, 128
Graphic matroid, 222
Graph isomorphism, 128
Graph morphism, 128
Graph of a mapping, 89
Greatest common divisor, 52
Greedy algorithm, 227
Ground set, 218
Group, 147, 158
Group automorphism, 159
Group endomorphism, 159
Group homomorphism, 154, 159
Group isomorphism, 159
Groupoid, 145
Groupoid homomorphism, 154
Groupoid word, 245

H

Hamilton graph/circuit/path, 134
Hamiltonian quaternions, 170

Hamming distance, 204
Hasse diagram, 77
Heterogeneous algebra, 187
Hexadecimal number, 58
Hilbert's basis theorem, 211
Homogeneous, 197
Homological algebra, 216
Homomorphism theorem for algebras, 157
Homomorphism theorem for graphs, 130
Homomorphism theorem for groups, 164
Homomorphism theorem for modules, 189
Homomorphism theorem for rings, 176
Homomorphism theorem for sets, 98
Hyperbolic geometry, 13
Hypercomplex conjugate, 170
Hypercomplex system, 169
Hypercube, 127
Hypergraph, 126
Hypothesis, 20

I
Ideal, 145
Ideal generated by A, 208
Ideal of a (semi)ring, 174
Idempotent, 180
Identical morphism, 234
Identical mapping, 86
Identity, 64, 70
Identity matrix, 172
If and only if, 8
Iff, 8
If..., then, 8
Image, 85
Image under a relation, 64
Imaginary, 55
Imaginary unit/number, 55
Implication, 8
Incidence mapping, 110
Incidence matrix, 120
Incident, 111
Inclusion, 30
Incomparable, 76
Indegree of a vertex, 112
Independent edges, 116
Independent events, 40
Independent mapping, 231
Independent set, 218
Indeterminate, 6, 173
Indirect proof, 22
Individual domain/variable, 17
Induced equivalence relation, 129
Induced congruence, 155, 163, 176, 188
Induced equivalence relation, 98

Induced surjection, 130, 163, 176, 188
Induction anchor/proof/step, 22
Induction hypothesis, 23
Inductive reasoning, 21
Infimum, 79
Information, 203
Inf-semilattice, 80
Inhomogeneous, 197
Initial condition of a recursion, 24
Initial object, 243
Injection, 87, 91, 92
Injection in category, 240
Injective, 85, 92
Injective composition, 93
Injective partial mapping, 92
Inner composition, 140
Input alphabet, 185
Integer, 52
Integer function, 86
Integral domain, 210
Interpretation, 17
Intersection, 31
Intersection of matroids, 228
Intersection of relations, 67
Interval nesting, 54
Inverse, 152
Inverse mapping, 94
Inverse relation, 67
Involution, 33
Irrational, 54
Irreducibility criteria, 201
Irreducible element, 209
Irreducible ideal, 209
Irreducible polynomial, 199
Irreflexive relation, 67
Isolated vertex, 111
Isomorphic, 92
Isomorphic matroids, 219
Isomorphism, 237
Isotone mapping, 105, 154

K
Kernel, 162, 175, 188
Kernel congruence, 155
K-homomorphism, 187
Klein four group, 160
K-linearity, 187
Königsberg bridge problem, 136
k-regular, 116
Kruskal algorithm, 115, 227
k-th injection, 87
k-th projection, 87
K-vector matroid, 218

L

Large category, 235
Largest element, 78
Largest lower bound, 79
Latin square, 151
Lattice, 80, 179
Laurent series, 172, 173
Law, 12
Left (near) (ring, etc.), 168
Left absorbing, 146
Left cancellable, 147
Left coset, 185
Left distributive, 167
Left group, 149, 153
Left ideal, 145, 174
Left inverse, 67, 94, 152
Left invertible, 67
Left neutral, 146
Left R-act, 184
Left simple, 149
Left solvable, 147
Left total, 67
Left unique, 67
Left zero, 146
Left zero in M^M, 146
Left zero semigroup, 147, 159
Left/right neutral, 146
Left/right one, 146
Left/right principal ideal, 208
Lemma, 20
Length of a vector, 144
Length of an edge sequence, 112
Length of the representation, 58
Lexicographic order, 77
Linear code, 205
Linear combination, 193
Linear equation system, 197
Linear extension, 195
Linear factor, 193
Linear hull, 193
Linearly (in)dependent, 194, 214
Linear mapping, 187
Linear order, 76
Linear space, 190
Line graph, 137
List notation, 161
Logical dual truth table, 10
Logical equivalence, 8
Logical equivalent, 9
Loop, 111, 147
Loop complement, 115
Lottery problem, 40
Lower Gauss bracket, 86
Lower semilattice, 80

M

Magma, 145
Main theorem, 20
Mapping, 84
Mapping onto, 92
Mapping rule, 85
Mapping theorem, 98, 99
Matching, 116
Matrix, 118
Matrix ring, 172
Matroid, 218
Maximal ideal, 209
Maximally cycle free, 117
Maximum, 78
Maximum element, 78
Maximum function, 142
Maximum matching, 116
Max-plus algebra, 187
Max-plus semifield, 169
Median, 143
Mengenknödel, 29
Message, 203
Message word, 203
Metric, 87, 115
Metric space, 87
Minimal element, 78
Minimally connected, 117
Minimal polynomial, 199
Minimum, 78
Minimum function, 142
Mixture problem, 41
Module over R, 184
Modulo, 72
Modus ponens, 13
Monoid, 146
Monoid homomorphism, 154
Monomorphic data type, 48
Monomorphism, 237
Morphism, 234
Moufang loop, 147
Multigraph, 110
Multiple edge, 110
Multiset, 30, 111
Multisorted system, 187

N

Nabla, 64
n-ary composition, 139
n-ary relation, 64
Natural language, 1
Natural numbers, 47, 48
n choose k, 43

Necessary/and sufficient, 8
Necessary for, 8
Negation, 8
Negation of statements, 19
Neighbors of a vertex, 114
Neutral element, 146
Neutral element for sets, 33
Neutral element in logic, 11
(n, m)-code, 204
Node set, 110
Noetherian, 210
Non-associative quasigroup, 152
Non-Fano matroid, 231
Non-Pappus matroid, 220
Non-set, 37
Normal subgroup, 163
Normalized polynomial, 172
Np-complete, 134
Number of partitions, 45

O
Object, 234
Octal number, 58
Octaves, 170
Octonions, 170
One, 146
One-to-one mapping, 92
One-to-one onto, 92
Open mapping, 231
Open set, 231
Operand domain, 184
Operation of relations, 67
Operation table, 145
Operator domain, 184
Oppositional, 67
Ordered algebraic structure, 153
Ordered field, 174
Ordered group, 153
Ordered groupoid, 153
Ordered pair, 31
Ordered ring, 174
Ordered r-tuple, 34
Ordered semigroup, 153
Ordered set, 76
Ordered triple, 34
Order on $\mathbb{N}$, 50
Order relation, 76
Ordinal number, 47
Origin, 190
Outdegree of a vertex, 112
Outer (external) operation, 184

P
Pairwise disjoint, 32
Parallel axiom, 12
Parallel edge, 110
Partial composition, 139
Partial inverse, 95
Partial mapping, 84
Partial order, 76
Partial transformation, 96
Partition matroid, 221
Partition of an equivalence relation, 71
Partition of a set, 35
Pascal's triangle, 46
Path, 112, 115
Path of length n, 115
Peano axioms, 48
Perfect matching, 116
Permutation, 96
Permutation group, 161
Permutation with/without repetition, 41
Placeholder, 6
Polynomial, 172, 191, 227
Polynomial ring, 171, 173
Poset, 76
Poset/partially ordered set, 76
Position vector, 190
Power of a matrix, 123
Power of a set, 34
Power series, 192
Power set, 32
Predecessor of a vertex, 112
Predicate/variable, 17
Preimage, 85
Preimage under a relation, 64
Premise, 8
Prime element, 208
Prime factorization, 57
Prime field, 177
Prime ideal, 209
Prime number, 52
Principal ideal, 208
Principal ideal domain, 210
Principal ideal ring, 211
Principle of inclusion and exclusion, 38
Principle of the excluded third, 7, 9
Product in categories, 240
Product of algebras, 165
Product of complex numbers, 55
Product of mappings, 88
Product of matrices, 121
Product of natural numbers, 49
Product of polynomials, 173
Projection, 87, 240

Projection edges to vertices, 111
Projection in categories, 240
Projection onto subset, 87
Projective geometry, 13
Proof, 20
Proof by contradiction, 22
Proof by contraposition, 22
Proof by mathematical induction, 22
Proper class, 38
Proper ideal, 174
Proper subset, 30

Q

Quantified variable, 18
Quantifier, 18
Quasigroup, 147
Quasiorder, 76
Quaternion group, 170
Quaternions, 170
Quotient field, 171
Quotient group, 163
Quotient module, 188
Quotient ring, 176
Quotient set, 71

R

R_{10}, 218
Range, 85
Rank, 226
Rank function, 226
Rational number, 53
Reachability matrix, 124
Real matroid, 219
Real number, 54
Recursion, 24
Recursive, 4
Recursive definition, 24
Recursively enumerable, 4
Reducible, 209
Reed Solomon code, 206
Reflexive relation, 67
Refutation of a conjecture, 21
Regular, 116
Regular matroid, 219
Relational system, 64
Relation graph, 65
Representation to base, 58
Representative independents, 142, 172
Representative system, 72
Representative system for residues modulo n, 72
Representing matrix, 196
Residue class field, 172

Residue class modulo n, 72
Residue class ring, 172
Restriction of the domain, 90
Restriction of the range, 90, 91
Retract/ion, 237
Reversal, 67
R-homomorphism, 187
Right (near) (ring, etc.), 168
Right group, 149
Right hand side, 197
Right inverse, 152
Right simple, 149
Right total, 67
Right unique, 67
Right zero semigroup, 147
Ring, 168
Ring homomorphism, 174
Ring of power series, 172
Ring of quotients, 176
R-linear algebra, 184
Roman number, 58
Root, 199
R-operand, 184
Rotation, 185
Rounding down, 86
Rounding up, 86
R-set, 184
Rule, 12
Russell's antinomy, 27
Russell's paradox, 37

S

Scalar, 190
Scalar multiplication, 186
Scalar product, 144, 187, 190
Schur-Hadamard multiplication, 122
Second component, 31
Section, 237
Semantics, 4
Semiautomaton, 185
Semi(group)automaton, 184
Semi-edge/ sequence/path/cycle/circuit, 112
Semifield, 168
Semigroup, 146
Semigroup homomorphism, 154
Semilattice, 80
Semilinear, 187
Semiring, 168
Set, 28
Set concrete, 235
Set dumplings, 29
Set identities, 32
Set of morphisms, 235

Signature of an algebra, 179
Simple graph, 110
Skew field, 168
Small category, 235
Smallest vector space, 190
Smallest field, 177
Smallests element, 78
Smallest upper bound, 79
Solution of a combinatorial optimization problem, 227
Solution of an algebraic equation, 56
Solution of an equation, 51
Solution of an equational system, 197
Solution of a recursion, 24
Source, 85
Source/domain of a morphism, 234
Source of an edge, 112
Spanning tree/forest, 115
Squaring the circle, 57
Standard basis, 192
Standard vector space, 190
Statement, 7
Statement form, 8
Statement variable, 7
State set, 185
Stirling numbers first/second kind, 45
Stirling numbers of the first kind, 45
Stirling's approximation formula, 41
Strong graph morphism, 128
Strong induction, 24
Strongly regular, 116
Strong mapping, 231
Structure preserving, 154, 179
Subalgebra/structure, 144
Subfield, 174
Subgroup, 162
Submodule, 188
Subobject in category, 237
Sub(semi)ring, 174
Subset, 2, 30
Sub(vector)space, 188
Successor, 48
Successor of a vertex, 112
Sudoku, 151
Sufficient for, 8
Sum of matrices, 121
Supremum, 79
Sup-semilattice, 80
Surjection, 92
Surjective, 85, 92
Surjective composition, 93
Surjective partial mapping, 92
Swallowing property, 145, 174
Symmetric difference, 31

Symmetric group, 161
Symmetric matrix, 119
Symmetric relation, 67
Syntax, 3
System of sets, 34

T
Target, 85
Target/codomain of a morphism, 234
Tarski's fix point theorem, 105
Target of an edge, 112
Tautology, 9, 16
Telephone number problem, 40
Tensor product, 241
Terminal object, 243
Ternary composition, 139
Ternary relation, 64
Tertium non datur, 7
Theorem, 20
Theorem of Bondy and Chvátal, 135
Theorem of Cantor, 107
Theorem of Cantor/Schröder/Bernstein, 104
Theorem of Dirac, 136
Theorem of Ore, 136
Theorem of Tarski (fix point theorem), 105
Theory formation, 21
There exists, 18
There exists exactly one, 18
There exists one and only one, 18
Topological space, 231
Torsion element, 215
Torsion free, 215
Torsion module, 215
Totally disconnected, 115
Total mapping, 84
Total order, 76
Transcendental number, 56
Transformation, 96
Transformation monoid, 146, 155
Transition function, 185
Transitive closure, 74, 124
Transitive relation, 67
Transposition, 162
Transversal, 221
Transversal matroid, 222
Travelling salesman problem, 134
Tree, 115, 123
Trisecting an angle, 57
Tropical semiring, 169, 187
Truth assignment of a propositional formula, 8
Truth table, 8
Type of an algebra, 179
Type of a system, 64

U

Unary composition, 139
Unary number representation, 60
Unary relation, 64
Uncountability of $\mathbb{R}$, 102
Uncountability of transcendental numbers, 102
Uncountable set, 100
Underlying mapping, 235
Underlying set, 235
Undirected graph, 110, 114
Uniform matroid, 221
Union, 31
Union of graphs, 241
Union of mappings, 88
Union of matroids, 228
Union of relations, 67
Unique factorization domain, 211
Uniquely solvable, 198
Uniqueness of free objects, 244
Uniqueness of One and Zero, 147
Unit, 208
Universal algebra, 178
Universally solvable, 198
Universal problem, 165, 240, 243, 244
Universal property, 165
Universal quantifier, 18
Universal relation, 64
Universal set, 28
Unknown, 6, 173, 197
Unsolvable problem, 57
Upper bound, 79
Upper semilattice, 80

V

Valency of a logic, 7
Valuation function, 211
Vámos matroid, 220, 226
Variable, 6, 51, 173
Variety of algebras, 180
Vector, 190
Vector matroid, 218
Vector product, 144
Vector space, 184, 190
Vector space homomorphism, 187
Venn diagram, 29
Vertex set, 110

W

Weakly/strongly connected, 113
Weak mapping, 231
Weight function, 211, 227
Well defined, 142
Well-ordering, 50
With one (ring, etc.), 168
Word, 2
Word semigroup/monoid, 245

Z

Zermelo/Fraenkel, 27, 103
Zero, 146
Zero divisor, 171, 177, 207
Zero matrix, 172
Zorn's Lemma, 193

The manufacturer's authorised representative in the EU is Springer Nature Customer Service Centre GmbH, Europaplatz 3, 69115 Heidelberg, Germany. If you have any concerns regarding our products, please contact ProductSafety@springernature.com

Printed and bound by CPI Group (UK) Ltd, Croydon, CR0 4YY
26/03/2026
02078964-0001